AGING AT THE MOLECULAR LEVEL

BIOLOGY OF AGING AND ITS MODULATION

VOLUME 1

AGING AT THE MOLECULAR LEVEL

Edited by

THOMAS VON ZGLINICKI

Professor of Cellular Gerontology
University of Newcastle
Henry Wellcome Laboratory for Biogerontology
Newcastle General Hospital
Newcastle upon Tyne
United Kingdom

SPRINGER-SCIENCE+BUSINESS MEDIA, B.V.

Library of Congress Cataloging-in-Publication Data

ISBN 978-90-481-6482-0 ISBN 978-94-017-0667-4 (eBook)
DOI 10.1007/978-94-017-0667-4

Printed on acid-free paper

Contents

About the series "Biology of aging and its modulation"

During the last 40 years, the study of the biological basis of aging has progressed tremendously, and it has now become an independent and respectable field of study and research. Several universities, medical institutes and research centers throughout the world now offer full-fledged courses on biogerontology. The interest of students taking such courses, followed by undertaking research projects for MSc and PhD studies, has also increased significantly. Cosmetic, cosmeceutical and pharmaceutical industry's ever increasing interest in aging research and therapy is also obvious. Moreover, increased financial support by the national and international financial agencies to biogerontological research has given much impetus to its further development.

This five-volume series titled "Biology of Aging and its Modulation" fulfills the demand for books on the biology of aging, which can provide critical and comprehensive overview of the wide range of topics, including the descriptive, conceptual and interventive aspects of biogerontology. The titles of the books in this series and the names of their respective editors are:

1. Aging at the molecular level (Thomas von Zglinicki, UK)
2. Aging of cells in and outside the body (S. Kaul and R. Wadhwa, Japan)
3. Aging of organs and systems (R. Aspinall, UK)
4. Aging of organisms (H. D. Osiewacz, Germany)
5. Modulating aging and longevity (S. Rattan, Denmark)

The target readership is both the undergraduate and graduate students in the universities, medical and nursing colleges, and the post-graduates taking up research projects on different aspects of biogerontology. We hope that these books will be an important series for the college, university and state libraries maintaining a good database in biology, medical and biomedical sciences. Furthermore, these books will also be of much interest to pharmaceutical, cosmaceutical, nutraceutical and health-care industry for an easy access to accurate and reliable information in the field of aging research and intervention.

Suresh I.S. Rattan, Ph.D., D.Sc.
Series Editor and Editor-in-Chief, Biogerontology
Danish Centre for Molecular Gerontology, Department of Molecular Biology,
University of Aarhus, Denmark

Preface

The essential cause of aging is molecular damage that slowly overwhelms cellular and organismic defense, repair and maintenance systems. In recent years, a wealth of highly sophisticated research has transformed this idea from a credible hypothesis not only to a major theory, but essentially to accepted knowledge. The present book describes some of the key elements in this transformation.

Whether oxygen-derived free radicals are the most important cause of molecular damage with relevance to aging or not, is still a matter of debate (see Chapters 10 and 11). However, oxygen free radicals are clearly the most thoroughly researched entity with respect to age-related molecular damage. Thus, the book starts off with a summary of what reactive oxygen species (ROS) are, what they do and how cells can defend themselves against it (Chapter 1). We then describe oxidative damage to DNA (Chapter 2), proteins (Chapter 3) and lipids (Chapter 4), the latter with special reference to comparative studies in mammals and birds, who have very different rates of ROS generation.

DNA damage, if unrepaired or mis-repaired, causes genomic instability and mutations. The inter-relationship between genomic instability and aging is best studied in human progerias, as shown in Chapter 5 by Bohr and Opresko. The role of nuclear mutations in cell aging is examined in Chapter 6 by Busuttil *et al.*, followed by the work of Turnbull and Barron (Chapter 7) on mitochondrial DNA mutations.

Aging is not only governed by damage. There is quite a number of confirmed or probable biological clocks at work, which seem to programme the aging process in a genetically determined fashion, and these are reviewed by Boukamp (Chapter 8). On the other hand, there is increasing evidence that the pace of these clocks is set in response to stress and damage, as shown by von Zglinicki (Chapter 9) for the telomere clock. Clearly, molecular damage, especially at low frequency, rarely promotes organismic aging directly on its own but triggers cellular response programmes, which then contribute to the aging process.

Recent work in transgenic mice very clearly demonstrates that oxidative damage and antioxidant defense might be important, but are by far not the sole determinants of aging (Chapter 10). Thus, non-oxidative modifications to DNA and protein (Chapter 11) as well as erroneous transcription and translation (Chapter 12) as it impacts on aging are summarized next. The present knowledge of the molecular links between environmental stress (especially caloric restriction), post-translational protein modification in chromatin, gene silencing, and aging are outlined in Chapter 13 by Cohen *et al.* Finally, the role that protein degradation plays in aging is considered in the chapters devoted to the proteasome (Chapter 14) and to the lysosome (Chapter 15).

My heartfelt gratitude goes to the chapter's authors that made this book possible. It is the quality and timeliness of their contributions that make this book exceptional. I also thank Suresh Rattan, the series editor, and the editors at Kluwer Academic Publishers for their support and patience. May this book contribute to further progress in the fascinating field of Biogerontology!

Thomas von Zglinicki
Newcastle, February 2003

Free Radical Production and Antioxidant Defense: A Primer

Nicolle Sitte[1] and Thomas von Zglinicki[2]
[1]Department of Anaesthesiology and Critical Care Medicine, CHARITÉ – University Medicine Berlin, Campus Benjamin Franklin, Berlin, Germany; [2]Henry Wellcome Laboratory for Biogerontology, Newcastle University, General Hospital, Newcastle, NE4 6BE, UK

Oxygen free radicals and aging

Aging is the progressive accumulation of changes over time that increases the probability of disease and death. There is good reason to believe that molecular damage, and the limited ability of cells, tissues, organs, and individuals to repair it or to maintain function despite it, is the principal driving force of the aging process. A major cause for molecular damage, although by far not the only one (see Chapters 11, 12 and 13, this volume), are highly reactive oxygen-derived free radicals, mostly, but not exclusively endogenously generated, as was first proposed in 1956 by Denham Harman in his "free radical theory of aging" [1]. Denham Harman was also the first to suggest in 1972 a prime role for mitochondria as the biological clock in aging, noting that the rate of oxygen consumption should determine the rate of accumulation of mitochondrial damage produced by free radical reactions [2]. These early suggestions have become a strong and vivid conceptual framework for aging research till today. Thus, a recapitulation of the essentials of the oxygen free radical/ antioxidant defense network in animal cells seems to be a prudent opening for a book dealing with the molecular mechanisms of aging. Figure 1 summarizes schematically the interplay between oxygen free radical generation, its damaging effects, and cellular antioxidant responses.

Sources of oxidants

What are free radicals?

A free radical is any molecular species capable of independent existence that contains one or more unpaired valence electrons not contributing to intramolecular bonding,

T. von Zglinicki (ed.), Aging at the Molecular Level, 1–10.

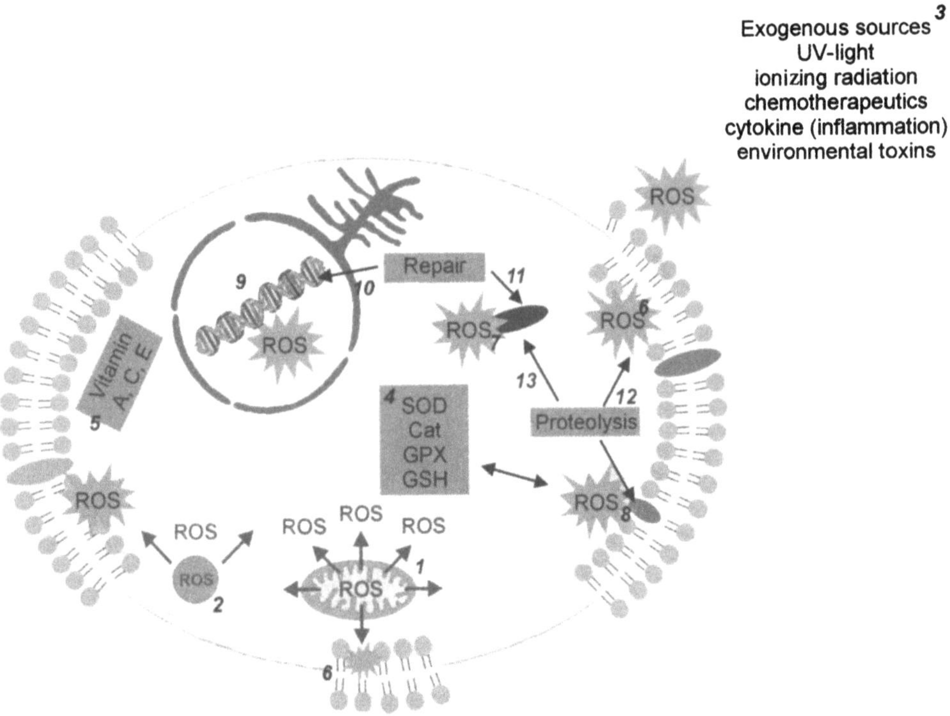

Figure 1. Sources of and cellular reactions to reactive oxygen species. Reactive oxygen species (ROS) are formed as a result of the intracellular metabolism in mitochondria (1) and peroxisomes (2) as well as of different cytosolic enzymatic systems. Additionally, production of ROS is enhanced by exogenous influences (3). The cell possesses antioxidant defense systems protecting against destructive effects of ROS that are responsible for physiological homeostasis. Most important enzymatic antioxidants (4) are superoxide dismutases (SOD), catalase (Cat), and glutathione peroxidases (GPX); non-enzymatic defense systems include glutathione (GSH), and vitamins like vitamin A, C and E (5). When the endogenous antioxidant defense is overwhelmed by high ROS production, membranes are damaged by lipid peroxidation (6), cytosolic enzymes (7) and membrane proteins (8) are functionally compromised and DNA is damaged (9). The secondary antioxidant defense system may partially repair damage of DNA (10) or proteins (11). In contrast, toxic products of lipid peroxidation are degraded (12) and oxidatively modified proteins are removed by proteolysis (13).

and is – in that sense – "free." Radicals are produced by oxidation/reduction reactions in which there is a transfer of only one electron at a time, or when a covalent bond is broken and one electron from each pair remains with each atom. Thus, a free radical has an unpaired electron. Many free radicals are highly reactive, owing to the tendency of electrons to pair – that is, to pair by the receipt of an electron from an appropriate donor or to donate an electron to an appropriate acceptor. Whenever a free radical reacts with a non-radical, a chain reaction is initiated until two free radicals react and then terminate the propagation with a 2-electron bond (each radical contributing its single unpaired electron). In biological systems free radicals have a range of transitory existences depending upon their reactivity.

The free radicals of special interest in aging are the oxygen free radicals. These free radicals often take an electron away from a "target" molecule to pair with their single free electron. This is called "oxidation." There are some closely related oxygen containing molecules that are not strictly free radicals but contribute to their production or are strong oxidants themselves, such as singlet oxygen and hydrogen peroxide. The term "reactive oxygen species" (ROS) is used to refer to these oxidants and the oxygen free radicals.

What are the major sources of free radical production?

There are many sources of free radicals both within and external (environmental) to cells. Many are produced by normal ongoing metabolism, especially from the electron transport system in the mitochondria and from a number of enzymatic reactions (e.g., xanthine oxidase, cytochrome P450, monoamine oxidase, nitric oxide synthase, lipoxygenases, cyclooxygenases, peroxidases, and other heme proteins and as a secretion of activated leukocytes [reviewed in ref. 3]). In the brain, free radicals are produced from the autoxidation of norepinephrine and dopamine. The autoxidation of catechols to quinones generates reduced forms of molecular oxygen, sources of free radicals (e.g., superoxide and hydrogen peroxide).

The activities of a number of key antioxidant enzymes, such as catalase, superoxide dismutase and glutathione peroxidase, may decrease with increasing age, though this is not unequivocal [4, 5]. As a result of this decline in the antioxidant defenses, it is possible that the impact of ROS increases with age. It is also possible their production increases, e.g., as a result of poorer coupling of electron transport components in mitochondria and other electron transport systems [4–7] and an increased level of redox-active metal ions which could catalyze oxidant formation [8, 9].

Bruce Ames and his colleagues claimed that oxidants generated by mitochondria are the major source of oxidative lesions that accumulate with age [6]. About 90% of the oxygen consumed in a cell is consumed within the mitochondria, predominantly via the inner mitochondrial membrane-located oxidative phosphorylation pathway. The mitochondria's main function is energy production. However, during oxidative phosphorylation in normal aerobic metabolism, the production of highly reactive oxygen radicals occurs continuously in the mitochondrial electron transport chain in which O_2 is reduced to H_2O. Mitochondrial electron transport is imperfect and results in the production of superoxide anion ($^{\cdot}O_2^-$) from the one-electron reduction of O_2. It has been estimated that about 2% of all the electrons traveling down the mitochondrial respiratory chain never make it to the end, but instead form superoxide [10].

Under normal metabolic conditions, each cell in our body is exposed to about 10^{10} molecules of superoxide each day. Once formed, superoxide is converted to other ROS. The two-electron reduction of O_2, with the addition of $2H^+$, generates hydrogen peroxide (H_2O_2). Most of the short-lived superoxide is metabolized by the mitochondrial Mn-SOD to H_2O_2 and oxygen. Compared to the extremely unstable superoxide or hydroxyl radicals, hydrogen peroxides are freely diffusible and relatively long-lived. Superoxide anions are also formed when molecular oxygen is subjected to ionizing radiation and acquires an additional electron, or when xanthine oxidase converts xanthine to uric acid and so produces superoxide radicals. Superoxide is also formed by the auto-oxidation of hydroquinones, catecholamines, and thiols.

The source of the highly toxic hydroxyl radical ($HO^{\cdot}$) is the Haber Weiss reaction between hydrogen peroxide and the superoxide radical anion:

$$\underset{\text{hydrogen peroxide}}{H_2O_2} + \underset{\text{superoxide anion}}{{}^{\cdot}O_2^-} \rightarrow \underset{\text{hydroxyl radical}}{{}^{\cdot}OH} + \underset{\text{hydroxyl ion}}{OH^-} + O_2$$

In the presence of small amounts of iron or copper, the Haber Weiss reaction is accelerated (Fenton reaction).

$$\underset{\text{hydrogen peroxide}}{H_2O_2} + Fe^{2+} \rightarrow \underset{\text{hydroxyl radical}}{{}^{\cdot}OH} + \underset{\text{hydroxyl ion}}{OH^-} + Fe^{3+}$$

Apart from the Fenton reaction, the reaction between hypochlorous acid and superoxide can effect the formation of hydroxyl radicals [11].

$$\underset{\text{hypochlorous acid}}{HOCl} + \underset{\text{superoxide anion}}{{}^{\cdot}O_2^-} \rightarrow O_2 + Cl^- + \underset{\text{hydroxyl radical}}{{}^{\cdot}OH}$$

In addition, nitric oxide ($NO^{\cdot}$) is also produced by mitochondria as a normal product of arginine metabolism [12]. The enzyme nitric oxide synthase (NOS), which produces the free radical gas $NO^{\cdot}$ from L-arginine, was found in mitochondria isolated from heart, skeletal muscle, kidney, and brain [13, 14]. Since $NO^{\cdot}$ react rapidly with ${}^{\cdot}O_2^-$ to form another very reactive oxidant, peroxynitrite ($ONOO^-$), it is clear that mitochondria are a major source of free radicals and oxidants.

$$\underset{\text{nitric oxide}}{NO^{\cdot}} + \underset{\text{superoxide anion}}{{}^{\cdot}O_2^-} \rightarrow \underset{\text{peroxynitrite}}{ONOO^-}$$

The permanent attack of cellular components by these radicals leads, for instance, to mutations in mitochondrial DNA that lead in turn to the production of less functional respiratory chain proteins, resulting in an increased free radical production and possibly more mitochondrial DNA mutations.

In addition to the mitochondrial respiratory chain, there are other endogenous sources of superoxide production. In particular, when leukocytes encounter micro-

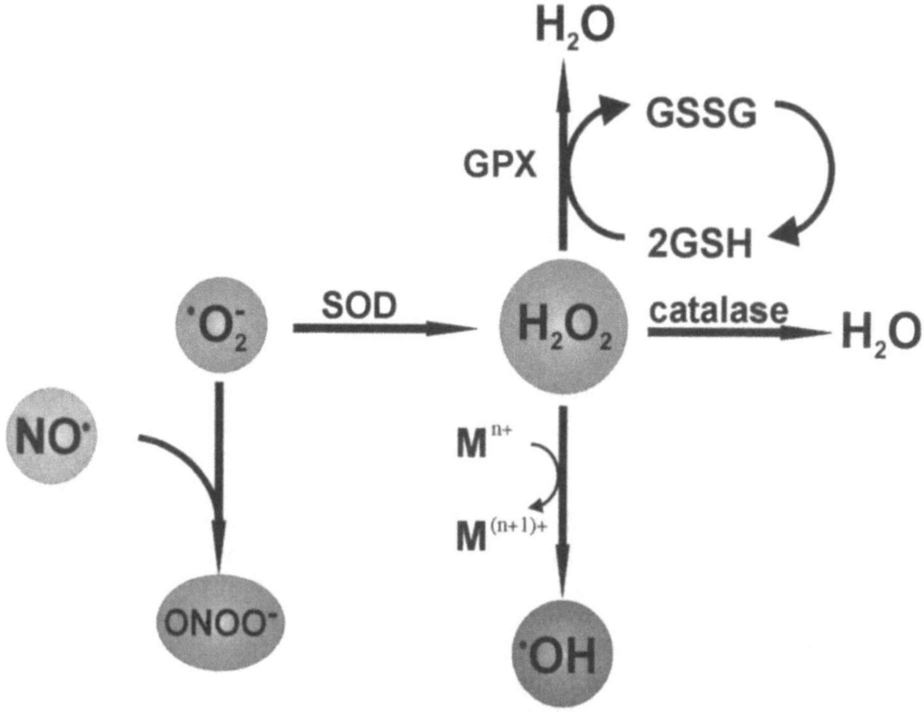

Figure 2. *Formation and metabolism of reactive oxygen species. Superoxide ($^{\cdot}O_2^-$) that is produced during normal cellular conditions mainly by the mitochondrial respiratory chain is dismutated to form hydrogen peroxide (H_2O_2) by superoxide dismutase (SOD). H_2O_2 is subsequently degraded to H_2O by antioxidants such as catalase and glutathione peroxidases. H_2O_2 also can be converted to the highly reactive hydroxyl radical ($^{\cdot}OH$) through transitional metal (M^{n+}) catalyzed reactions. Nitric oxide ($NO^{\cdot}$) can react rapidly with $^{\cdot}O_2^-$ to form highly reactive peroxynitrite ($ONOO^-$).*

organisms or other pathogens invading our bodies, they start to generate large amounts of superoxide. The chemical basis for respiratory burst is the activation of the plasma membrane-bound NADPH oxidase from neutrophils [15], which catalyzes the monoelectronic reduction of oxygen to the superoxide anion:

$$2O_2 + NADPH \rightarrow 2^{\cdot}O_2^- + NADP^+ + H^+$$

Phagocytes are another potent source of oxidant production and produce H_2O_2, $HO^{\cdot}$, $NO^{\cdot}$, and hypochlorous acid (HOCl), besides $^{\cdot}O_2^-$. HOCl [10, 16, 17] is an inflammatory mediator and a strongly oxidizing and chlorinating compound that can generate other reactive metabolites, such as nitryl chloride (NO_2Cl) and nitrogen dioxide ($NO_2^{\cdot}$) in the presence of nitrite [18]. The potentially deleterious effects of reactive oxygen, reactive nitrogen, and chlorinating species can affect the aging process.

Additionally, many environmental stimuli including cytokines, ultraviolet light, nuclear radiation, chemotherapeutic agents, hyperthermia and even growth factors generate high levels of ROS that can perturb the normal redox balance and shift cells into a state of oxidative stress. During infection and chronic inflammation, massive amounts of nitric oxide and superoxide radicals form in immune cells to fight off invading bacteria and viruses. Made in excess, these oxidants can harm and combine to form another toxic chemical that produces further damage in DNA and other cellular components.

Antioxidants

When reactive oxygen species (ROS) are generated in living systems, a wide variety of antioxidants come into play. Under normal conditions the damaging actions of ROS are minimized and controlled to a certain level by abundant protective and repair mechanisms that cells possess, including many enzymes, and redox active molecules. The relative importance and efficacy of these various antioxidants in the body depends on which of the ROS is involved, how it is generated, where it is generated, and which target of damage is selected. Thus, an antioxidant may protect against free radicals in one system but fails to protect in other systems. In certain circumstances, an antioxidant may not only have a protective action but may even aggravate the situation. The chemistry *in vivo* is highly complex, and antioxidants under unusual circumstances can become pro-oxidants. This is possible in conditions where iron or copper are not in their normal non-catalytic state, as occurs in some diseases (e.g., hemachromatosis, and iron overload) or following trauma.

Antioxidants, both enzymatic and non-enzymatic, limit oxidative damage to biological molecules by various mechanisms. Antioxidant activity may be accomplished by three different mechanisms: (1) by inhibiting the generation of ROS; (2) by directly scavenging free radicals by means of anti-radical scavenging enzymes such as SOD, catalase, and GSH-PX; and (3) by raising the endogenous antioxidant defenses, that is, up-regulated expression of the genes that encode the enzymes SOD, catalase, or GSH-PX.

Enzymatic antioxidants

Superoxide dismutase (SOD)

The dismutation of superoxide ($^{\cdot}O_2^-$) by SOD [19] to hydrogen peroxide (H_2O_2) (see Figure 2) is generally considered to be the primarily antioxidant defense of the body because this enzyme prevents the further generation of free radicals. SOD exists in virtually every oxygen-respiring organism and its major function is to catalyze the dismutative reaction.

$$\underset{\text{superoxide anion}}{2\,O_2^-} + 2H^+ \rightarrow \underset{\text{hydrogen peroxide}}{H_2O_2} + O_2$$

The superoxide dismutases are divided into three distinct classes depending on the metal ion content, namely, Cu/ZnSOD, MnSOD, and FeSOD. Some of the SOD activity is extracellular, but the bulk of the activity is intracellular, where it is divided between the mitochondria (MnSOD) and the cytosolic compartments (Cu/ZnSOD). The SOD level varies according to the part of the body. The highest levels of SOD are found in the liver, adrenal gland, kidney, and spleen.

Catalase

Catalase is one of the major primary antioxidant defense mechanisms. This enzyme is primarily found in the peroxisomes, but is also located, less abundantly, in mitochondria. It works primarily to catalyze the decomposition of hydrogen peroxide to oxygen and water, a function shared with glutathione peroxidase. Both enzymes detoxify oxygen-reactive radicals by catalyzing the formation of hydrogen peroxide derived from superoxide (see Figure 2). The liver, kidney and red blood cells possess high levels of catalase.

$$\underset{\text{hydrogen peroxide}}{2H_2O_2} \rightarrow 2H_2O + O_2$$

Glutathione peroxidases

The enzyme glutathione peroxidase catalyzes the reduction of hydrogen peroxide and organic hydroperoxides in reactions requiring glutathione (see Figure 2). Both selenium dependent and selenium independent glutathione peroxidases protect against free radical damage by reducing peroxides. Glutathione peroxidases are intracellular, located in the mitochondria and the cytosol of the cell. Glutathione peroxidases play an important role in the prevention of lipid peroxidation to maintain the structure and function of biological membranes.

$$\underset{\text{lipid peroxide}}{ROOH} + \underset{\text{reduced glutathione}}{2GSH} \rightarrow \underset{\text{oxidized glutathione}}{GSSG} + \underset{\text{lipid}}{ROH} + H_2O$$

$$\underset{\text{hydrogen peroxide}}{H_2O_2} + \underset{\text{reduced glutathione}}{2GSH} \rightarrow \underset{\text{oxidized glutathione}}{GSSG} + 2H_2O$$

Non-enzymatic antioxidants

In addition to the antioxidant enzymes, there are several small-molecule antioxidants that play important roles in antioxidant defense systems. These small-molecule antioxidants are particularly important in blood and the fluids present in the extracellular space, where antioxidant enzymes are absent or present only in small quantities. The small-molecule antioxidants include lipid-soluble and water-soluble components. The lipid-soluble antioxidants are localized to cellular membranes and lipoproteins, whereas the water-soluble antioxidants are present in aqueous fluids, such as blood and the fluids within cells and surrounding them.

Vitamin E

The generic term "vitamin E" refers to at least eight structural isomers of tocopherol, among which is alpha-tocopherol, the best known isomer and the one processing the most potent antioxidant activity. Vitamin E is the most widely distributed antioxidant in nature and is found in both the plant and animal kingdoms. As the major lipid soluble antioxidant present in all cellular membranes, vitamin E protects against lipid peroxidation. Vitamin E acts directly with a variety of oxy radicals, superoxide radical, and singlet oxygen. Vitamin E can protect against some of the symptoms of selenium deficiency and vice versa. Vitamin E is the major free radical chain terminator in the lipophilic environment. The lipid solubility of tocopherol may explain why high levels are found in adrenal gland, heart, gonadal, and liver tissues. By reacting with lipid peroxyl and alkoxyl radicals, vitamin E is able to effectively protect against membrane lipid peroxidation. The sparing and synergistic actions appear to be the result of the ability of both tocopherol and selenium-dependent glutathione to decrease the production of lipid peroxidation products [20, 21].

Carotenoids

Other lipid-soluble antioxidants are beta-carotene (a vitamin A precursor) and related substances called carotenoids, such as alpha-carotene, lycopene (the red colour in tomatoes), lutein, and zeaxanthine. Beta-carotene is considered to be the most efficient quencher of singlet oxygen. Although beta-carotene is the major carotenoid precursor of vitamin A, vitamin A does not quench singlet oxygen and is said to have a very small capability of scavenging free radicals. Carotenoids protect lipids against peroxidation by quenching free radicals and other reactive species. Beta-carotene traps free radicals through its inhibition of lipid peroxidation induced by the xanthine oxidase system.

Beta-carotene, like vitamin C, seems to function as both an antioxidant and a prooxidant. Under low O_2 partial pressures, beta-carotene exhibits excellent radical-scavenging activity. At higher oxygen partial pressures, its capacity to trap free radicals shows autocatalytic prooxidant effects, with concomitant loss of its antioxidant activity [22, 23].

Coenzyme Q-10

Coenzyme Q-10 is a lipid-soluble powerful reducing agent that prevents lipid peroxidation. It resides in the inner membrane of mitochondria and plays an important role in the respiratory chain.

Vitamin C (ascorbic acid)

In contrast to vitamin E, vitamin C is water soluble and functions more effectively in an aqueous environment. Vitamin C and E work synergistically in quenching free radicals and singlet oxygen. Vitamin C also regenerates the reduced antioxidant form of vitamin E; one of its major functions may well be to recycle oxidized vitamin E. As a reducing and antioxidant agent, ascorbic acid reacts directly with superoxide and hydroxyl free radicals and various lipid hydroperoxides. Vitamin C, when compared with other water-soluble antioxidants, provides the most effective protection against plasma lipid peroxidation. Ascorbic acid acts as both an antioxidant and a prooxidant. As an antioxidant, vitamin C has a sparing effect on the antioxidant actions of vitamin E and selenium. Excessive amounts of vitamin C may act as a prooxidant in the presence of the transition metals iron (Fe^{2+}) and copper (Cu^{2+}) by generating cofactors of activated oxygen radicals during the promotion of lipid peroxidation. For further information on vitamin C [24, 25].

Glutathione (GSH)

The tripeptide-reduced glutathione (γ-L-glutamyl-L-cysteinyl-glycine) is the most abundant low-molecular-weight thiol present in almost all mammalian cellular systems. Reduced glutathione is characterized by its reactive thiol group and its γ-glutamyl bond, which makes it resistant to peptidase attack. The versatility of glutathione is due to its chemical properties that allow it to serve as both a nucleophile and an effective reductant by interacting with numerous electrophilic and oxidizing compounds such as hydrogen peroxide, superoxide, and hydroxyl free radicals. Glutathione plays an effective reductant role in a variety of detoxification processes. Glutathione readily interacts with free radicals, especially hydroxyl and carbon radicals, by donating a hydrogen atom. Reactions of this type provide protection by neutralizing reactive hydroxyl radicals which are a major source of free radical pathology, including cancer. Glutathione affects radiation sensitivity by its ability to scavenge radiation-induced free radicals as well as to restore radiation-damaged molecules by hydrogen donation. Additionally, GSH reduces disulfide linkage of proteins and of other molecules and thus maintains the reduced state of glycolytic and antioxidant enzymes.

Uric acid

Studies have demonstrated that uric acid is a potent physiological antioxidant, playing a major role in both extracelluar and intracellular defense mechanisms. Although the precise biochemical mechanism is not completely determined, uric acid appears to have a sparing action for plasma ascorbate, probably by complexing transition metals such as iron and copper. Furthermore, uric acid may serve as an important protective agent of the overall defense systems, thus helping to increase the life span of the individual [22].

References

1. Harman D (1956). Aging: a theory based on free radical and radiation chemistry. *J Gerontol.* 2: 298–300.
2. Harman D (1972). The biological clock: the mitochondria? *J Am Geriatr Soc.* 20: 145–7.
3. Halliwell B, Gutteridge JMC (1999). *Free Radicals in Biology and Medicine*, 3rd edn. Oxford: Oxford University Press.
4. Sohal RS, Brunk UT (1992). Mitochondrial production of pro-oxidants and cellular senescence. *Mutat Res.* 275: 295–304.
5. Sohal RS, Orr WC (1992). Relationship between antioxidants, prooxidants, and aging process. *Ann NY Acad Sci.* 663: 74–84.

6. Ames BN, Shigenaga MK, Hagen TM (1995). Mitochondrial decay in aging. *Biochim Biophys Acta* 1271: 165–70.
7. Sohal RS, Sohal BH, Orr WC (1995). Mitochondrial superoxide and hydrogen peroxide generation, protein oxidative damage, and longevity in different species of flies. *Free Rad Biol Med.* 19: 499–504.
8. Garland D (1990). Role of site-specific, metal-catalyzed oxidation in lens aging and cataract: a hypothesis. *Exp Eye Res.* 50: 677–82.
9. Rikans LE, Cai Y (1993). Diquat-induced oxidative damage in BCNU-pretreated hepatocytes of mature and old rats. *Toxicol Appl Pharmacol.* 118: 263–70.
10. Chance B, Sies H, Boveris A (1979). Hydroperoxide metabolism in mammalian organs. *Physiol Rev.* 59: 527–605.
11. Folkes LK, Candeias LP, Wardman, P (1995). Kinetics and mechanisms of hypochlorous acid reactions. *Arch Biochem Biophys.* 323: 120–6.
12. Giulivi C, Poderoso JJ, Boveris A (1998). Production of nitric oxide by mitochondria. *J Biol Chem.* 273: 11038–43.
13. Bates TE, Loesch A, Burnstock G, Clark JB (1995). Immunocytochemical evidence for a mitochondrially located nitric oxide synthase in brain and liver. *Biochem Biophys Res Comm.* 213: 896–900.
14. Bates TE, Loesch A, Burnstock G, Clark JB (1996). Mitochondrial nitric oxide synthase: a ubiquitous regulator of oxidative phosphorylation? *Biochem Biophys Res Comm.* 218: 40–4.
15. Laporte F, Doussiere J, Vignais PV (1990). Respiratory burst of rabbit peritoneal neutrophils. Transition from an NADPH diaphorase activity to an $^{\cdot}O_2(-)$-generating oxidase activity. *Eur J Biochem.* 194: 301–8.
16. Klebanoff SJ (1980). Oxygen metabolism and the toxic properties of phagocytes. *Ann Int Med.* 93: 480–9.
17. Hurst JK, Barrette WC Jr (1989). Leukocytic oxygen activation and microbicidal oxidative toxins. *Crit Rev Biochem Molec Biol.* 24: 271–328.
18. Eiserich JP, Hristova M, Cross CE, *et al.* (1998). Formation of nitric oxide-derived inflammatory oxidants by myeloperoxidase in neutrophils. *Nature* 391: 393–7.
19. Fridovich I (1995). Superoxide radical and superoxide dismutases. *Annu Rev Biochem.* 1995; 64: 97–112.
20. Machlin LJ and Bendich A (1987) Free radical tissue damage: protective role of antioxidant nutrients. *FASEB J.* 1: 441–5.
21. Machlin LJ, Gabriel E (1980). Interactions of vitamin E with vitamin C, vitamin B12, and zinc. *Ann NY Acad Sci.* 355: 98–108.
22. Ames BN (1983). Dietary carcinogens and anticarcinogens. Oxygen radicals and degenerative diseases. *Science* 221: 1256–64.
23. Yu MW, Zhang YJ, Blaner WS, Santella RM (1994). Influence of vitamin A, C, and E and beta-carotene on aflatoxin B1 binding to DNA in woodchuck hepatocytes. *Cancer* 73: 596–604.
24. Pauling L (1970). Evolution and the need for ascorbic acid. *Proc Natl Acad Sci USA* 67: 1643–8.
25. Benzie IF (2000). Evolution of antioxidant defence mechanisms. *Eur J Nutr.* 39: 53–61.

Oxidative DNA Damage and Repair – Implications for Aging

Erling Seeberg
Centre of Molecular Biology and Neuroscience, and Institute of Medical Microbiology, University of Oslo, The National Hospital, NO-0027, Oslo, Norway

Introduction

Reactions of DNA with oxygen radical species represent an important cause of cellular deterioration and have long been considered a major component of aging in various biological systems. All aerobic organisms are faced with a load of unwanted side reactions caused by the incomplete reduction of oxygen, for example during electron transport and oxidative phosphorylation. Major oxygen radical species are formed such as superoxide and hydroxyl radicals as well as hydrogen peroxide (Figure 1). Among these, hydroxyl radicals are clearly the most reactive component. However, its extremely short half-life makes it unlikely that it can diffuse from its major origin of production in the mitochondria into the nucleus and induce damage to the chromosomal DNA. However, hydrogen peroxide may do so and will be converted to hydroxyl radicals by the Fenton reaction. This requires Fe^{2+} that is normally tightly associated with DNA thus leading to the formation of hydroxyl radicals if peroxide also is present. To prevent this cascade of events two important scavenger mechanisms have evolved; superoxide dismutase that converts superoxide to peroxide and catalase that hydrolyzes peroxide to water. In addition, glutathione inside cells will also function as a scavenger of radicals and will be recycled through the action of glutathione peroxide (see Chapter 1, this volume). Nevertheless, these proteins and scavenger molecules cannot completely remove all ROS and oxidative DNA damage will be induced, to the DNA of the nucleus but more extensively to the mitochondrial DNA. It has been estimated that more than 10 000 DNA lesions per cell per day will be generated by such reactions.

ROS induce a wide variety of structurally different lesions in DNA and more than 50 different types of structural changes have been described, which affect either the base or the sugar residues of the DNA [1, 2]. The purpose of this review is to consider some of the most important types of such lesions, their biological consequences and

T. von Zglinicki (ed.), Aging at the Molecular Level, 11–25.

the repair systems involved in the removal of such lesions. Roughly described, the lesions can be divided into five distinct groups depending on their origin and/or biological consequence for the cell. These comprise oxidized purines, oxidized pyrimidines, abasic sites, single-strand breaks and double-strand breaks. Attack on the sugar moiety will in most cases result in a strand break with ends that need to be tailored by nucleases and gaps that must be filled by polymerases in order for rejoining by ligase to occur. Attack on the base residues results in a variety of modifications that for the pyrimidines often involve reactions at 5,6 double bond and for the purines modifications of the immidazole rings. In addition, the amino groups of the nucleobases are generally unstable and a certain amount of deamination will also be associated with oxygen radical attack. Deamination also occurs spontaneously in DNA through hydrolysis and the reader is referred to recent reviews for a comprehensive description of the repair of deaminated bases such as uracil and hypoxanthine [3]. In the following, repair mechanisms for the different subgroups of damage referred to above will be described separately.

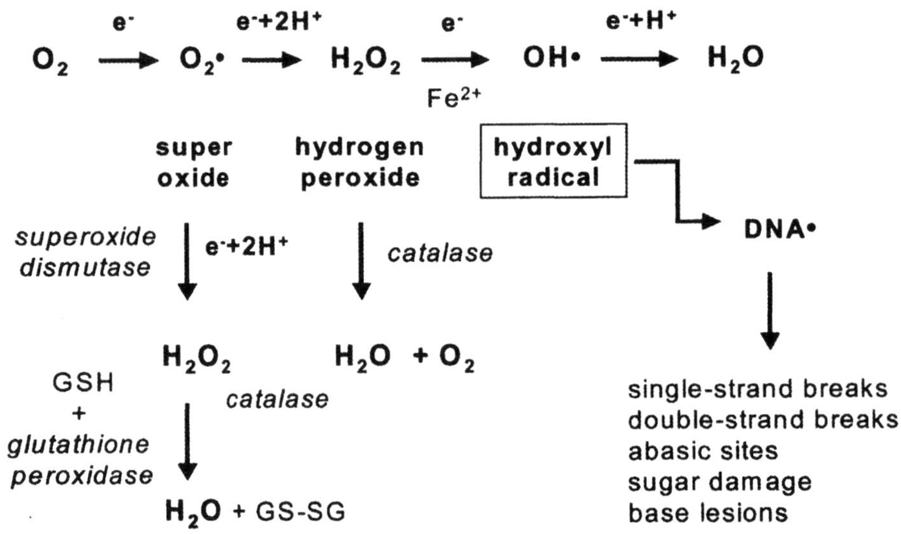

Figure 1. *Steps in the formation of reactive oxygen species and scavenger reactions operating to eliminate and prevent these molecules from reacting with cellular macromolecules.*

Base excision repair

The most important mechanism involved in the repair of oxidative DNA damage is base excision repair (BER). BER is universally present in all organisms from bacteria to man and has probably evolved to handle particularly frequent types of spontaneous DNA damage [1, 4]. BER is initiated by a DNA glycosylase that recognizes the modified base, which is then removed in a free form by enzymatic cleavage of the N-glycosylic bond (Figure 2). An abasic site is formed (AP-site, apurinic or apyrimidinic site), which is acted upon by two types of enzymes; an AP lyase that cleaves the DNA strand 3′ to the deoxyribosephosphate (dRP) residue or by an AP endonuclease hydrolysing the phosphodiester bond 5′ to dRP.

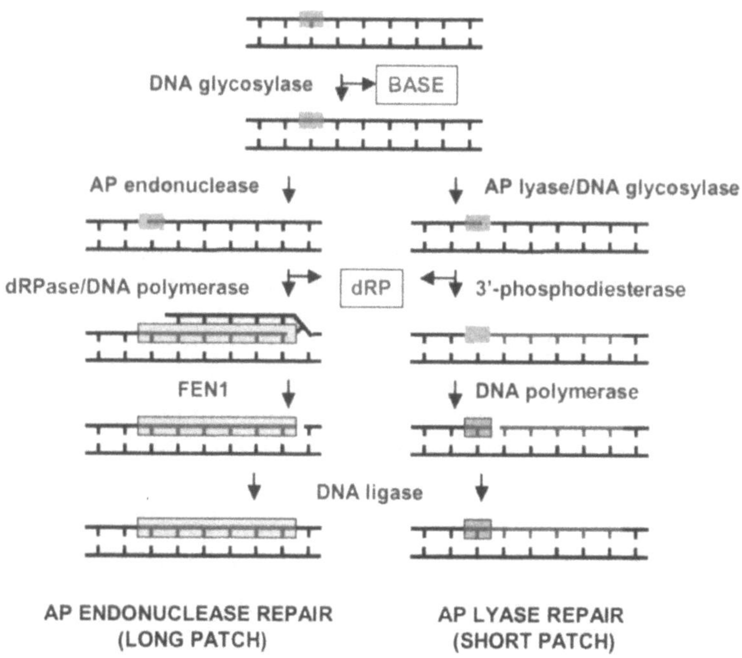

***Figure** 2. Base excision repair of DNA damage. The model implies that single nucleotide replacement occurs predominantly after lyase mediated strand cleavage and long patch repair after an initial AP endonuclease cleavage.*

The choice between these two alternative steps will be discussed further below. However, many DNA glycosylases are equipped with an intrinsic AP-lyase activity and are therefore termed bifunctional. Once the break is formed, the remaining deoxyribosephosphate residue will be removed either by a 3′–5′ phosphodiesterase or by a 5′–3′ dRPase, respectively, depending on the site of the initial strand cleavage. A one-nucleotide gap is formed, which may be further extended by the action of Flap Endonuclease I (FEN1) producing a longer gap [5]. A DNA polymerase fills the resulting gap and the strand is finally joined by DNA ligase. *In vitro* the entire repair reaction can be reconstituted with four different purified proteins from human cells; a DNA glycosylase (e.g., hNTH1), AP-endonuclease I (HAP1/APE), DNA polymerase β (Polβ) and DNA ligase I or III. However, *in vivo* several other accessory factors are also involved, e.g., XRCC1, which are probably required to execute the repair in coordinated way without detrimental side reactions such as extensive DNA degradation potentially originating from the site of strand breaks.

Removal of oxidized purines

Oxidative agents induce two major types of purine lesions; 8-oxopurines and fomamidopyrimidines (Figure 3). For several reasons, 8-oxoguanine (8-oxoG) has been subject to extensive investigations. First, it is one of the most abundant lesions induced by ROS and therefore can be used as a marker for exposure. Second, it is easy to quantify since HPLC followed by electrochemical detection can be applied with high sensitivity. However, the yield of 8-oxoG in DNA has been grossly overestimated due to artefactual introduction of 8-oxoG during DNA isolation. Early reports have indicated a steady state level of up to three per 10^4 guanines whereas the real levels are probably below one per 10^6 [6]. Conclusions made on the effects of oxidations in earlier studies should therefore be cited with caution. Third, 8-oxoG is a highly premutagenic lesion that introduces GC to TA transversions with high frequencies due to its ability of forming stable base pairs with adenine. Forth, an intricate repair system for preventing mutations occurring from 8-oxoG lesions was discovered in bacteria and was shown to be present also in mammalian cells. In human cells, this involves a DNA glycosylase, OGG1, for removal of 8-oxoG across

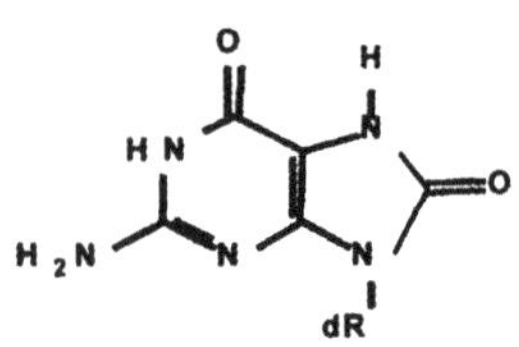

8-oxo-7,8-dihydroguanine
(8-oxoG)

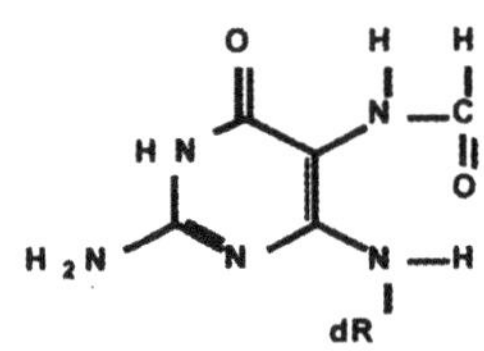

2,6-diamino-4-hydroxy
5-formamidopyrimidine
(faPyG)

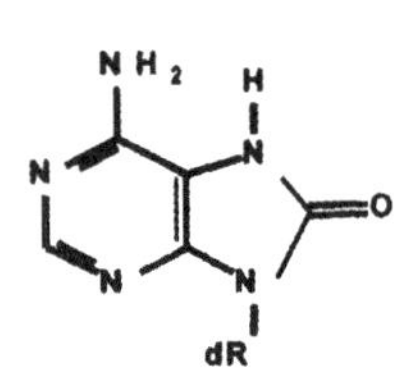

8-oxo-7,8-dihydroguanine
(8-oxoA)

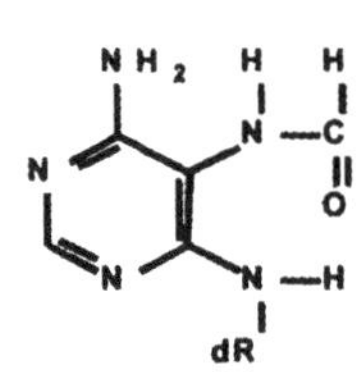

4,6-diamino-
5-formamidopyrimidine
(faPyA)

Figure 3. *Frequent oxidative modifications of purines.*

C in a standard BER pathway (Figure 4; ref. 7; see Table 1 for a list of human genes involved in the repair of oxidative DNA damage). In addition, another glycosylase, MYH, will remove adenine that is misincorporated across 8-oxoG during replication [8]. This re-establishes the 8-oxoG:C base pair which is then subject to further repair by a DNA glycosylase different from the one involved in prereplicative repair [9]. We do not know yet what enzyme this is but a likely candidate is a member of the newly discovered family of Fpg like DNA glycosylases in mammalian cells denoted NEIL1 [10]. Steps succeeding the glycosylase action are thought to be common for the three BER reactions. However, the role of accessory factors may well be different and the MYH removal of A appears to be associated with DNA replication and also tightly coordinated with the subsequent removal of 8-oxoG. A source of oxidative DNA damage not originating from the direct reaction of DNA with ROS is incorporation of oxidized bases from the nucleotide pool. In the case of 8-oxoG this is prevented by the human MTH1 function that specifically hydrolyzes 8-oxoGTP to 8-oxoGMP thus minimizing the availability of oxidized nucleotides and the formation of A:T to C:G transversions [11].

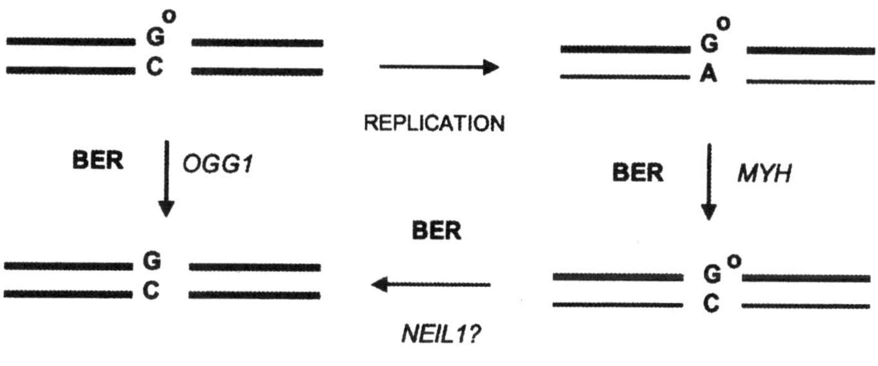

Figure 4. Repair reactions to prevent mutations from occurring at the site of 8-oxoG in DNA.

Removal of oxidized pyrimidines

The major human DNA glycosylase involved in the removal of oxidized pyrimidines is NTHL1, which is highly conserved from bacteria to man. Typical lesions handled by NTHL1 are thymine glycols, 5,6-dihydrouracil, 5-hydroxycytosines, urea and 5-formyluracil (see Figure 5). Some of these, e.g., thymine glycol and urea, pose a strong block to replication and are strongly cytotoxic lesions. Others, such as dihydrocytosine and formyluracil, are both potent premutagenic lesions. Human NTHL1 belongs to the Helix-hairpin helix family of DNA glycosylases that was recognized on the basis of sequence similarity to similar genes in bacteria (nth) and yeast (NTG1, NTG2) [12, 13]. Like OGG1, NTHL1 is a bifunctional DNA glycosylase with an associated AP lyase activity, implying that strand breaks will be

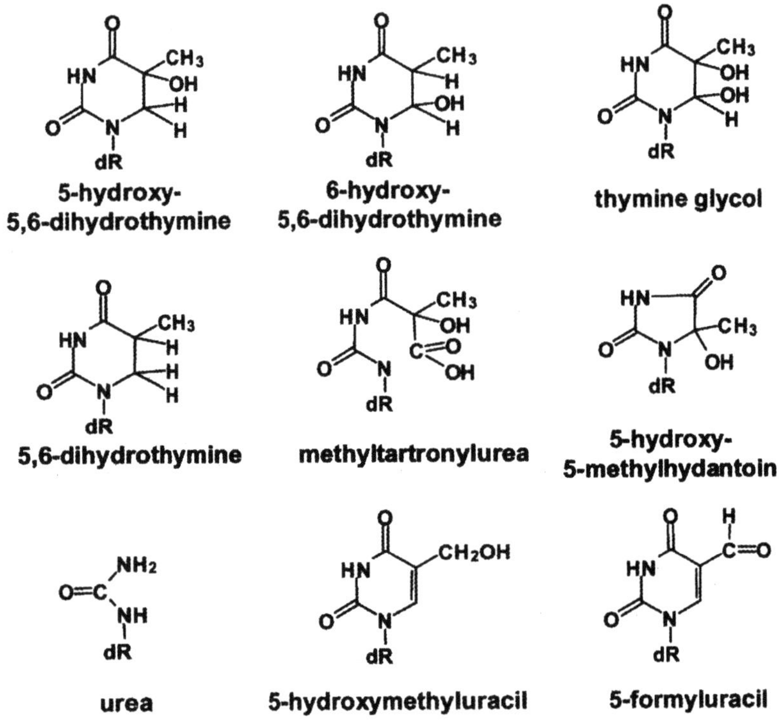

Figure 5. *Oxidative DNA damage at thymine residues.*

formed at the site of the oxidized nucleotide with NTHL1 alone. However, whereas the AP lyase activity of OGG1 is weak, has a strong preference for C in the opposite strand and is uncoupled from the base removal event, the lyase of NTH1 is much stronger and appears largely associated with the glycosylase reaction.

Repair of abasic sites

Abasic sites are probably the most frequent type of DNA damage formed. It has been calculated that purine bases will be lost spontaneously by hydrolysis at a rate that approaches 10 000 per cell per day [1]. Additionally, abasic sites are induced by many different DNA damaging agents including ROS and by DNA glycosylase mediated release of base lesions. In many cases, abasic sites represent more severe damage than modified base residues and base excision is therefore often going from bad to worse. However, despite that abasic sites can assume different chemical forms (e.g., natural, oxidized and reduced; see ref. 14) it nevertheless unifies the damage into one particular class towards which many repair pathways have evolved. AP endonucleases are probably the most important enzymes for this purpose and the repair follows the path outlined above for BER. Human cells contain one major (APE/

HAP1) and one minor AP endonuclease (APE2). Null mutant mice of APE are embryonic lethal indicating the essential nature of this gene function. Furthermore, heterozygous APE mice are hypersensitive to increased oxidative stress [15]. APE2 is sequence related to APE1 and Exonuclease III in bacteria, but relative to these is mutated at specific residues, which makes it a very poor endonuclease [16]. It is presently unclear if this protein has any role in the repair of abasic sites *in vivo*. However, a similar function exists in yeast and mutant analysis of *S. cerevisiae* indicates a specific role in the repair of alkylation damage as a back up function when/if the primary AP endonuclease is mutated [17, 18]. The situation could be similar in mammalian cells, however, the presence of APE2 is apparently not sufficient to alleviate embryonic lethality of APE1.

A second important function in the handling of abasic sites is provided by the DNA glycosylases themselves. The glycosylases will contribute to the repair of abasic sites either through their associated AP lyase activity or by tight binding to the AP site as such. Such binding is important to prevent the chemically reactive AP sites from forming covalent links to proteins or other cellular molecules. Most of the DNA glycosylases have a very low turnover number due to high affinity for the product of their own reaction. This is probably a characteristic part of the coordinated action of the BER pathway and it has been shown that the human AP endonuclease stimulates the OGG1 mediated removal of 8-oxoG by greatly enhancing its turnover number.

Accessory factors required for accurate BER in vivo

Cell biology studies and mutant analysis have shown that factors other than those described above are important for BER *in vivo*. In particular, the role of XRCC1 in BER was established years ago without finding any enzymatic or cofactor role of this protein. However, interaction studies have shown that XRCC1 binds several other proteins with a known function in BER and therefore might serve as a scaffold protein. XRCC1 has been shown to interact with ADPRT (Poly (ADP-ribose) polymerase, PARP; [19]), DNA ligase III [20] and DNA polymerase β [21], thus linking the function of Polβ and LIGIII together. It also suggests an important role of ADPRT in BER and this has been confirmed by studies monitoring BER in cell free extracts. ADPRT has long been considered a sensor of nicks in DNA and is clearly involved in the repair of DNA single-strand breaks. When ADPRT is activated a gross consumption of NAD takes place and energy deprivation will occur. After ischemia and exposure to high levels of ROS this reaction plays an important role in the induction of cell death and ADPRT inhibitors have a significant therapeutic role after stroke conditions [22].

BER enzymes also associate with components normally involved in DNA replication. Co-localization studies have indicated an interaction between the DNA glycosylases MYH [23] and NEIL3 [10], and the Proliferating cell nuclear antigen (PCNA) and replication factor RPA. This suggests a link between BER of oxidative DNA damage and DNA replication as has been previously shown for the repair of uracil in DNA. This further indicates that A misincorporated across 8-oxoG in the DNA template is removed by MYH in a process associated with replication fork progression.

Repair of strand breaks in DNA

ROS also induce strand breaks in DNA mostly through attack on the sugar residues. To some extent, ROS produce the same pattern of DNA damage as ionizing radiation. More than half of the radiation effects are indirect and caused by hydroxyl radicals induced by radiolysis of water. From previous radiobiological studies much is known about the repair of both single- and double-strand breaks in DNA. Single-strand break repair is an efficient and rapid process that is completed within 1–2 h after a challenging dose of X-rays in mammalian cells. Repair of single-strand breaks induced by ROS are handled in a similar way and is often considered part of the BER pathway. This is not entirely correct as strand breaks represent a very different signal in the cells than base damage. Strand breaks are only transiently formed during repair of base damage. Nevertheless, the proteins involved are overlapping with a requirement for XRCC1, POLB and LIG3. ADPRT appears to be more important for the repair of single-strand breaks than for BER, which is consistent with the nick sensing property of this protein. A protein complex has been isolated that constitutes POLB, XRCC1, ADPRT, LIGIII and Polynucleotide kinase (PNK), which appears essential for the single-strand break repair process [24]. PNK has so far not been implicated in BER, however, it appears to be required after breaks induced by ROS or X-rays.

Double-strand breaks are produced at a frequency of about one tenth of that of single-strand breaks after ionizing radiation. This is partly due to the penetrating effects of high energetic direct radiation and the frequency will be less in the case of ROS. However, damage caused by ROS also tends to cluster and multiple lesions at one site affecting both strands will occasionally also lead to the formation of double-strand breaks. Repair of double-strand breaks proceeds either by homologous recombination involving RAD50/51/52/54 and other proteins involved in DNA recombination, or by nonhomologous end rejoining involving DNA-dependent protein kinase, KU70, KU80, LIG4 and other accessory proteins. It is beyond the scope of this review to discuss these important repair pathways in further detail. However, it should be emphasized that recombinational repair clearly is very important in the repair of oxidative DNA damage as judged from genetic evidence in yeast and partly also in mammalian systems.

Repair of oxidative DNA damage in mitochondrial DNA

The mitochondrial DNA is highly prone to oxidative DNA damage because of its close proximity to the mitochondrial inner membrane and the electron transport machinery. It has been estimated that as much as 5% of the oxygen consumed by the electron transport chain may by converted to ROS by incomplete oxygen reduction [25]. The mitochondrial genome does not carry any DNA repair genes by itself and repair of mitochondrial DNA depends on the transport into the mitochondria of repair proteins synthesized in the cytoplasm. However, surprisingly many of the BER functions are also found in the mitochondria and the mitochondrial forms are generally produced from alternative splicing of the nuclear genes yielding proteins containing mitochondrial localization signals. For example, the *OGG1* gene has an additional exon 8 that is located far downstream of the other coding parts [18]. The mitochondrial form harbours sequences derived from exon 8 including a mitochondrial localization signal instead of a sequence from exon 7, which encodes the nuclear signals. More often, the mitochondrial form results from an alternative exon at the N-terminus of the protein that dominates the nuclear localization signal that is often residing towards the C- terminal part. This is the case for NTHL1. BER proteins shown to be present in the mitochondria include OGG1, NTHL1, MYH, LIGIII, MPG, APE, APE2 and POLG. The latter is the only DNA polymerase shown to be present in the mitochondria and can serve the same functions as POLB in the nucleus having both dRPase and gap filling activities [26]. The presence of most BER proteins in the mitochondria clearly emphasizes the importance of the BER pathway for the repair of oxidative DNA damage. Furthermore, there is little evidence so far for a role of other repair pathways in the repair of mitochondrial DNA but further work is required before this can be firmly concluded.

Other pathways involved in the repair of oxidative DNA damage

Nucleotide excision repair (NER) is normally active only in the repair of bulky DNA lesions and not in the removal of smaller oxidative lesions, which do not cause major distortions of the helix structure. Nevertheless, both genetic and biochemical evidence have indicated that NER also may be active against abasic sites in DNA, at least in yeast [27, 28]. In addition, certain types of oxidative lesions will be causing major structural distortions such as 5′,8-cyclo-2-deoxyadenosine in which the adenine is attached to the sugar residue at two different positions. This lesion cannot be removed by the BER pathway and appears to be a preferred substrate for the NER machinery [29]. Accumulation of this lesion, for example in the brain tissue, might account for neurological disorders that are associated with the NER defect in patients with Xeroderma pigmentosum (XP). XP patients are extremely sensitive to UV-light exposure due to the defect in the repair of UV-lesions such as pyrimidine dimers, which cause severe forms of malignant melanoma at an early age. However, a high incidence of internal tumors cannot be ascribed to a defect in the repair of UV damage and the accumulation of certain type of oxidative lesions may be important for other types of diseases in the XP patients.

It was early recognized that NER is more active in transcribed genes than in the genome overall [30]. The standard NER machinery requires XPA through XPG, ERCC1, PCNA, Replication factor C, Polymerase ε and LIG1 [31]. In addition, transcription coupled repair of UV damage requires the gene products deficient in patients with Cockayne syndrome, CSA and CSB, and possibly also proteins involved in mismatch repair. Together with seven other proteins, both XPB and XPD are subunits of the general transcription initiation factor TFIIH, which is essential for transcription as well as for transcription-coupled repair. More recent studies have indicated that transcription-coupled repair also is active in the removal of oxidative DNA damage including thymine glycols and 8-oxoG [32]. This process has so far been shown to require CSB, XPG and TFIIH, but it is presently unclear if all the

Table 1. *Human genes known to be involved in the repair of oxidative DNA damage. The list is not exhaustive and does not include genes involved in recombination and nucleotide excision repair, which are also handling certain types of oxidative DNA damage*

Gene	Protein function	Size (kD)	Mouse mutant	Chromosomal map position
OGG1	DNA glycosylase 8-oxoG, faPy	38	viable	3p26
NTHL1	DNA glycosylase ThyGlyc, 5-ohC	35	viable	16p13
MYH	DNA glycosylase A opposite 8-oxoG	61	viable	1p33
NEIL1	DNA glycosylase 5-ohC, ThyGlyc, 8-oxoG, faPy	44	?	15q22
NEIL2	DNA glycosylase Cytosine damage	35	?	8p23
NEIL3	DNA glycosylase FaPy	68	?	4q34
MPG	DNA glycosylase ethenoA, ethenoG	33	viable	16p13
TDG	DNA glycosylase ethenoC	46	?	12q24
SMUG1	DNA glycosylase 5-oh*me*U	31	?	12q13
APE1	AP endonuclease	35	embryonic lethal	14q11
APE2	AP endonuclease?	59	?	Xq11
FEN1	Flap endonuclease	43	embryonic lethal	11q12
POLB	DNA polymerase β	38	embryonic lethal	8p11
POLG	DNA polymerase γ	140		15q25
LIG1	Single-strand break ligase	102	mid-term embryonic viability	19q13
LIG3	BER ligase	103	embryonic lethal	17q11
XRCC1	Scaffold protein	70	embryonic lethal	19q13
ADPRT	Poly(ADP-ribose) polymerase	113	viable	1q41
PCNA	Replication protein	29		20p13

other NER proteins are also required for transcription-coupled repair of oxidative DNA damage. The mechanism behind transcription-coupled repair remains obscure but is thought to involve recognition of the damage by RNA polymerase stalling at the lesion followed by polymerase release to allow repair from an open protein/DNA complex.

Evidence for the role of oxidative DNA repair in the prevention of aging

DNA repair mechanisms are probably the most important cellular protection against the development of cancer and are also essential for living cells to survive in an environment that would otherwise cause unacceptable mutation frequencies. DNA repair disorders are often associated with elevated cancer frequencies and heterozygosity of defects in mismatch repair genes are known to be autosomal dominant in promoting nonpolyposis colon cancer in humans. The role of DNA repair in the prevention of aging is partly supported by repair genes being mutated in patients with premature aging syndromes such as Werner syndrome (WRN, RecQ helicase) and Cockayne syndrome [see Chapter 5, this volume]. Perhaps the best recent evidence for the role of oxidative DNA repair in the prevention of aging comes from the study of mouse models with targeted defects in XPD and XPA. Both double mutants and single mutant XPD mice show many signs of accelerated aging and high sensitivity to oxidative DNA damage [33]. The particular XPD mutation studied in this case was the same as that known to promote trichothiodystrophy in man; a disease associated with brittle hair, reduced size, mental retardation and premature aging, but not cancer. Sometimes, mutations in the XPD gene, which causes a repair deficiency, will not have a cancer phenotype probably due to a serious defect of particular XPD alleles in transcription caused by a defective XPD subunit of TFIIH.

Several mutant mouse models have been generated with targeted deletions of genes in the BER pathway. The general picture is that DNA glycosylase defective mutants are viable and without obvious phenotypic abnormalities. In contrast, mutations leading to defects in later stages of the BER pathway are embryonic lethal indicating an essential role of the BER pathway for normal life existence. Some embryonic lethal knock-outs include *APE, POLB, LIG1, FEN1 and XRCC1.* The fact that glycosylase defective mutants lacking NTHL1 and OGG1 are normally viable and without apparent abnormal phenotypes has been ascribed to the presence of back-up repair pathways. This is also consistent with the observation that these animals still have ample abilities of removing the oxidized base residues from DNA. The mouse OGG1 mutant shows a modest increase of 10–15 fold in the steady state level of 8-oxoG in the nuclear DNA and a 20–50 fold increase in the amount of 8-oxoG in mitochondrial DNA in nonproliferative tissues and essentially normal levels in other tissues [34, 35]. When mutated mouse embryonic fibroblasts are exposed to oxidative agents inducing 8-oxoG lesions then repair is slower than in wild type but still quite efficient [34, 36]. Further analyses with plasmid assays have indicated that the $mogg1^{-/-}$ mutation causes a defect of global genome repair but that the mutated animals are proficient in transcription coupled repair [36]. Furthermore, another glycosylase for repair of 8-oxoG has recently been discovered, NEIL1, that appears to represent a backup repair activity in these animals [10]. This enzyme may be upregulated in the mutant cells, especially in younger animals [Eide, Seeberg and Klungland, unpublished]. Similarly, two different back up DNA glycosylases for removal of thymine glycols have been detected in mice mutants of NTHL1 [37]. One of these appears to be NEIL1, which has activities both against oxidized purines and pyrimidines [10, 37–40]. In fact, four different DNA glycosylases, OGG1, NTHL1, NEIL1 and NEIL3 appear to remove formamidopyrimidines (FaPy) from DNA thus emphasizing the redundancy in the glycosylase mediated removal of oxidized base residues in DNA [10]. We must await the construction and characterization of mice with multiple defects in glycosylase genes before firm conclusions can be made about the importance of removing oxidized bases to prevent disease and accelerated aging.

Concluding remarks

Sophisticated mechanisms have evolved to repair oxidative DNA damage of which base excision repair appears to be the most efficient and elaborate. The function of this repair pathway is likely to be essential as indicated from embryonic lethality of mouse with targeted disruptions of specific genes involved in this pathway. Surprisingly, mouse mutants with defects in the enzymes recognizing and removing the modified base residues are essentially without phenotypic abnormalities as if these functions are less important than those involved in the repair of abasic sites and strand breaks in DNA. However, this can most likely be ascribed to a large redundancy in the gene functions involved in base damage removal.

ROS are formed spontaneously inside cells as a by product of the electron transport chain and by other important processes such as stress and inflammatory processes, such as the oxidative burst operating to kill infectious agents. In the brain, excessive stimulation of glutamergic neurons will result in receptor activation and Ca^{2+} influx, formation of nitric oxide by activation of nitric oxide synthase, and conversion of nitric oxide by superoxide to peroxynitrite ($ONOO^-$). Peroxynitrite is an extremely reactive oxidant that produces a complex pattern of oxidative DNA damage independent of its ability to generate hydroxyl radicals. It thus appears that repair of oxidative DNA damage may be particularly important in the brain and could be essential to avoid neurological deterioration associated with Alzheimer's and Parkinson's disease, typical aging diseases. DNA repair disorders in man are almost always associated with neurological disorders thus substantiating this notion. Further work is required to monitor DNA repair of oxidative DNA damage in the elderly population in order to get more evidence for the role of DNA repair for longevity and, if applicable, to identify individuals prone to dementia prior to onset for early administration of preventive measures.

References

1. Lindahl T (1993). Instability and decay of the primary structure of DNA. *Nature* 362: 709–15.
2. Demple B, Harrison L (1994). Repair of oxidative damage to DNA: enzymology and biology. *Annu Rev Biochem.* 63: 915–48.
3. Krokan HE, Nilsen H, Skorpen F, Otterlei M, Slupphaug G (2000). Base excision repair of DNA in mammalian cells. *FEBS Lett.* 476: 73–7.
4. Seeberg E, Eide L, Bjørås M (1995). The base excision repair pathway. *Trends Biochem Sci.* 20: 391–7.
5. Klungland A, Lindahl T (1997). Second pathway for completion of human DNA base excision-repair: reconstitution with purified proteins and requirement for DNase IV (FEN1). *EMBO J.* 16: 3341–8.
6. ESCODD (2003). Measurement of DNA oxidation in human cells by chromatographic and enzymic methods. *Free Radic Biol Med.* 34: 1089–99.
7. Bjørås M, Luna L, Johnsen B, *et al.* (1997). Opposite base-dependent reactions of a human base excision repair enzyme on DNA containing 7,8-dihydro-8-oxoguanine and abasic sites. *EMBO J.* 16: 6314–22.
8. Yang H, Clendenin WM, Wong D, *et al.* (2001). Enhanced activity of adenine-DNA glycosylase (Myh) by apurinic/apyrimidinic endonuclease (Apel) in mammalian base excision repair of an A/GO mismatch. *Nucleic Acids Res.* 29: 743–52.
9. Dantzer F, Bjørås M, Luna L, Klungland A, Seeberg E (2003). Comparative analysis of 8-oxoG:C,8-oxoG:A,A:C and C:C DNA repair in extracts from wild type or 8-oxoG DNA glycosylase deficient mammalian and bacterial cells. *DNA Repair*, in press.
10. Morland I, Rolseth V, Luna L, Rognes T, Bjoras M, Seeberg E (2002). Human DNA glycosylases of the bacterial Fpg/MutM superfamily: an alternative pathway for the repair of 8-oxoguanine and other oxidation products in DNA. *Nucleic Acids Res.* 30: 4926–36.
11. Egashira A, Yamauchi K, Yoshiyama K, *et al.* (2002). Mutational specificity of mice defective in the MTH1 and/or the MSH2 genes. *DNA Repair* (Amst) 1: 881–93.
12. Aspinwall R, Rothwell DG, Roldan-Arjona T, *et al.* (1997). Cloning and characterization of a functional human homolog of Escherichia coli endonuclease III. *Proc Natl Acad Sci USA* 94: 109–14.
13. Luna L, Bjoras M, Hoff E, Rognes T, Seeberg E (2000). Cell-cycle regulation, intracellular sorting and induced overexpression of the human NTH1 DNA glycosylase involved in removal of formamidopyrimidine residues from DNA. *Mutat Res.* 460: 95–104.
14. Doetsch PW, Cunningham RP (1990). The enzymology of apurinic/apyrimidinic endonucleases. *Mutat Res.* 236: 173–201.
15. Meira LB, Devaraj S, Kisby GE, *et al.* (2001). Heterozygosity for the mouse Apex gene results in phenotypes associated with oxidative stress. *Cancer Res.* 61: 5552–7.
16. Hadi MZ, Ginalski K, Nguyen LH, Wilson DM III (2002). Determinants in nuclease specificity of Apel and Ape2, human homologues of *Escherichia coli* exonuclease III. *J Mol Biol.* 316: 853–66.
17. Johnson RE, Torres-Ramos CA, Izumi T, Mitra S, Prakash S, Prakash L (1998). Identification of APN2, the Saccharomyces cerevisiae homolog of the major human AP endonuclease HAP1, and its role in the repair of abasic sites. *Genes Dev.* 12: 3137–43.
18. Seeberg E, Luna L, Morland I, *et al.* (2000). Base removers and strand scissors: different strategies employed in base excision and strand incision at modified base residues in DNA. *Cold Spring Harbor Symp Quant Biol.* 65: 135–42.

19. Masson M, Niedergang C, Schreiber V, Muller S, Menissier-de Murcia J, de Murcia G (1998). XRCC1 is specifically associated with poly(ADP-ribose) polymerase and negatively regulates its activity following DNA damage. *Mol Cell Biol.* 18: 3563–71.
20. Caldecott KW, McKeown CK, Tucker JD, Ljungquist S, Thompson LH (1994). An interaction between the mammalian DNA repair protein XRCC1 and DNA ligase III. *Mol Cell Biol.* 14: 68–76.
21. Wilson SH (1998). Mammalian base excision repair and DNA polymerase beta. *Mutat Res.* 407: 203–15.
22. Herceg Z, Wang ZQ (2001). Functions of poly(ADP-ribose) polymerase (PARP) in DNA repair, genomic integrity and cell death. *Mutat Res.* 477: 97–110.
23. Matsumoto Y (2001). Molecular mechanism of PCNA-dependent base excision repair. *Prog Nucl Acid Res Mol Biol.* 68: 129–38.
24. Whitehouse CJ, Taylor RM, Thistlethwaite A, *et al.* (2001). XRCC1 stimulates human polynucleotide kinase activity at damaged DNA termini and accelerates DNA single-strand break repair. *Cell* 104: 107–17.
25. Bohr VA (2002). Repair of oxidative DNA damage in nuclear and mitochondrial DNA, and some changes with aging in mammalian cells. *Free Radic Biol Med.* 32: 804–12.
26. Bogenhagen DF, Pinz KG, Perez-Jannotti RM (2001). Enzymology of mitochondrial base excision repair. *Prog Nucl Acid Res Mol Biol.* 68: 257–71.
27. Torres-Ramos CA, Johnson RE, Prakash L, Prakash S (2000). Evidence for the involvement of nucleotide excision repair in the removal of abasic sites in yeast. *Mol Cell Biol.* 20: 3522–8.
28. Guillet M, Boiteux S (2002). Endogenous DNA abasic sites cause cell death in the absence of Apn1, Apn2 and Rad1/Rad10 in *Saccharomyces cerevisiae. EMBO J.* 21: 2833–41.
29. Kuraoka I, Bender C, Romieu A, Cadet J, Wood RD, Lindahl T (2000). Removal of oxygen free-radical-induced 5′,8-purine cyclodeoxynucleosides from DNA by the nucleotide excision-repair pathway in human cells. *Proc Natl Acad Sci USA* 97: 3832–7.
30. Bohr VA, Smith CA, Okumoto DS, Hanawalt PC (1985). DNA repair in an active gene: removal of pyrimidine dimers from the DHFR gene of CHO cells is much more efficient than in the genome overall. *Cell* 40: 359–69.
31. Hoeijmakers JH (2001). Genome maintenance mechanisms for preventing cancer. *Nature* 411: 366–74.
32. Le Page F, Klungland A, Barnes DE, Sarasin A, Boiteux S (2000a). Transcription coupled repair of 8-oxoguanine in murine cells: the oggl protein is required for repair in nontranscribed sequences but not in transcribed sequences. *Proc Natl Acad Sci USA* 97: 8397–402.
33. de Boer J, Andressoo JO, de Wit J, *et al.* (2002). Premature aging in mice deficient in DNA repair and transcription. *Science* 296: 1276–9.
34. Klungland A, Rosewell I, Hollenbach S, *et al.* (1999). Accumulation of premutagenic DNA lesions in mice defective in removal of oxidative base damage. *Proc Natl Acad Sci USA* 96: 13300–5.
35. de Souza-Pinto NC, Eide L, Hogue BA, *et al.* (2001). Repair of 8-oxodeoxyguanosine lesions in mitochondrial dna depends on the oxoguanine dna glycosylase (OGG1) gene and 8-oxoguanine accumulates in the mitochondrial dna of OGG1-defective mice. *Cancer Res.* 61: 5378–81.
36. Le Page F, Kwoh EE, Avrutskaya A, *et al.* (2000b). Transcription-coupled repair of 8-oxoguanine: requirement for XPG, TFIIH, and CSB and implications for Cockayne syndrome. *Cell* 101: 159–71.

37. Takao M, Kanno S, Kobayashi K, *et al.* (2002). A back-up glycosylase in Nth1 knock-out mice is a functional Nei (endonuclease VIII) homologue. *J Biol Chem.* 277: 42205–13.
38. Takao M, Kanno S, Shiromoto T, *et al.* (2002). Novel nuclear and mitochondrial glycosylases revealed by disruption of the mouse Nth1 gene encoding an endonuclease III homolog for repair of thymine glycols. *EMBO J.* 21: 3486–93.
39. Hazra TK, Izumi T, Boldogh I, *et al.* (2002). Identification and characterization of a human DNA glycosylase for repair of modified bases in oxidatively damaged DNA. *Proc Natl Acad Sci USA* 99: 3523–8.
40. Bandaru V, Sunkara S, Wallace SS, Bond JP (2002). A novel human DNA glycosylase that removes oxidative DNA damage and is homologous to *Escherichia coli* endonuclease VIII. *DNA Repair* (Amst) 1: 517–29.
41. Rosenquist TA, Zaika E, Fernandes AS, Zharkov DO, Miller H, Grollman AP (2003). The novel DNA glycosylase, NEIL1, protects mammalian cells from radiation-mediated cell death. *DNA Repair* (Amst) 2: 581–91.

Oxidative Damage to Proteins

Nicolle Sitte
Department of Anaesthesiology and Critical Care Medicine, CHARITÉ – University Medicine Berlin, Campus Benjamin Franklin, Berlin, Germany

Introduction

Protein oxidation *in vivo* is a natural consequence of aerobic life. Oxygen radicals and other reactive oxygen species that are generated as by-products of cellular metabolism or from environmental sources, cause modifications to amino acids of proteins. These generally result in loss of protein function and/or enzymatic activity. Living in an oxygenated environment has required the evolution of effective cellular strategies to detect and detoxify metabolites of molecular oxygen known as reactive oxygen species (ROS). The appropriate and inappropriate production of oxidants, together with the ability of organisms to respond to oxidative stress, is intricately connected to aging and life span.

Research has shown that the levels of oxidized proteins increase with age, and certain unknown oxidation pathways may be partly responsible for protein oxidation. There are several factors that may influence the accumulation of oxidized proteins. For example, species-specific differences in metabolic rate may exist, which could influence the formation of oxidized macromolecules. Also, less efficient removal of oxidized proteins through proteolytic cleavage may promote the accumulation of oxidized protein with aging. Furthermore, several proteolytic enzymes responsible for degrading oxidized proteins may decline with age in tissues. In addition, defenses either enzymatic or non-enzymatic may decline with age in specific tissues. Little is known about the complex biochemistry of the accumulation and removal of oxidized amino acids *in vivo*. However, it is reasonable to predict that the accumulation of oxidized proteins is dependent upon the balance between pro-oxidant, antioxidant, and proteolytic activities.

T. von Zglinicki (ed.), Aging at the Molecular Level, 27–45.

Post-translational modifications of protein structures by oxidative stress

An inescapable side-product of oxidative metabolism is the production of ROS, which can damage lipids, nucleic acids, and proteins. Under normal physiological conditions, there exists a balance between the production of oxidizing species and antioxidants in the cell. Nevertheless, cellular macromolecules are attacked permanently by ROS. When the ROS-concentration exceeds the cellular ability to remove them and repair cellular damage – as occurring both in the aging process and under pathological conditions – oxidative stress arises, which results in the widespread oxidation of biomolecules, including proteins. Additionally, many environmental stimuli including cytokines, UV radiation, chemotherapeutic agents, and hyperthermia generate high levels of ROS. In a complex structure like a protein, in addition to modification of amino acid side chains, oxidation reactions are able to destroy, at least partially, the secondary and tertiary structure of the protein molecule. These protein modifications by oxidants lead generally to loss of biological function of the protein. Biological effects of protein oxidative damage depend on the acting oxidant itself, the reactivity, localization, the proximity of the radical to the target, and the presence of readily oxidisable amino acid residues in the protein molecule. On the other hand, the actual oxidizing capacity of an oxidant in a cell depends on the nature, function and activity of the antioxidative defenses, the ratio of radical/oxidant and available repair systems. Mammalian cells exhibit only limited direct repair mechanisms and, for the most part, oxidized proteins appear to undergo selective proteolysis. Protein oxidation induces structural changes, which results in protein unfolding and consequently in an increase of the hydrophobicity of the protein surface. Surface hydrophobicity is *in vitro* the key factor for the recognition and degradation of the substrate by several proteases, especially the proteasome [1]. Oxidized forms of proteins are generally degraded more rapidly than their counterparts. If oxidatively modified proteins are not rapidly removed by proteolytic enzymes, they can undergo direct chemical fragmentation, or can form large aggregate and covalent cross-links, which are actually poor substrates for the proteasome. The final step of the cross-linking process is the formation of a water insoluble material called age-pigment or lipofuscin (see section on "Accumulation and degradation of oxidized proteins during aging"), which can not be degraded and accumulates in senescent cells (see Chapter 15, this volume).

Mechanisms of oxidative modifications of protein structures

Oxidation of proteins normally is caused by free radicals, and this process, from a chemical thermodynamics standpoint, is an exothermic event. It is today well established that various reactive oxygen species are able to damage proteins. Among other modifications, ROS can convert proteins to carbonyl derivatives, either directly, by oxidation of some amino acid residue side chains and by oxidative cleavage of peptide bonds. That means, there are various possible oxidative pathways that can occur: (a) oxidation of the protein backbone, (b) formation of protein-protein cross-linkages, (c) oxidation of amino acid side chains and (d) protein fragmentation. Additionally, ROS can damage proteins indirectly by the oxidation of lipids and carbohydrates to derivatives that can form protein carbonyl adducts or by oxidation of glycated proteins.

Oxidation of the protein backbone

Oxidants are able to cleave the polypeptide backbone chemically. Therefore, the result of an oxidative cleavage of a polypeptide are modified peptides and not peptides, like in the case of degradation by proteolytic enzymes [2].

The cleavage is occurring at the C_{α}-atom of the polypeptide chain. Backbone oxidation is initiated by the ˙OH-dependent abstraction of the α-hydrogen atom of an amino acid residue leading to the formation of a carbon-centered radical. (Figure 1, reaction a).

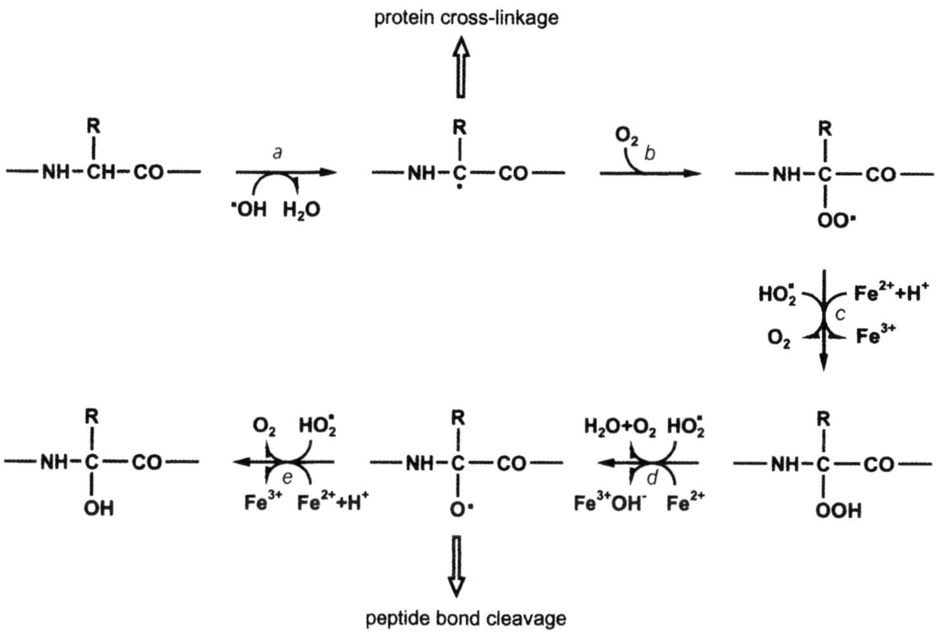

Figure 1. Oxygen free radical-mediated protein oxidation.

The carbon-centered radical reacts rapidly with oxygen to form an alkylperoxyl radical intermediate (Figure 1, reaction b), which can give rise to alkylperoxide reaction (Figure 1, reaction c), followed by formation of an alkoxyl radical (Figure 1, reaction d), which can be converted to a hydroxyl protein derivative (Figure 1, reaction e) [3]. These reactions can be mediated by Cu^{+}, Fe^{2+} or the hydroperoxyl radical (HOO˙).

The alkyl, alkylperoxyl, and alkoxyl radical intermediates may undergo side reactions with other amino acid residues in the same or different protein molecule to produce another carbon-centered radical (reaction f) capable of reacting similarly to the reactions illustrated in Figure 1.

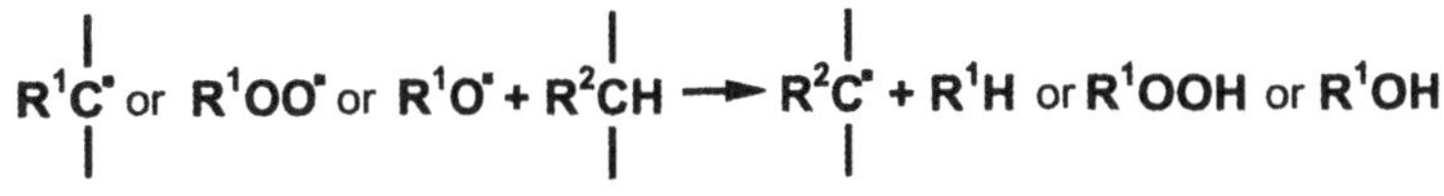

(reaction f)

Alkoxyl derivates of proteins are also capable of undergoing peptide bond cleavage. This cleavage can occur by either of two mechanisms [3]: the diamide pathway or the α-amidation pathway (Figure 2). Furthermore, peptide bond cleavage can occur also by ROS-mediated oxidation of glutamyl side chains. In this mechanism, oxalic acid is generated and the N-terminal amino acid of one peptide fragment exists as the pyruvyl derivative [3]. Cleavage of the peptide bond will result in formation of carbonyl groups that are often used as a marker of protein oxidation.

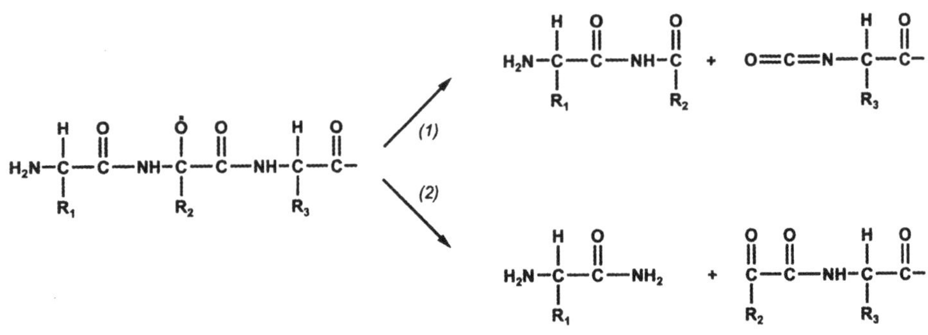

***Figure 2**. Peptide bond cleavage by the (1) diamide and (2) α-amidation pathway.*

Oxidation of amino acid side chains

The oxidative modifications of amino acid residues depend on the presence of oxidation-susceptible groups and on steric availability of these groups for oxidant attack. All amino acid residues of proteins are potential targets for oxidation by ·OH generated upon exposure to ionizing radiation [4–6] or by high hydrogen peroxide concentration in the presence of Cu^{2+} or Fe^{2+} [7, 8]. Although most amino acid side-chains are susceptible to oxidation, only some oxidations products have been fully characterized [3]. Common products of oxidation of amino acid residues are shown in Table 1. Several authors used the determination of oxidation products of specific amino acids as marker of protein oxidation.

Table 1. Oxidation of amino acid residue side chains [reviewed in ref. 10]

Amino acid	Products
arginine	Glutamic semialdehyde
cysteine	Disulfides (Cys-S-S-Cys, Cys-S-S-R); CySOH; CySOOH; $CySO_2H$
glutamic acid	oxalic acid; pyruvate adducts
histidine	2-oxohistidine; aspartic acid; asparagine
leucine	3-, 4-, and 5-OH-leucine
lysine	α-aminoadipic semialdehyde
methionine	methionine sulfoxide; methionine sulfone
phenylalanine	2-, 3-, and 4-OH-phenylalanine; 2,3-dihydroxy-phenylalanine
proline	glutamylsemialdehyde; 2-pyrrolidone; 4- and 5-OH-proline; pyroglutamic acid
threonine	2-amino-3-ketobutyric acid
tryptophan	2-, 4-, 5-, 6-, and 7-OH-tryptophan; formylkynurenine, 3-OH-kynurenine; nitrotryptophan
tyrosine	3,4-dihydroxyphenylalanine; Tyr-Tyr cross-links; 3-nitrotyrosine
valine	3-hydroxyvaline

Aliphatic side chains such as those of Val, Leu, Ile, Lys, Glu, Arg and Pro) can be oxidized to hydroperoxides, alcohols and carbonyl compounds. The levels of these oxidative markers in normal samples and various pathologies have been reviewed [9].

The residues of *sulfur-containing amino acids* cysteine and methionine in proteins are by far the most sensitive to oxidation by almost all kinds of ROS [3, 5]. The nucleophilic role of sulhydryls is well documented in the literature and it is consequentially that active site cysteines are important catalytic entities in many reactions (glutathione S-transferases, glyceraldehyde 3-phosphate dehydrogenase, glutathione reductase, thioredoxin/glutaredoxin), and in peptidase activities (calpains, caspase, papain) [for review see ref. 11]. Protein sulfhydryls can be oxidized to several states including reversibly oxidized forms such a protein disulfide, S-thiolated, S-nitrosylated, and sulfenic acid as well as more highly oxidized states such as the sulfinic and sulfonic acid forms of protein cysteines (Table 1). The sulfur-containing amino acids are easily and reversibely oxidized under even mild conditions, leading to disulfides and methionine sulfoxide, respectively. These are the only oxidative modifications of proteins that can be repaired. Most of the biological systems contain disulfide reductases and methionine sulfoxide reductases (Figure 3) [12] that convert the oxidized forms of cysteine and methionine residues back to their unmodified forms. Reduced glutathione normally functions to maintain reduced thiols on proteins. Many reactive cysteines found on protein surfaces have an antioxidant or protective function. Based on experimental evidence, it has been suggested that the cyclic methionine oxidation/reduction is a protective mechanism that serves as a ROS scavenger system to protect such proteins from more serious oxidative damage [13]. This is further supported by the existence of methionine sulfoxide reductases [12].

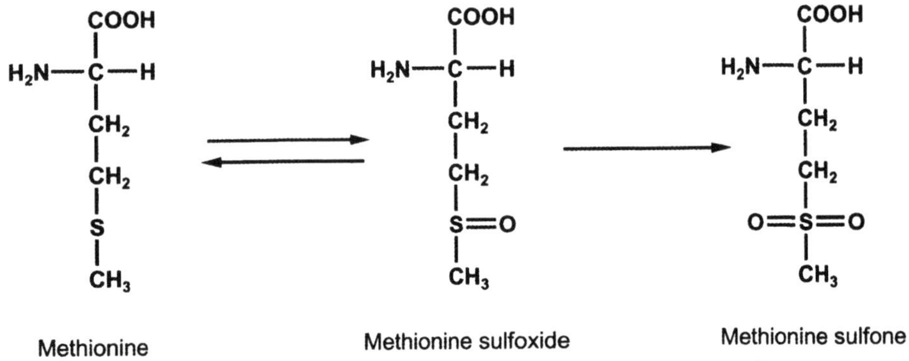

Figure 3. *Methionine oxidation. The first step of methionine oxidation to methionine sulfoxide is reversible owing to methionine sulfoxide reductase; further oxidation to methionine sulfone is irreversible.*

Examples for specific *oxidation of aromatic amino acid residues* (see Table 1) are the conversion of phenylalanine to *o*-tyrosine, *m*-tyrosine and 3-nitrophenylalanine, respectively (Figure 4).

Another amino acid residue present in proteins that is susceptible to oxidation is the indole moiety of tryptophan. The reducing potential is considerably less than that of cysteine and methionine, so oxidation of tryptophanyl residues usually does not occur until all exposed thiol residues are oxidized. Tryptophan residues are the only chromophoric moieties in proteins which can be photooxidized to tryptophanyl radicals by UV radiation [14]. Tryptophan residues readily react with all reactive oxygen species and reactive nitrogen species (Figure 5) and are oxidized to kynurenine compounds and to various hydroxy derivates. Histidine residues are converted to 2-oxohistidine, asparagine, and aspartic acid residues [reviewed in ref. 9].

Hydroxyl radical acting on proteins also induces tyrosyl radical formation, which may form intra- or interprotein cross-linked derivatives through dityrosine bridges (Figure 6) [15]. An important oxidative process with profound functional and structural consequences involves the irreversible nitration of tyrosin residues by peroxynitrite ($ONOO^-$) [16] (see Figure 6). The formation of nitrotyrosine is an irreversible process and therefore can interfere with intracellular signaling pathways as it blocks protein activation/deactivation via preventing the phosphorylation/dephosphorylation of tyrosines [17] and locks the proteins into a relatively inactive

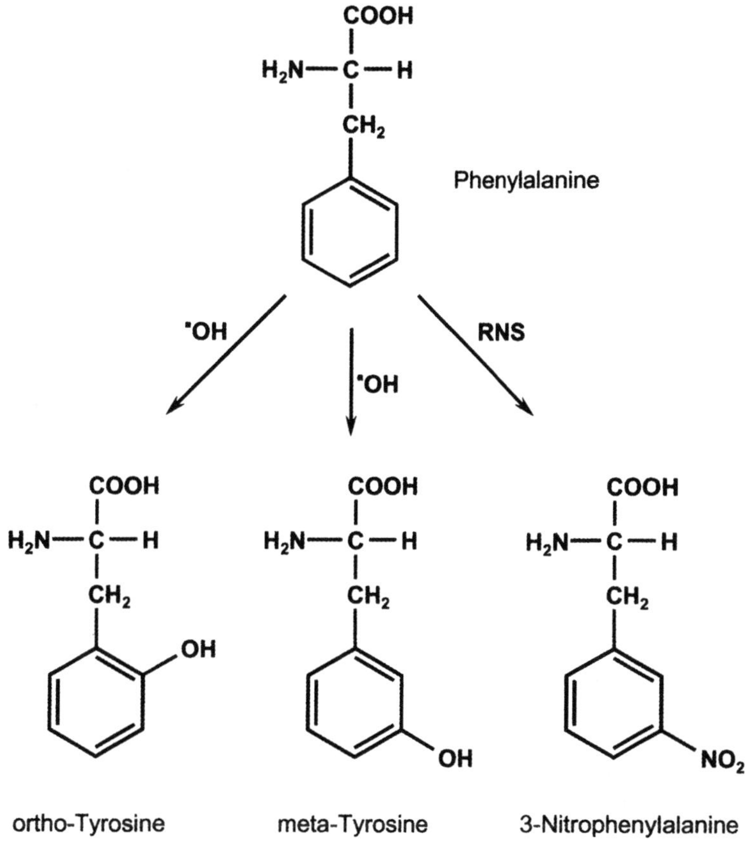

Figure 4. *Oxidation of phenylalanine by reactive oxygen and reactive nitrogen species (RNS).*

configuration. Likewise, nitrotyrosine formation within structural proteins (e.g., actin and neurofilaments) can interfere with their polymerization and leads to disruption of cytoskeletal elements [18, 19].

Generation of protein carbonyl derivatives

Lysine, arginine, threonine, and proline residues of proteins are particularly sensitive to metal-catalyzed oxidation, leading in each case to the formation of carbonyl derivatives [20]. As described above, peptide carbonyl compounds are also obtained as fragmentation products of peptide bond cleavage. In addition, carbonyl groups may be introduced into proteins by reactions with lipid peroxidation products (Figure 7) [21] (Michael addition) including 4-hydroxy-2-nonenal (HNE), 2-propeneal (acrolein) [22] or malondialdehyde [22, 23]. These reactive alkenals react with nucleophilic side-chains of Cys, His, and Lys residues, resulting in covalent addition

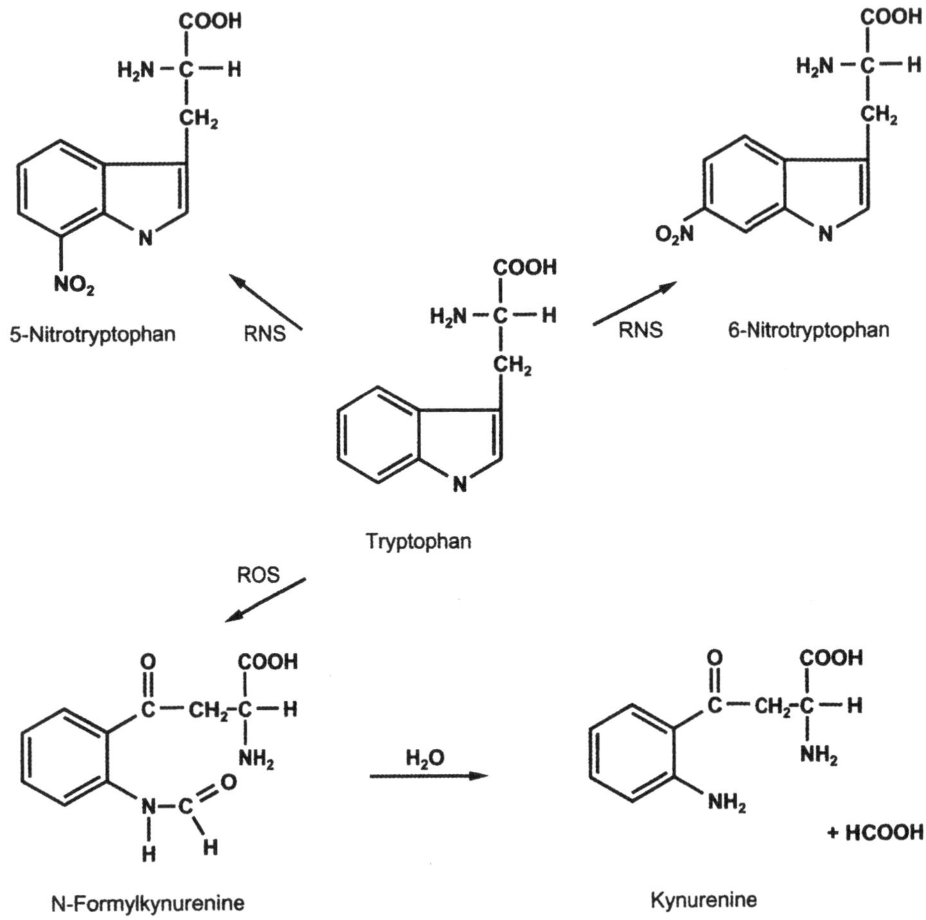

Figure 5. *Oxidative modification of tryptophan by reactive oxygen species (ROS) and reactive nitrogen species (RNS).*

of an aldehyde carbonyl group to the peptide chain. In some cases, protein cross-linking can occur via Schiff base formation between the carbonyl functionality on the alkenal and an amine on an adjacent protein [24]. Furthermore, carbonyl compounds are also formed with reactive carbonyl derivatives (ketoamines, ketoaldehydes, deoxysones) generated as a result of the reactions of reducing sugars or their oxidation products with the amino group of lysine residues of proteins (by mechanisms referred to as glycation and glycoxidation) [25–27] (Figure 7).

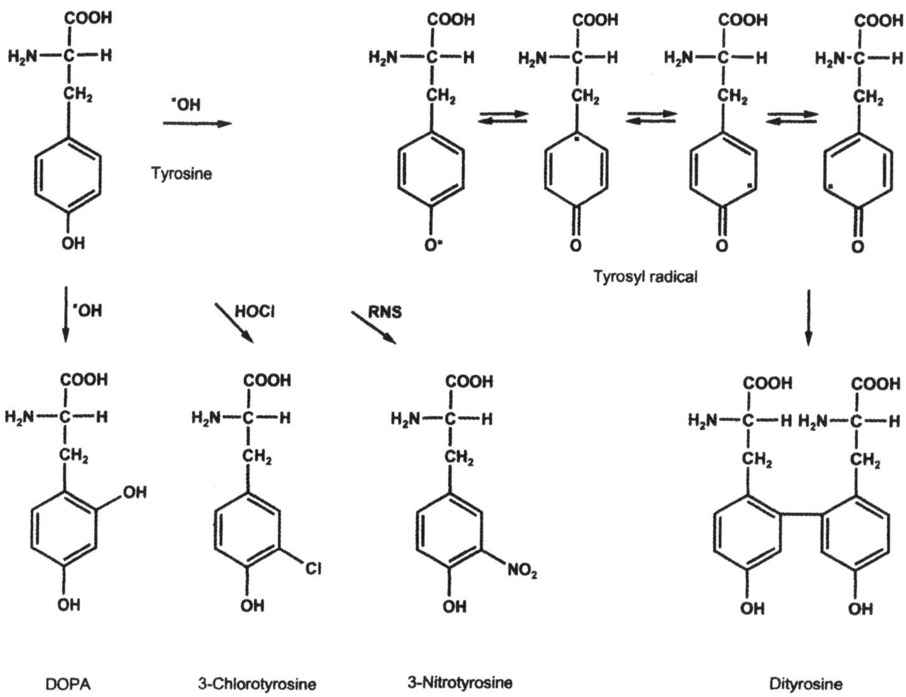

Figure 6. Oxidative modification of tyrosine by reactive oxygen species, reactive nitrogen species (RNS), and hypochlorite.

In view of the fact that carbonyl compounds are generated by many different oxidative processes, the detection and quantification of protein carbonyl groups [28] are used as a marker of ROS-mediated protein damage during oxidative stress, aging, and a number of diseases.

Formation of protein-protein cross-linkages

Oxidative modification of a protein can lead to the formation of several different kinds of protein-protein cross-linked derivatives [for review see ref. 29].

(a) In the absence of O_2, two different carbon-centered radicals generated by OH abstraction of H atoms may react with each other to form (–C–C–) protein-cross-links, thus propagating the free radical initiated oxidation across and between proteins.

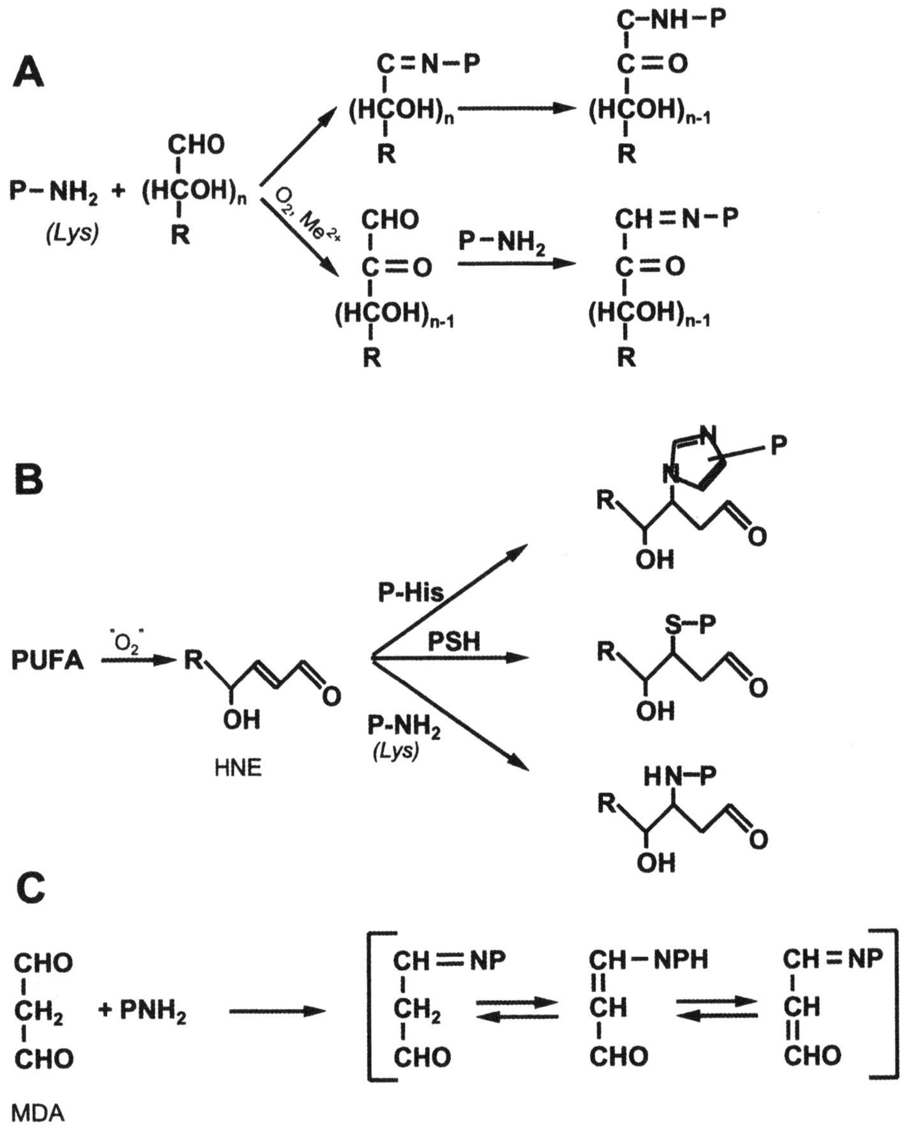

***Figure** 7. Protein carbonyl formation by glycation, glycoxidation, and by reactions with peroxidation products of polyunsaturated fatty acids (PUFA) **A**: reactions of sugars with protein lysyl amino groups ($P\text{-}NH_2$). **B**: Michael addition of 4-hydroxy-2-nonenal (HNE) to protein lysine ($P\text{-}NH_2$), histidine (P-His), or cysteine (PSH) residues. **C**: reaction of protein amino groups (PNH_2) with the lipid peroxidation product malondialdehyd (MDA).*

(b) The oxidation of sulfhydryl groups can generate disulfide- (–S–S–) cross-linked proteins.

(c) The tyrosine oxidation can lead to dityrosine- (–Tyr–Tyr–) cross-linked compounds (see Figure 6).

(d) The carbonyl group formed in the direct oxidation of amino acid side chains of one protein may react with the lysine amino groups of another protein obtaining Schiff-base cross-linked products.

(e) The Michael addition reaction of either cysteine sulfhydryl groups, lysine amino groups, or histidine imidazole groups of proteins with the double bonds of 4-hydroxynonenal and other α,β-unsaturated fatty aldehydes (e.g., acrolein), formed during lipid peroxidation, leads to the generation of a protein carbonyl compound, which upon further reaction with the lysine amino group of the same or a different protein will lead to Schiff-based cross-linked derivatives (see Figure 7).

(f) Reaction of both aldehyde groups of the lipid peroxidation product malondialdehyde with lysine amino groups within the same or two different proteins will lead to Schiff-based cross-linkages (see Figure 7).

Repair systems
Oxidative damage repair mechanisms are a critical component in maintaining intracellular homeostasis. Cellular repair systems may be classified as either direct or indirect [30]. Until now only little is known about direct repair of oxidatively modified proteins.

Direct repair systems. The re-reduction of oxidized sulfhydryl groups is one important direct repair process (see section "Oxidation of amino acid side chains"). The cysteine residues, which are highly susceptible for oxidation, often form a disulfide bond, changing the tertiary structure by producing a more rigid protein molecule. If disulfide bonds are generated between two proteins, the formation of cross-linked compounds can be initiated. Another important sulfhydryl oxidation process is the oxidation of methionine residues to methionine sulfoxide (see section "Oxidation of amino acid side chains"; Figure 3). It was suggested that this oxidation serves as a protective mechanism for further oxidative modification of other amino acid residues in the protein. Methionine sulfoxide reductase can regenerate methionine residues within such oxidized protein molecules and restore the biological function.

Indirect repair systems. Indirect repair involves two steps: first, the damaged protein molecule must be recognized, removed, and completely degraded to amino acids. Next, a replacement of the removed protein must be synthesized *de novo* [30]. It appears probable that most of the amino acids from an oxidized and degraded protein are reutilized for *de novo* protein synthesis.

Oxidized forms of proteins are generally degraded more rapidly than their natural counterparts. Indeed, the normal functions of a cell involve the regular elimination of these altered molecules. In extensive studies it was demonstrated that after oxidant attack towards proteins the hydrophobicity of the protein surface increases. Surface hydrophobicity serves as the key factor for recognition and degradation of the substrate by several proteases, especially by the proteasome [1]. The proteasome complex, which constitutes up to 1% of total cellular protein, seems to be responsible for the degradation of oxidized proteins in the cytoplasm and nucleus of eukaryotic cells. The proteasome recognizes hydrophobic amino residues, aromatic residues, and bulky aliphatic residues that are exposed during the oxidative rearrangement of secondary and tertiary protein structure. The recognition of such normally shielded hydrophobic residues is the suggested mechanism by which the proteasome catalyzes the selective removal of oxidatively modified cellular proteins. The proteasome-specific inhibitor lactacystin was able to inhibit the elimination of protein-bound carbonyl groups, thus demonstrating a direct correlation between protein turnover and removal of protein-bound carbonyl groups [31]. With increasing radical exposure of proteins the proteasome becomes more degraded [1, 32]. Because strongly oxidized proteins tend to aggregate and form covalent cross-links, which are actually poor substrates for the proteasome, cells prevent the accumulation of oxidized proteins by rapid degradation of moderately oxidized proteins. With age there is less efficient removal of oxidized proteins through proteolytic cleavage, which may cause the accumulation of protein carbonyls with aging (see sections on "Protein oxidation and aging" and "Accumulation and degradation of oxidized proteins during aging").

Protein oxidation and oxidative stress

The intracellular level of oxidized proteins reflects the balance between the rate of protein oxidation and the rate of oxidized protein degradation. This balance is a complex function of various circumstances that lead to the generation of reactive oxygen species, on the one hand, and of multiple factors determining the concentrations and/or activities of proteases involved in degradation of oxidatively modified proteins, on the other. There is a large body of literature documenting studies on protein oxidation in different models. A majority of these studies used protein-bound carbonyls as a marker for protein oxidation. Studies exposing animals or cell cultures to various conditions of oxidative stress yielded similar results by showing increases in the levels of protein carbonyls [reviewed in ref. 10]. Thus e.g., the carbonyl content of proteins has been shown to rise in hepatocytes and lung tissue of rats and whole body proteins of houseflies during exposure of these organisms to hyperoxia, in heart and lung of rats exposed to ischemia-reperfusion; in neutrophils and macrophages during periods of oxidative burst, in human plasma and hearts of rats exposed to ozone; in human plasma exposed to cigarette smoke; in flight muscle mitochondria of houseflies and heart and brain of mice during exposure to X-rays.

Protein oxidation and aging

It is known for a long time that aging is associated with the accumulation of altered forms of a number of enzymes [33]. At least one of every three proteins in the cell of older animals carries a carbonyl group [34] and those modified proteins are likely dysfunctional either as enzymes or structural proteins and more sensitive to heat inactivation and to proteolysis. Many studies using various models of aging have shown that the intracellular level of oxidized proteins, measured as protein carbonyls, increases exponentially as a function of age. These investigations were performed in cell culture as a function of age of the fibroblast donor [35], as well as in flies and other species like e.g., rats, mice, and gerbils [36–38]. In all these animals a correlation between life span and amount of oxidized proteins was demonstrated. A increase in the amount of oxidized proteins has been also reported in a study comparing different aging models of proliferating human fibroblasts [39, 40] and aging of nondividing human fibroblasts [41].

Manipulations which change life span also alter the carbonyl content of tissues in affected animals. Extension of life span can be obtained by reduction of the rate of reactive oxygen species production as well as by increasing the antioxidant defense. Interestingly, flies which are prevented from flying had an average life span of 48 days compared to 21 days for those allowed normal flight. The maximum life span increased even from 40 to 80 days [42]. After two-thirds of average life span of flyers, the carbonyl content in flyers' protein was already 55% higher than in those restricted from flying. Moreover, administration of the spin-trapping agent N-tert-butyl-α-phenylnitrone [43–45] or limiting food intake (caloric restriction) [46, 47] has been shown to extend life span and to decrease the oxidative damage to protein in the brains of rodents, with a concurrent improvement in age-related behavioral deficits. Besides, genetic manipulation of life span were achieved by simultaneous overexpression of Cu/Zn superoxide dismutase and catalase genes in transgenic *Drosophila* [48], resulting in an extended mean and maximum life span. These transgenic flies exhibited a delayed loss of physical performance and less oxidative damage to proteins.

Age-related changes have been observed in other markers of protein oxidation as well. Some studies detected the products of specific amino acid modifications, such as dityrosine [49], *o*-tyrosine or nitrotyrosine [50], or methionine sulfoxide [50], to name only a few. Alterations in overall protein thiol content are also good indicators of the age-related changes in protein structure and oxidant status [37]. The level of oxidized methionine (MetO) in rat liver was found to rise with animal age. This age-related increase in MetO correlated with an age-related increase in protein surface hydrophobicity and aggregation and has also been used as marker for rising protein oxidation [50, 51]. Several studies have verified that irreversible oxidation of protein sulfhydryls is more extensive in aged tissue samples. If these damaged proteins are not sufficiently removed by proteolysis, aged cells may accumulate increased amounts of protein containing damaged sulfhydryls. An adequate pool of glutathione as a protectant is essential to prevent this increase in sulfhydryl oxidation. However, changes in the glutathione redox status during the aging process indicate that there is only a limited potential for regeneration of oxidized protein sulfhydryls left [reviewed in ref. 11].

Accumulation and degradation of oxidized proteins during aging
The carbonyl content of tissues from various species has been shown to increase dramatically in the last third of life span, reaching such a level that on average one out of every three protein molecules carries the modification [34]. It was observed that old animals are more susceptible than young ones to oxidative stress-induced protein oxidation. Thus, exposure of old houseflies to X-irradiation led to greater increase in the protein carbonyl content and to greater loss of glucose-6-phosphate dehydrogenase activity compared to young flies [52]. The age-related accumulation of oxidized proteins may result from an increase in protein oxidation or a decline in the degradation of oxidized proteins. It was suggested that both processes actually contribute to an age-related accumulation of oxidized compounds.

There is a strong indication that protein turnover in cells and tissues tends to decline with age, and we now have good evidence demonstrating an actual decrease in the activity of the major cytosolic protease – the proteasomal system [reviewed in ref. 53]. In comparative investigations of protein turnover during cellular senescence in fibroblasts in proliferating [40] as well as nondividing cells [41, 54] a good correlation between accumulation of oxidized or cross-linked proteins and the decline of proteasome activity and overall cellular protein turnover during *in vitro* senescence was revealed. This decline in overall protein turnover was accompanied by a dramatic increase in the levels of oxidized and cross-linked proteins. In both models of cellular senescence, the proteasome content and the transcription of proteasome subunits were unchanged, despite significantly decreased intracellular proteasome activities. The clear conclusion was that the proteasomal system was being inhibited by the accumulated oxidized or cross-linked protein aggregates, leading to a progressively diminishing cellular ability to degrade oxidized proteins [40, 41, 54]. During aging, when proteolytic capacity may decline below a critical threshold of activity required to cope with normal oxidative stress levels, oxidized proteins may not undergo appropriate proteolysis and may instead cross-link with each other or form extensive hydrophobic bonds. Such large aggregates of damaged proteins are detrimental to normal cell functions and viability. It was suggested that the final step of the cross-linking process in the cell is the formation of a water insoluble self-fluorescent material, which accumulates in senescent cells [55]. This material is termed differently: age-pigment-like fluorophores, lipofuscin or ceroid. Furthermore, it could be demonstrated that lipofuscin accumulates in cellular proliferative senescence [39, 40] and in postmitotic cells with age [41, 51, 56]. The involvement of reactive oxygen and nitrogen species was hypothesized as one of the initial steps responsible for the formation of this material. Since lipofuscin was found mainly in lysosomes, it was postulated that the origin of this material may be at least partially also the result of a dysfunction of lysosomes in senescent cells [56].

Comprising, the proteasome function is very important in controlling the level of modified proteins in eukaryotic cells. Because the steady-state level of oxidized proteins reflects the balance between the rate of protein oxidation and the rate of protein degradation, age-related accumulation of altered protein can be due to a combination of circumstances like increasing oxidant production, decline in primary

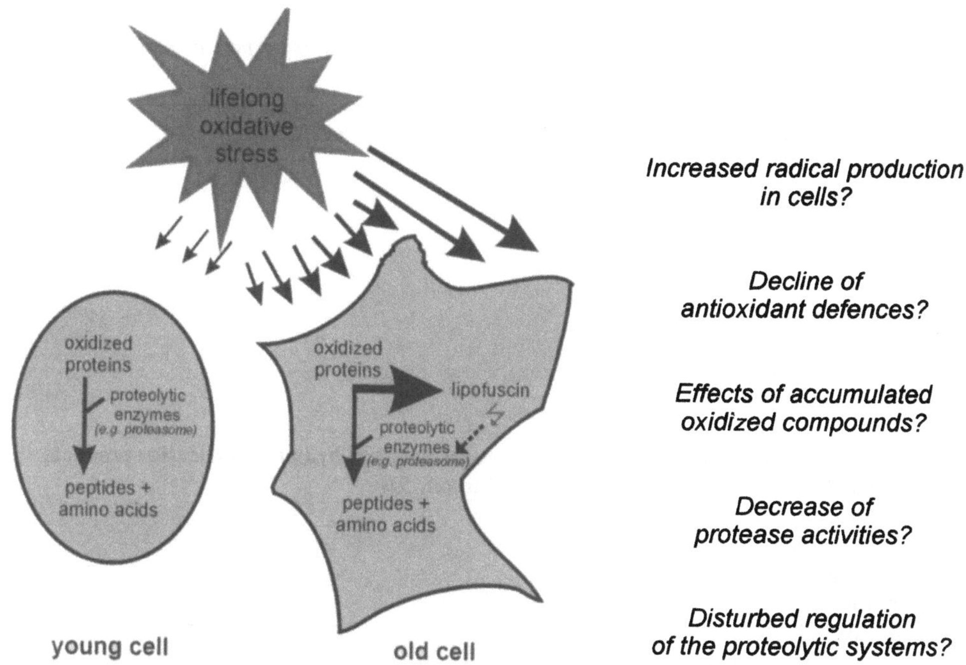

Figure 8. *Accumulation of oxidized proteins in aging cells. Increased production of reactive oxygen species via mitochondria combined with a decline of the cellular antioxidant systems leading to enhanced protein oxidation in old cells. Whereas in younger cells proteolytic systems are able to degrade oxidized or damaged proteins rapidly, some oxidatively modified and unfolded proteins tend to cross-link and are accumulated in older cells resulting in lipofuscin formation. Oxidized proteins and lipofuscin can be potent inhibitors of the proteasomal system leading to decreased degradation of oxidized proteins. The figure suggests some circumstances which could be the possible reasons for the accumulation of oxidized proteins in older cells.*

antioxidant defense, and a decrease of repair and progressive decline in proteasome activity (Figure 8).

Protein oxidation and diseases

We are constantly exposed to ROS generated from endogenous and some exogenous sources. These ROS react with biological molecules, such as DNA, proteins, and lipids, causing structural and functional damage. Oxidative damage accumulates in human tissues with age and can causally contribute to a number of degenerative diseases. Based on various markers for protein oxidation, it has been shown that a number of age-related diseases listed in Table 2, is associated with elevated levels of oxidatively modified proteins [reviewed in refs. 10, 57]. The described injuries alter

Table 2. *Diseases associated with elevated levels of different markers for protein oxidation*

Diseases associated with increased levels of oxidized protein [reviewed in refs. 10, 57]
- Alzheimer's disease
- Amyotrophic lateral sclerosis
- Cataractogenesis
- Rheumatoid arthritis
- Muscular dystrophy
- Respiratory distress syndrome
- Progeria
- Parkinson's disease
- Werner' syndrome

Diseases associated with elevated content of proteins modified by lipid peroxidation products (malondialdehyde, 4-hydroxynonenal) [reviewed in ref. 57]
- Parkinson' disease
- Cardiovascular disease
- Atherosclerosis
- Alzheimer's disease
- Iron-induced renal carcinogenesis
- Dextran sulfate-induced colitis
- Experimental pancreatitis

Diseases associated with increased level of protein glycation/glycoxidation end products [reviewed in ref. 57]
- Diabetes mellitus
- Atherosclerosis
- Alzheimer's disease
- Down' syndrome
- Parkinson's disease

Diseases associated with an elevated content of protein nitrotyrosine damage [reviewed in ref. 57]
- Alzheimer's disease
- Multiple sclerosis
- Lung injury
- Endotoxemia
- Atherosclerosis

cell functions and increase the risk of heart disease, stroke, cancer, and brain disease. Oxidation of low density lipoprotein (LDL), the "bad" cholesterol, causes it to stick more easily to blood vessel walls, facilitating the formation of plaques in arteries, leading to atherosclerosis. If plaques detach as clots and travel in the circulation they can block vessels in the heart, causing a heart attack, or in the brain, causing stroke. Vascular damage and other forms of oxidant damage to brain cells are associated with Alzheimer's disease. Free radical injury increases the risk of wrinkles, cataracts, blindness and arthritis. Antioxidants, both enzymatic and non-enzymatic, limit oxidative damage to biological molecules by various mechanisms. Dietary antioxidants, such as vitamins C and E, significantly contribute to antioxidant defense systems in humans and may help protect us from certain age-related degenerative disease.

References

1. Pacifici RE, Kono Y, Davies, KJ (1993). Hydrophobicity as the signal for selective degradation of hydroxyl radical modified hemoglobin by the multicatalytic proteinase complex, proteasome. *J Biol Chem.* 268: 15405–11.
2. Stadtman ER (1993). Oxidation of free amino acids and amino acid residues in proteins by radiolysis and by metal-catalyzed reactions. *Annu Rev Biochem.* 62: 797–821.
3. Garrison WM (1987). Reaction mechanisms in the radiolysis of peptides, polypeptides, and proteins. *Chem Rev.* 87: 381–98.
4. Garrison WM, Jayko ME, Bennet, W (1962). Radiation-induced oxidation of proteins in aqueous solution. *Radiat Res.* 16: 487–502.
5. Swallow AJ (1960). Effect of ionizing radiation on proteins, RCO groups, peptide bond cleavage, inactivation, -SH oxidation In: AJ Swallow, ed. *Radiation Chemistry of Organic Compounds.* New York: Pergamon Press, pp. 211–24.
6. Schuessler H, Schilling K (1984). Oxygen effect in radiolysis of proteins. Part 2. Bovine serum albumin. *Int J Radiat Biol.* 45: 267–81.
7. Huggins TG, Wells-Knecht MC, Detorie NA, Baynes JW, Thorpe SR (1993). Formation of *o*-tyrosine and dityrosine in proteins during radiolytic and metal-catalyzed oxidation. *J Biol Chem.* 268: 12341–7.
8. Neuzil J, Gebiki JM, Stocker, R (1993). Radical-induced chain oxidation of proteins and its inhibition by chain-breaking antioxidants. *Biochem J.* 293: 601–6.
9. Davies MJ, Fu S, Dean RT (1999). Stable markers of oxidant damage to proteins and their application in the study of human disease. *Free Radic Biol Med.* 27: 1151–63.
10. Stadtman ER, Berlett BS (1997). Reactive oxygen-mediated protein oxidation in aging and diseases. *Chem Res Toxicol.* 10: 485–94.
11. Thomas JA, Mallis RJ (2001). Aging and oxidation of reactive protein sulfhydryls. *Exp Gerontol.* 36: 1519–26.
12. Stadtman ER, Berlett BS (1998). Reactive oxygen-mediated protein oxidation in aging and disease. *Drug Metab Rev.* 30: 225–43.
13. Levine RL, Mosoni L, Berlett BS, Stadtman ER (1996). Methionine residues as endogenous antioxidants in proteins. *Proc Natl Acad Sci USA* 93: 15036–40.
14. Bazin M, Patterson LK, Santus R (1983). Direct observation of monophotonic photoionization in tryptophan excited by 300-nm radiation. A laser photolysis study. *J Phys Chem.* 87: 189–90.

15. Leeuwenburgh C, Rasmussen JE, Hsu FF, Mueller DM, Pennathur S, Heinecke JW (1997). Mass spectrometric quantification of markers for protein oxidation by tyrosyl radical, copper, and hydroxyl radical in low density lipoprotein isolated from human atherosclerotic plaques. *J Biol Chem.* 272: 3520–6.
16. Beckman JS (1996). Oxidative damage and tyrosine nitration from peroxynitrite. *Chem Res Toxicol.* 9: 836–44.
17. Martin BL, Wu D, Jakes S, Graves DJ (1990). Chemical influences on the specifity of tyrosine phosphorylation. *J Biol Chem.* 265:7108–111.
18. Crow JP, Ye YZ, Strong M, Kirk M, Barnes S, Beckman JS (1997). Superoxide dismutase catalyzes nitration of tyrosine by peroxynitrite in the rod and head domains of neurofilament-L. *J Neurochem.* 69: 1945–53.
19. Eiserich JP, Estevez AG, Bamberg TV, *et al.* (1999). Microtubule dysfunction by post-translational nitrotyrosination of alpha-tubulin: a nitric oxide-dependent mechanism of cellular injury. *Proc Natl Acad Sci USA* 96: 6365–70.
20. Amici A, Levine RL, Tsai L, Stadtman ER (1989). Conversion of amino acid residues in proteins and amino acid homopolymers to carbonyl derivatives by metal-catalyzed reactions. *J Biol Chem.* 264: 3341–6.
21. Refsgaard H, Tsai L, Stadtman ER (2000). Modifications of proteins by polyunsatured fatty acid peroxidation products. *Proc Natl Acad Sci USA*, 97: 611–16.
22. Esterbauer H, Schaur RJ, Zollner H (1991). Chemistry and biochemistry of 4-hydroxynonenal, malonaldehyde and related aldehydes. *Free Radic Biol Med.* 11: 81–128.
23. Burcham PC, Kuhan YT (1996). Introduction of carbonyl groups into proteins of the lipid peroxidation product, malondialdehyde. *Biochem Biophys Res Commun.* 220: 996–1001.
24. Butterfield DA, Stadtman ER (1997). Protein oxidation processes in aging brain. *Adv Cell Aging Gerontol.* 2: 161–91.
25. Kristal BS, Yu BP (1992). An emerging hypothesis: synergistic induction of aging by free radicals and Maillard reactions. *J Gerontol.* 47: B107–14.
26. Baynes JW (1991). Role of oxidative stress in development of complications in diabetes. *Diabetes* 40: 405–12.
27. Monnier V, Gerhardinger C, Marion MS, Taneda S (1995). In: Cutler RG, Packer L, Bertram J, Mori A, eds. *Oxidative Stress and Aging*. Basel, Switzerland: Birkhauser Verlag, pp. 141–9.
28. Levine RL, Williams JA, Stadtman ER, Shacter E (1994). Carbonyl assays for determination of oxidatively modified proteins. *Methods Enzymol.* 233: 346–57.
29. Stadtman ER (1997). Free radical mediated oxidation of proteins. In: Özben T, ed. *Free Radics, Oxidative Stress, and Antioxidants. Pathological and Physiological Significance.* NATO ASI Series, Series A: Life Sciences, 296. New York: Plenum Press, pp. 51–65.
30. Davies KJ (2000). Oxidative stress, Antioxidant defenses, and damage removal, repair, and replacement systems. *IUBMB Life* 50: 279–89.
31. Sitte N, Merker K, Grune T (1998). Proteasome-dependent degradation of oxidized proteins in MRC-5 fibroblasts. *FEBS Lett.* 440: 399–402.
32. Grune T, Reinheckel T, Davies KJ (1997). Degradation of oxidized proteins in mammalian cells. *FASEB J.* 11: 526–34.
33. Rothstein M (1984). Changes in enzymatic proteins during aging. In: Roy AK, Chatterjee, eds. *Molecular Basis of Aging*. New York: Academic Press, pp. 209–32.
34. Stadtman ER, Levine RL (2000). Protein oxidation. *Ann NY Acad Sci.* 899: 191–208.
35. Oliver CN, Ahn BW, Moerman EJ, Goldstein S, Stadtman ER (1987). Age-related changes in oxidized proteins. *J Biol Chem.* 262: 5488–91.

36. Agarwal S, Sohal RS (1994). Aging and proteolysis of oxidized proteins. *Arch Biochem Biophys.* 309: 24–8.
37. Agarwal S, Sohal RS (1994). Ageing and protein oxidative damage. *Mech Ageing Dev.* 75: 11–19.
38. Agarwal S, Sohal RS (1996). Relationship between susceptibility to protein oxidation, aging, and maximum life span potential of different species. *Exp Gerontol.* 31: 365–72.
39. Sitte N, Merker K, von Zglinicki T, Grune T (2000). Protein oxidation and degradation during proliferative senescence of human MRC-5 fibroblasts. *Free Radic Biol Med.* 28, 701–8.
40. Sitte N, Merker K, von Zglinicki T, Grune T, Davies KJA (2000). Protein oxidation and degradation during cellular senescence of human BJ fibroblasts: part I – effects of proliferative senescence. *FASEB J.* 14(15): 2495–502.
41. Sitte N, Merker K, von Zglinicki T, Davies KJA, Grune T (2000). Protein oxidation and degradation during cellular senescence of human BJ fibroblasts: part II – aging of nondividing cells. *FASEB J.* 14(15): 2503–10.
42. Sohal RS, Agarwal S, Dubey A, Orr WC (1993). Protein oxidation damage is associated with life expectancy of houseflies. *Proc Natl Acad Sci USA* 90: 7255–9.
43. Smith CD, Carney JM, Starke-Reed PE, *et al.* (1991). Excess brain protein oxidation and enzyme dysfunction in normal aging and in Alzheimer disease. *Proc Natl Acad Sci USA* 88: 10540–3.
44. Stadtman ER (1992). Protein oxidation and aging. *Science* 257: 1220–4.
45. Dubey A, Forster MJ, Sohal RS (1995). Effect of spin-trapping compound N-tert-butyl-α-phenylnitrone on protein oxidation and life span. *Arch Biochem Biophys.* 324: 249–54.
46. Sohal RS, Agarwal S, Candas M, Forster MJ, Lal H (1994). Effect of age and caloric restriction of DNA oxidative damage in different tissues of C57BL/6 mice. *Mech Ageing Dev.* 76: 215–24.
47. Masoro EJ (2000). Caloric restriction and aging: an update. *Exp Gerontol.* 35: 299–305.
48. Orr WC, Sohal RS (1994). Extension of life span by overexpression of superoxide dismutase and catalase in *Drosophila melanogaster*. *Science* 263: 1128–30.
49. Giulivi C, Davies KJ (1994). Dityrosine: a marker for oxidatively modified proteins and selective proteolysis. *Methods Enzymol.* 233: 363–71.
50. Chao CC, Ma YS, Stadtman ER (1997). Modification of protein surface hydrophobicity and methionine oxidation by oxidative systems. *Proc Natl Acad Sci USA*, 94: 2969–74.
51. Nakano M, Oenzil F, Mizuno T, Gotoh S (1995). Age-related changes in the lipofuscin accumulation of brain and heart. *Gerontology* 41(2): 69–79.
52. Agarwal S, Sohal RS (1993). Relationship between aging and susceptibility to protein oxidative damage. *Biochem Biophys Res Commun.* 194: 1203–6.
53. Grune T, Shringarpure R, Sitte N, Davies KJA (2001). Age-related changes in protein oxidation and proteolysis in mammalian cells. *J Gerontol A Biol Sci Med Sci.* 56: B459–67.
54. Sitte N, Huber M, Grune T, *et al.* KJA (2000). Proteasome inhibition by lipofuscin/ceroid during postmitotic aging of fibroblasts. *FASEB J.* 14(11): 1490–8.
55. Yin D (1996). Biochemical basis of lipofuscin, ceroid, and age pigment-like fluorophores. *Free Radic Biol Med.* 21: 871–88.
56. Brunk UT, Terman A (1998). The mitochondrial-lysosomal axis theory of cellular aging. In: Cadenas E, Packer L, eds. *Understanding the Process of Aging*. Basel: Marcel Dekker, pp. 229–50.
57. Stadtman ER (2001). Protein oxidation in aging and age-related diseases. *Ann NY Acad Sci.* 928: 22–38.

Aging Rate, Mitochondrial Free Radical Production, and Constitutive Sensitivity to Lipid Peroxidation: Insights From Comparative Studies

Reinald Pamplona[1] and Gustavo Barja[2]
[1]Department of Basic Medical Sciences, Faculty of Medicine, University of Lleida, Lleida 25198, Spain; [2]Department of Animal Biology-II (Animal Physiology), Faculty of Biology, Complutense University, Madrid 28040, Spain

Introduction

There is a most fundamental question in gerontology: what are the mechanisms determining the rate of animal aging? Why human beings, for instance, can reach 122 years whereas rats only last at best 4 years? Fortunately, some answers to that fundamental question are emerging in recent years. Many different "theories" of aging have been proposed in the past, mainly because gerontological knowledge was seriously limited. The mitochondrial oxygen radical theory of aging [1], is now the one more supported by available information, and has one additional relevant value, it can integrate inside a common theoretical body various of the mechanistic theories of aging proposed in the past: the mutation accumulation theory, the rate of living theory, the cross-linking theory, and the waste-product accumulation theory (e.g., lipofuscin).

The idea that oxygen radicals, specially those of mitochondrial origin [1], are causally related to the basic aging process is increasingly receiving support from scientific studies [2–4]. That support comes from several lines of evidence including, most importantly, comparisons between animal species with different aging rates. Any theory of aging should fit with four main characteristics of this natural process: aging is progressive, endogenous, irreversible and deleterious for the individual (in the sense that it damages the soma). The progressive character of aging means that causes of aging must be present during the whole life span, both at young and old age, at approximately the same levels. Furthermore, aging is an endogenous process. Therefore, exogenous factors can not be causes of the intrinsic aging process (e.g.,

T. von Zglinicki (ed.), Aging at the Molecular Level, 47–64.

UV rays, dietary antioxidants, etc.). The endogenous character of aging means that the rate of aging of different animal species, and thus their maximum lifespan potentials (MLSPs), is mostly determined by their genes, not by the environment. This explains why different animal species age at widely different rates in similar environments. Conversely, the mean life span, frequently and wrongly called "longevity" (which is calculated from the amount of time that each individual lives) is mainly determined by the environment and to a lesser extent by the genotype (heritability of mean life span in humans is commonly agreed to be around 30%). This is the reason why many environmental factors like smoking, amount of saturated fat, unbalanced diets, sedentary life, and possibly antioxidants are so important for the determination of age of death. Conversely, no matter what an elephant eats or does, it will never age in two years like a healthy rat. Thus, mortality should not be confused with aging, even though aging increases the probability of death. In relation to this, the interindividual variation in the time lived inside a given species (mainly environmentally determined) should not be confused with interspecies variation longevity (which is strongly genetically determined). Confusion in this, as well as between mean and maximum life span, can delay gerontological advance and should be avoided.

In order to understand the rate of aging, its fundamental causes, which are finally responsible for maximum longevity, must be unravelled. The mitochondrial oxygen radical theory of aging fulfils the four criteria stated above: ROS (reactive oxygen species) are *endogenously* produced at mitochondria under normal physiological conditions, they are produced continuously throughout life (and can thus lead to *progressive* aging changes), and their *deleterious* effects on mtDNA (mitochondrial DNA) are *irreversible* due to accumulation of somatic DNA mutations during aging in post-mitotic tissues. The first three characteristics are at least applicable to the second known parameter relating aging and oxidative stress, the degree of fatty acid unsaturation of cellular membranes also in post-mitotic tissues. In this article comparative studies of oxidative stress-related parameters performed in animals with different maximum longevities will be summarized. These subjects have been previously reviewed by us in the past [4–6].

Antioxidants and longevity

In the past, most investigations trying to clarify the relationship between oxidative stress and aging studied antioxidant rather than endogenous prooxidant factors, mainly because their measurement was much more easy, and also because it was assumed (most probably wrong) that only longevity determinant genes, not pro-aging genes, existed. The idea that aging could be due to a decrease in antioxidants in old age was soon discarded when it was observed that the changes in endogenous antioxidants as a function of age did not follow a consistent pattern, showing decreases, increases, or lack of changes in old animals depending on the particular antioxidant, tissue or animal strain [7–8]. Early comparative studies, however, led some authors to suggest that antioxidants were longevity determinants [9, 10]. This was based on comparative studies of mammals with widely different MLSPs, which showed that some tissue antioxidants like CuZnSOD (superoxide dismutase) were positively correlated with MLSP, but this was true only after dividing their value by the metabolic rate of the whole animal [9, 10]. Since the animals compared greatly differed in body size [9, 10], and larger mammals have lower metabolic rates, the positive correlations observed with MLSP were mainly due to the lower oxygen consumption per kg of the long-lived animals (which is well known since the early work of Max Rubner almost one century ago), not to higher levels of antioxidants. When tissue antioxidants were directly studied as a function of MLSP without that mathematical transformation, 10 out of the 12 independent investigations performed in seven different laboratories showed that endogenous antioxidant enzymes and low molecular weight antioxidants are negatively correlated with maximum longevity [6], while in two studies no correlation was evident [9, 11]. These negative correlations constitute the strongest (although indirect) evidence available to date that the rate of oxygen radical generation in tissues *in vivo* in normal conditions must be lower in long-lived than in short-lived species [6, 12, 13].

On the other hand, what are the effects of increasing tissue antioxidants *in vivo*? Generally in agreement with the results of the comparative investigations described above, out of 16 available studies on life-long experimental modification of antioxidant levels, performed in large samples of animals using dietary supplementation, pharmacological induction, or transgenic techniques to increase the tissue antioxidants, four investigations found some increase in MLSP [14–17] whereas in the twelve remaining studies MLSP did not change [18–29]. The general trend for a lack of effect is even more evident in vertebrate animals, where out of eight investigations, only one described a small increase (12%) in MLSP in mice [16]. while the seven remaining studies found no effect in frogs, mice, or rats [18, 19, 21, 22, 24–26]. Furthermore, 1.5- to 4-fold overexpression of CuZnSOD in various tissues of transgenic mice produced a strongly pathological phenotype affecting many vital organs and led to higher resistance to stress without modifying or even decreasing MLSP in relation to controls [30]. In another investigation, 4- to 13-fold CuZnSOD overexpression in mice generated an array of neurodegenerative changes including swelling and vacuolization of mitochondria in neurons, axonal degeneration and loss of spinal motoneurons during aging [31]. A lack of positive effect of GSH-reductase overexpression on MLSP together with increased resistance to stress has been described in *Drosophila melanogaster* [32], while overexpression of MnSOD in mice increased [33] or did not change [34] resistance to hyperoxia. Laboratories that previously supported the concept that antioxidants were important longevity determinants also agree nowadays with earlier proposals [12, 13] that mitochondrial ROS production, rather than antioxidants, seems to be the relevant parameter related to aging [32, 35].

In the case of mean life span, an increase in its value in antioxidant-treated or antioxidant-induced animals was a much more frequent finding in the studies described above. Those increases in mean life span suggest that antioxidants can non-specifically protect against many causes of early death – they can increase survival – especially when the experiments are performed under suboptimum conditions. These protective effects can be very important to avoiding early death in human populations living in suboptimum environments and having non-rectangular survival curves. But the general lack of effect of antioxidants on MLSP indicates that they do not slow the intrinsic aging process, which also agrees with their presence at much lower (instead of higher) levels in long-lived than in short-lived species. Many experiments recently performed in SOD knockout mice also support this concept. Thus, aging rate does not seem to change in homozygous CuZnSOD [36, 37], extracellular SOD [38], GSH-peroxidase [39], or heterozygous MnSOD [40], knock-out mutant mice; the effects are limited to lack of modification [40, 41] or increased susceptibility to higher than normal stress [36, 38, 42, 43]. In the case of homozygous MnSOD knockout mice, the lack of this enzyme produces a strongly pathological phenotype (a most interesting observation) that leads to death with dilated cardiomyopathy a few days after birth [44, 45]. But this phenotype is clearly different from normal aging.

The rate of mitochondrial oxygen radical production: a main cause of aging rate?

In the previous section it was shown that antioxidants do not control the endogenous rate of aging and MLSP. What is then the connection between aging and oxidative stress? Various studies point to two parameters that can fulfill such a role. One parameter that correlates with aging rate in the appropriate sense is the rate of generation of reactive oxygen species (ROS) by mitochondria. All the investigations performed to date have shown that the rate of ROS production of mitochondria isolated from post-mitotic tissues (the ones more relevant to aging) is lower in long-lived than in short-lived species [4, 46, 47]. This occurs in all kinds of long-lived homeothermic animals independently of their rates of oxygen consumption, low in mammals of large body size and high in birds of small size. This key observation can explain why endogenous tissue antioxidants correlate negatively with MLSP: long-lived animals have constitutively low levels of antioxidants simply because they produce ROS at a low rate. The studies in birds are specially illustrative since these animals live much longer than mammals of similar body size and metabolic rate [48, 49] which strongly disagrees with the rate of living theory of aging when applied to comparisons between these two phylogenetic classes (while it still holds essentially true when comparing between different birds, or comparing between different mammals). However, in spite of their high rates of oxygen consumption, birds have low rates of mitochondrial free radical production, in many cases because their free radical leak (the percentage of O_2 converted to ROS at their respiratory chain) is lower than in rodents [4, 47], A lower rate of mitochondrial ROS production has been found in pigeons (MLSP = 35 years) than in rats (MLSP = 4 years) as well as in canaries (MLSP = 24 years) and parakeets (MLSP = 21 years) when compared to mice (MLSP = 3.5 years), thus extending the observation to three different bird families: *Columbiformes, Passeriformes, and Psitacciformes* [4]. Both small birds and large mammals have low rates of ROS production, in agreement with their slow aging rates, whereas metabolic rate is slow in the second but high in the first kind of animals. Thus, the mitochondrial rate of ROS generation correlates better than the metabolic rate with maximum longevity. The site in the respiratory chain responsible for the longevity-related difference in ROS production has been studied in rats and pigeons, and has been found to be Complex I. It is interesting that the only known manipulation that is able to decrease aging rate, caloric restriction, does not change antioxidants in a consistent way [50] but decreases mitochondrial ROS generation both in mice [50] and rats [51–53], and the difference in ROS production between ad libitum fed and restricted rats is is also localized at Complex I, not at Complex III [51–53]. This suggests that decreasing mitochondrial oxygen radical generation at Complex I is a common mechanism used to slow down aging both across animal species, and inside a given species during caloric restriction.

mtDNA oxidative damage and animal longevity

What is the consequence of the low rate of ROS production of long-lived animals at the level of DNA? This is specially relevant since once a postmitotic cell has lost all the copies of a given gene it can not be recuperated, at variance with detrimental changes in proteins or lipids, molecules that can be resynthesized if the corresponding genes are still intact. It has been recently found that the level of oxidative damage measured as 8-oxo,7,8-dihydro-2′-deoxyguanosine (8-oxodG) in heart and brain mitochondrial DNA (mtDNA) of homeothermic animals (mammals or birds) also correlates negatively with MLSP whereas this does not happen in the case of nuclear DNA (nDNA) [54, 55]. Most interestingly, long-term caloric restriction in rodents also decreases heart and liver 8-oxodG in mtDNA but not in nDNA [51, 53]. Furthermore, agreeing with an early proposal coming from studies limited to rat liver [56], it has been observed that steady-state oxidative damage is higher in mtDNA than in nDNA in the heart and brain of all the 11 species of mammals or birds studied [54, 55].

The above described findings indicate that the flux of oxidative damage through DNA is higher in the mtDNA of short-lived than in that of long-lived animals, as well as in the mtDNA when compared to nDNA in all species, mammals or birds [54, 57]. Thus, it is the rate of oxygen radical attack to mtDNA what seems to differ between animals with different longevity [57]. Why would this be important if there is repair of 8-oxodG? (it is now known that 8-oxodG is actively repaired not only in the nucleus, but also, and probably even more intensely, in the mitochondria).

It would be relevant because ROS can also cause many other kinds of DNA damage in addition to 8-oxodG generation, and some of these changes are not efficiently repaired and accumulate during aging. These irreversible changes – somatic mutations – increase strongly during aging in postmitotic cells like those of skeletal muscle, heart, or neurons. The relatively higher rate of mitochondrial ROS production of short-lived animals can be an important cause of their much faster rate of accumulation of mtDNA mutations during aging, which occurs after 70–100 years in humans, but after only 2–3 years in mice. Furthermore, since 8-oxodG is mutagenic [58], the higher 8-oxodG level present in the mtDNA of short-lived animals in relation to that of long-lived ones would additionally contribute to their higher rate of accumulation of mtDNA mutations. The higher 8-oxodG level of mtDNA in comparison with nDNA, or of mtDNA in ad libitum-fed versus caloric restricted animals, would have a similar consequence: a higher rate of accumulation of somatic mutations in both cases.

Aging rate and sensitivity to lipid peroxidation

Studies summarized in the previous sections indicate that MLSP is inversely related to mitochondrial free radical generation and mtDNA oxidative damage. Nevertheless, additional factors can also lead to a low level of oxidative damage in long- vs. short-lived animal species. Oxygen radicals attack all cellular macromolecules, not only DNA, and these other cellular components may be also an important source of damage. The polyunsaturated fatty acid (PUFA) residues of phospholipids are extremely sensitive to oxidation. Every phospholipid in every membrane of the cell contains an unsaturated fatty acid residue esterified to the 2-hydroxyl group of the glycerol moiety of the phospholipid. Many of these are polyunsaturated and the presence of a methylene group between two double bonds renders the fatty acid sensitive to ROS-induced damage, their sensitivity to oxidation exponentially increasing as a function of the number of double bonds per fatty acid molecule [59]. Consequently, the high concentration of PUFAs in phospholipids not only makes them prime targets for reaction with oxidizing agents but also enables them to participate in long free radical chain reactions.

A low degree of fatty acid unsaturation in cellular membranes, and particularly in the inner mitochondrial membrane, may be advantageous by decreasing their sensitivity to lipid peroxidation. This would also protect other molecules against lipoxidation-derived damage. In agreement with this, it has been found that long-lived animals have a lower degree of total tissue and mitochondrial fatty acid unsaturation (low Double Bond Index, DBI) than short-lived ones. Thus, an early study found that the DBI of liver mitochondrial phospholipids was lower in pigeons (MLSP, 35 years) than in rats (MLSP, 4 years; Figure 1A), and was also lower in humans (MLSP, 122 years) than in pigeons [60]. These results were later confirmed by another independent laboratory [61]. Furthermore, also comparing rat with pigeon, the same result was obtained (Figure 1) in heart mitochondria [62] and skeletal muscle (unpublished results). Nevertheless, the possibility remained that the low DBI found in pigeon (Order *Columbiformes)* has a physiological significance unrelated to MLSP. To discern whether a low DBI is a general characteristic of these highly long-lived animals, additional bird species were studied. It was then observed that the heart DBI of both canary (MLSP, 24 years; Order *Passeriformes)* and parakeet (MLSP, 21 years; Order *Psittaciformes*) was lower than that of mice (MLSP, 3.5 years) (Figure 1B) [63]. A negative correlation between DBI and MLSP in mammals with different longevities was also obtained in liver mitochondrial phospholipids [64], and heart [65] (Figure 2, upper left) and liver [66] total phospholipids. Thus, a low degree of unsaturation of cellular membranes in postmitotic tissues seems to be a characteristic of long-lived vertebrate homeotherms, both birds and mammals.

What is the physiological significance of that characteristic? Various possibilities have been proposed. Some authors [67–69] have suggested that mammals of large body size have a low DBI in order to decrease their metabolic rates, because the

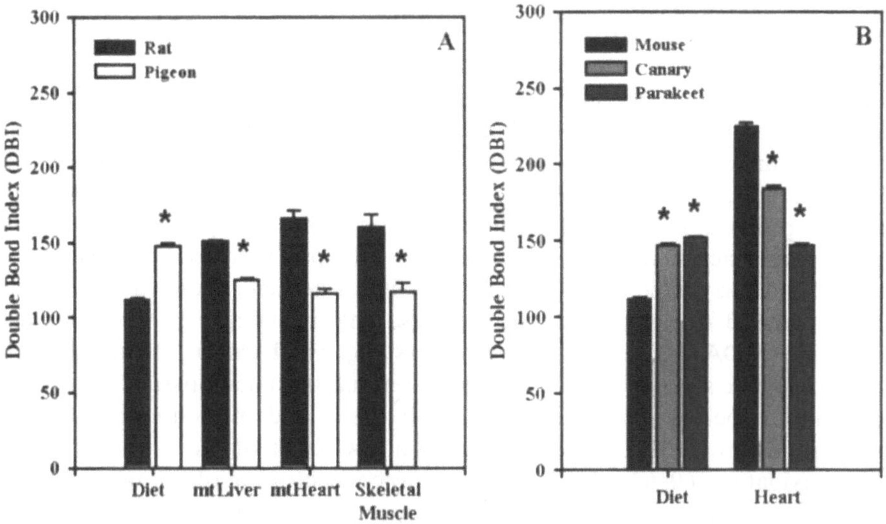

Figure 1. *(A) Double bond index (DBI) of fatty acid present in the diet and tissue lipids (liver and heart mitochondria and skeletal muscle) of rats (MLSP, 4 years) and pigeons (MLSP, 35 years). (B) DBI of fatty acids present in the diet and heart lipids of mice (MLSP, 3.5 years), canaries (MLSP, 24 years) and parakeets (MLSP, 21 years). DBI=[(Σmol% Monoenoic × 1) + (Σmol% Dienoic × 2) + (Σmol% Trienoic × 3) + (Σmol% Tetraenoic × 4) + (Σmol% Pentaenoic × 5) + (Σmol% Hexaenoic × 6)].*

lower the DBI of a membrane, the lower would be its permeability to ions (maintenance of various trans-membrane gradients is one of the main determinants of metabolic rate), suggesting that membranes can act as pacemakers for overall metabolic activity. In agreement with this hypothesis, the permeability to Na^+ and K^+ in liver hepatocytes [70] and to H^+ in the inner mitochondrial membranes [71] correlates negatively with body size. While this could be true for mammals of different sizes, it cannot explain the low DBI of birds because they have a metabolic rate similar or even higher than that of mammals of similar size. But the studied birds and the mammals of large body size share a common trait: their MLSP is very high (they age slowly) [72]. Thus, it can be hypothesized that the low DBI of long-lived homeotherms (either mammals or birds) could have been selected during evolution to decrease membrane lipid peroxidation and its lipoxidative consequences to other cellular macromolecules including proteins [73] and DNA [74]. Thus, the low DBI of long-lived mammals of large body size would protect their tissues against oxidative damage and, simultaneously, it could contribute to lower their metabolic rate, whereas in birds only the first of these two functions can be operative. Therefore, the general relationship in all homeotherms (either mammals or birds) is the negative association between DBI and MLSP, not between DBI and metabolic rate. The low DBI of birds does not fit with their very high metabolic rates, whereas it does fit with their high longevity. Thus, the high specific metabolic activity of birds must be due to other factors different from the low degree of unsaturation of their cellular membranes.

Lipid peroxidation generates hydroperoxides and endoperoxides, which undergo fragmentation to produce a broad range of reactive intermediates, like alkanals, alkenals, hydroxyalkenals, glyoxal, and malondialdehyde (MDA) [75]. These carbonyl compounds, and possibly their peroxide precursors, react with nucleophilic groups in proteins, resulting in chemical modification of the protein. The modification of amino acids in proteins by products of lipid peroxidation results in the chemical, nonenzymatic formation of a variety of adducts grouped in a family of products termed Advanced Maillard Products, that includes, e.g., the malondialdehyde-lysine (MDA-lys) adducts, among others [76], which can be useful indicators of protein oxidative stress *in vivo*. In this context, it has been demonstrated that in long-lived animal species a low degree of total tissue and mitochondrial fatty acid unsaturation (low DBI) is accompanied by a low sensitivity to *in vivo* and *in vitro* lipid peroxidation [60, 62, 63, 65, 66] and a low concentration of lipoxidation-derived adducts in several tissues and mitochondrial proteins [62, 66] (see Figure 2 for heart phospholipids). A negative correlation between sensitivity to lipid autoxidation and MLSP in homogenates from mammalian kidney and brain has also been described by others [77].

However, the occurrence of correlation does not necessarily means that a cause-effect relationship is operative. In order to clarify whether the low DBI of long-lived animals protects their mitochondria from lipid oxidation and lipoxidation-derived protein modification, an experimental dietary study of *in vivo* modification of the DBI of rat heart mitochondria was performed [78]. The diets used were specially designed to partially circumvent the homeostatic system of compensation of dietary-induced changes in DBI which operates at tissue level. The analysis of heart mitochondria showed that the dietary manipulation was successful, since the DBI was lower in the SAT than in UFA group. The decrease in the DBI significantly lowered *in vivo* levels of lipid peroxidation and MDA-lysine in heart mitochondria [78]. Further studies (unpublished) showed us that experimentally decreasing the DBI in brain tissue also lowered MDA-lys, N^{e}-(carboxymethyl)lysine, N^{e}-(carboxyethyl)-lysine and aminoadipic semialdehyde (a component of protein carbonyls) in proteins and 8-oxodG in mtDNA. These observations demonstrate that lowering the DBI of cellular membranes can protect post-mitotic tissues against lipid peroxidation and lipoxidation-derived protein damage.

The membrane acyl composition of the mammals and birds studied indicate that their biological membranes maintain a similar fatty acid average chain length (around 18 carbon atoms), and a similar ratio of saturated vs unsaturated fatty acids irrespective of animal longevity. The low DBI observed in long-lived species is obtained by modulating the type of unsaturated fatty acid that participates in membrane composition. So, there is a systematic redistribution between the types of PUFAs present from the highly unsaturated arachidonic (20:4n-6, AA) and docosahexaenoic (22:6n-3, DHA) acids in short-lived animals to the less unsaturated linoleic acid (18:2n-6, LA), and, in some cases, linolenic acid (18;3n-3, LNA) in the

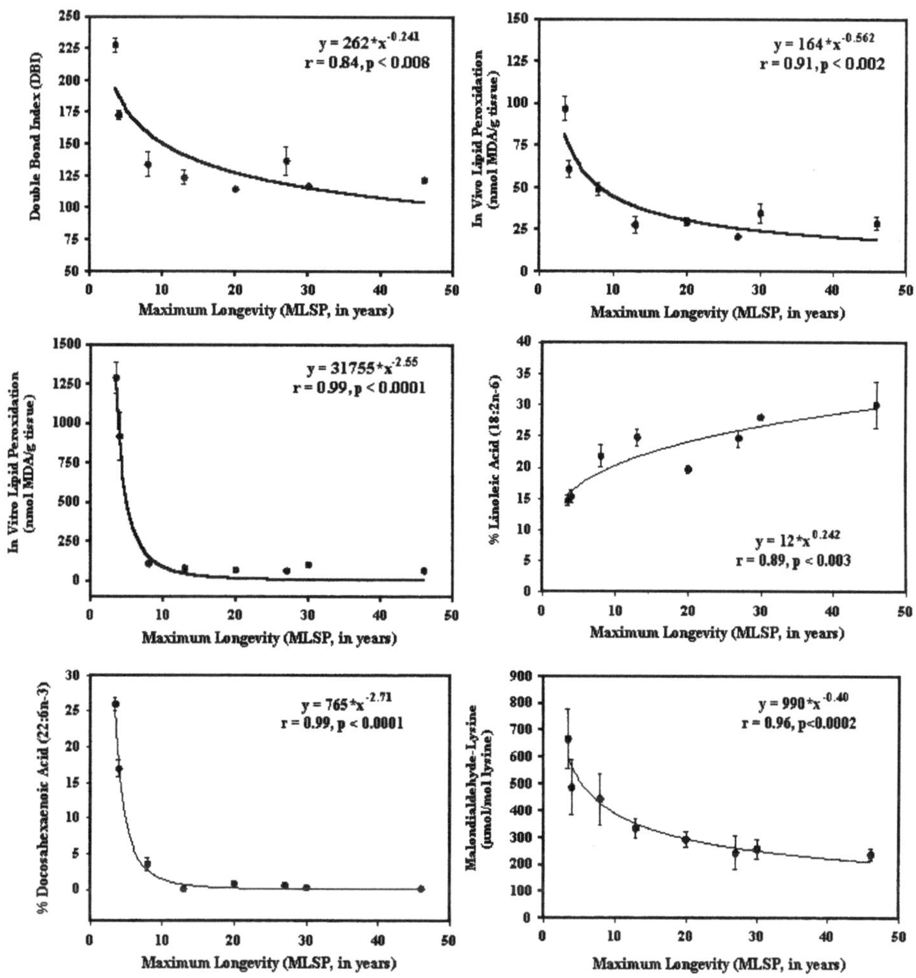

Figure 2. *Relationship between maximum longevity (MLSP) and double bond index (DBI, upper left), in vivo lipid peroxidation (upper right), in vitro lipid peroxidation (middle left), linoleic acid (18: 2n-6) content (middle right), docosahexaenoic acid (22: 6n-3) content (lower left) and the lipoxidation-derived protein damage marker malondialdehyde-lysine (lower right) in heart phospholipids of eight mammalian species. The MLSP of the selected species are: mouse, 3.5 years (Mus musculus); rat, 4 years (Rattus norvegicus); guinea pig, 8 years (Cavia porcellus); rabbit, 13 years (Oryctolagus caniculus); sheep, 20 years (Ovis aries); pig, 27 years (Sus scrofa); cow, 30 years (Bos taurus) and horse, 46 years (Equus caballus). Values are means ± SEM.*

long-lived ones, at mitochondrial and tissue level [60, 62–66] (Figure 2). Furthermore, the DBI of the different diets did not correlate with MLSP. This indicates that the inter-species differences could not be due to variations in the degree of unsaturation of dietary fats.

While the findings described above had not been previously related to MLSP, two comparative reports relating fatty acid unsaturation to mammalian body size existed. In the first study it was observed that DHA acutely decreased as body size increased in the order mouse-rat-rabbit-man-whale [79], which is also an order of increasing MLSP, although this was not commented by the authors. In the second report [67] it was found that the DBI was negatively correlated with body size in the heart, skeletal muscle and kidney cortex of five species, mouse-rat-rabbit-sheep-cattle, whereas in the liver the negative trend did not reach statistical significance and in the brain no correlation was observed. The fatty acids mainly responsible for these differences were again DHA and AA that decreased as body size increased, and LA which showed progressively larger levels in animals of larger size. This finding has recently been extended to a wide variety of mammals and birds species collating data from the literature regarding the acyl composition of total tissue phospholipids [80].

In relation to the mechanisms responsible for the particular fatty acid composition of homeothermic vertebrates with different longevities, two are possible: the fatty acid desaturation pathway, and the deacylation-reacylation cycle. The estimation of delta-5 and delta-6 desaturase activities indicated that they were various fold lower in long-lived species compared with short-lived ones [60, 62, 66]. This can explain why DHA and AA decrease and LA and/or LNA increase from short- to long-lived animals. The same was also postulated in the studies referred above since desaturases are the limiting enzymes in the pathways of n-3 and n-6 synthesis of the highly unsaturated PUFAs AA (20:4n-6) and DHA (22:6n-3) from their dietary precursors, LA (18:2n-6) and LNA (18:3n-3) respectively. The authors concluded that the main factor responsible for the different DBIs was the possession of low delta-5/-6 desaturase activities in animals of large body size [67, 80] Thus, desaturation pathways would make available *in situ* the n-6 and n-3 fatty acids to phospholipid acyltransferases in order to remodel the phospholipid acyl groups, postulating the presence of constitutively low species-specific desaturase activities in long-lived animals. The acyltransferase/n-6 desaturase activity ratio is about 10:1 in tissues [81], which suggests that the desaturases are the main limiting factor responsible for the observed DBI-longevity relationship. However, a relevant role for a phospholipid-specific deacylation-reacylation system can not be discarded since it has recently been observed that this redistribution affects particularly phosphatidylcholine and phosphatidyletanolamine fractions in liver mitochondria, although not cardiolipin [82]. The presence of constitutively low desaturase activities in long-lived animals can explain why feeding corn oil (rich in LA) to primates increases mainly LA (to 30% of total fatty acids) instead of AA (only to 10% of total) in their tissues [83], whereas in short-lived rodents dietary LA leads to strong increases in AA. Similarly to those primates, human monastic communities which chronically consume only corn oil as the main dietary fat source (67% rich in LA) have lipid profiles with around 30% LA but only 9% AA in their lipoproteins [84].

Animals with high MLSP have a low degree of membrane fatty acid desaturation based in a redistribution between types of PUFAs without any alteration in the total (%) PUFA content or in the average chain length. This is an elegant evolutionary strategy, since it can allow to decrease the sensitivity to lipid peroxidation and lipoxidation-derived damage to cellular macromolecules without strongly altering fluidity/microviscosity, a fundamental property of cellular membranes for the proper function of receptors, ion pumps, and transport of metabolites, among others. This occurs because membrane fluidity is known to increase acutely with the formation of the first and less with the second double bond due to their introduction of "kinks" in the fatty acid molecule, whereas additional (the third and following) double bonds cause few further variations in fluidity [85]. This is so because the kink has a larger impact on fluidity when the double bond is situated near the center of the fatty acid chain (first double bond) than when it is situated progressively nearer to its extremes (next double bond additions). In the case of sensitivity to lipid peroxidation, however, double bonds increase it irrespective of being situated at the center or laterally on the fatty acids. Thus, by substituting fatty acids with 4 or 6 double bonds by those having only 2 (or sometimes 3) double bonds, the sensitivity to lipid peroxidation is strongly decreased in long-lived animals, whereas the fluidity of the membrane is essentially maintained. This phenomenon, reminiscent of membrane acclimation to temperature at PUFA level in poikilotherms, has been named *homeoviscous longevity adaptation* [5]. The adjustment of the DBI of each organ and species independently of the diet also indicates that it is an endogenous trait under genome control. This occurs through PUFA-induced repression of the expression of genes controlling PUFA synthesis, whereas PUFA-deficient diets increase the expression of those genes [86]. These genome-based mechanisms are responsible for the decreases in PUFAn-6 induced by diets rich in PUFAn-3 as well as for the reverse, which operate mainly trough variations in delta-5/-6 activities [87].

Strongly unsaturated fatty acids like AA and DHA can have detrimental effects *in vivo*. Examples of this include decreases in respiratory control and increases in proton leak in mitochondria, increased mitochondrial breakage and dysfunction, peroxisome proliferation, fatal ventricular fibrillation in rats, neurological damage, increased lipid peroxidation in association with various diseases, increased incidence of death from apoplexy, or sudden cardiac death in humans. Increases of more than one order of magnitude in AA (to 500 μM) occur in the brain during ischemia and even concentrations of AA and eicosapentaenoic acid (20:5n-3, EPA) in the much lower 20–40 μM range uncouple mitochondria and cause tissue edema [88]. Hypermetabolic uncoupling effects of thyroid hormones on rat liver mitochondria are due to a great extent to increased AA/LA ratios caused by increases in desaturase activities induced by the hormone, whereas LA is considered a "proton plug" or coupler [see ref. 89 for review]. Furthermore, the largest amounts of unsaturated fats in the healthy human diet must be present as fatty acids with low degrees of unsaturation like oleic acid and LA, whereas beneficial levels of dietary n-3 PUFAs (the n-3 "paradox") occur only at low 1% optimum dietary levels. Furthermore, these beneficial effects, which seem mainly related to avoidance of blood coagulation and perhaps to promotion of apoptosis of heavily damaged cells [90], are observed in

***Table 1**. Known characteristics of long-lived animals related to oxidative stress*

1. Low levels of endogenous enzymatic and non-enzymatic antioxidants
2. LOW LEVELS OF MITOCHONDRIAL ROS GENERATION
3. Low steady-state levels of 8-oxodG specifically in mitochondrial DNA
4. Low rate of accumulation of somatic mtDNA mutations
5. LOW DEGREES OF FATTY ACID UNSATURATION IN CELLULAR MEMBRANES
6. Low sensitivity to lipid peroxidation and lipoxidation-derived protein damage

The characteristics shown in bold letters can be mechanistically involved in the determination of the low rate of aging of long-lived animals. Those shown in capital letters (points 2 and 5) can be causal in relation to others: point 2 causing 3, 4 and 6; point 5 causing 6, and perhaps 4? Caloric restricted animals share with long-lived species characteristics 2, 3 and 4. Characteristic 1 is an evolutionary consequence of characteristics 2 and 5, and is obviously not involved in maximum life span determination. For references and further explanation, see text.

humans probably because their conversion of dietary LNA to highly unsaturated fatty acids like DHA is strongly limited thanks to their low delta-5/-6 desaturase activities.

Other observations also suggesting that fatty acid desaturation influences aging rate include: (i) Decreases in the less unsaturated LA and LNA and increases in the highly unsaturated AA, 22:4n-3, 22:5n-3 and DHA membrane fatty acids have been described during aging in rat liver or heart [91, 92], while fasting decreases delta-5/-6 desaturases, and caloric restriction avoids the age-related increases in DBI by increasing oleic acid and LA and decreasing 22:4n-3, 22:5n-3, DHA and the peroxidizability index in rat liver mitochondrial phospholipids [91] and decreases 22:6n-3 in rat heart [93]. (ii) The senescent accelerated prone mouse (SAM-P) has higher levels of the very unsaturated AA and DHA and peroxidizability index and lower levels LA than SAM-resistant controls [94]. (iii) Most interestingly, Eskimos are human populations showing unusually low incidence of coronary heart disease, psoriasis, rheumatoid arthritis and asthma and have very low levels of AA in plasma phospholipids due to a genetic lack of delta-5 desaturase activity which persists even after changing them to a LA-rich diet [95].

Final conclusions

Nowadays only two traits controlling the level of oxidative stress are known to correlate with the maximum longevity of animals in the appropriate sense: the rate of mitochondrial oxygen radical generation and the degree of unsaturation of membrane fatty acids (see Table 1). Both can lead to a lower endogenous generation of damage in long-lived species. These two molecular traits are significantly lower in all the relatively long-lived homeothermic vertebrates so far studied, may be main causes of the low rate of aging of long-lived mammals or birds, and are genetically determined characteristics as it is expected for causes of aging and maximum longevity. They fit with the concept that causes of aging must be endogenous, must operate progressively, and lead to the accumulation of irreversible detrimental changes (like mtDNA mutations). Both work through a common mechanism: they decrease the rate of generation of endogenous damage. This makes sense since that kind of mechanism is less energetically expensive and more efficient than, for instance, increasing antioxidants or repair in order to achieve a high MLSP [57]. Both determine rate processes (rates of ROS production and rates of lipid peroxidation or lipoxidation-derived damage), which is consistent their hypothesized role: to control a rate process, aging.

Acknowledgments

The results from our laboratories described in this review were supported by FIS grants to G. Barja 90/0013, 96/1253, and 99/1049, and to R. Pamplona 98/0752 and 00/0753.

References

1. Harman D (1972). The biological clock: the mitochondria? *J Am Geriatr Soc.* 20: 145–7.
2. Beckman KB, Ames B (1998). The free radical theory of aging matures. *Physiol Rev.* 78: 547–81.
3. Weindruch R, Sohal RS (1997). Caloric intake and aging. *N Engl J Med.* 337: 986–94.
4. Barja G (1999). Mitochondrial free radical generation: sites of production in states 4 and 3, organ specificity and relationship with aging rate. *J Bioenerg Biomembr.* 31: 347–66.
5. Pamplona R, Barja G, Portero-Otín M (2002). Membrane fatty acid unsaturation, protection against oxidative stress, and maximum life span: a homeoviscous-longevity adaptation. *Ann NY Acad Sci.* 959: 475–90.
6. Pérez-Campo R, López-Torres M, Cadenas S, Rojas C, Barja G (1998). The rate of free radical production as a determinant of the rate of aging: evidence from the comparative approach. *J Comp Physiol B* 168: 149–58.
7. Barja de Quiroga G, López-Torres M, Pérez-Campo R (1992). Relationship between antioxidants, lipid peroxidation and aging. In: Emerit I, Chance B, eds. *Free Radicals and Aging*. Basel: Birkhäuser, pp. 109–23.
8. Benzi G, Moretti A (1995). Age- and peroxidative stress-related modifications of the cerebral enzymatic activities linked to mitochondria and the glutathione system. *Free Radic Biol Med.* 12: 77–101.
9 Tolmasoff JM, Ono T, Cutler RG (1980). Superoxide dismutase: correlation with life-span and specific metabolic rate in primate species. *Proc Natl Acad Sci USA* 77: 2777–81.
10. Cutler RG (1986). Aging and oxygen radicals. In: Taylor AE, Matalon S, Ward P, eds. *Physiology of Oxygen Radicals*. Bethesda, MD: American Physiological Society, pp. 251–85.
11. Sohal, RS Sohal, BH, Brunk UT (1990). Relationship between antioxidant defenses and longevity in different mammalian species. *Mech Ageing Dev.* 53: 217–27.
12. López-Torres M, Pérez-Campo R, Rojas C, Cadenas S, Barja G (1993). Maximum life span in vertebrates: correlation with liver antioxidant enzymes, glutathione system,

ascorbate, urate, sensitivity to peroxidation, true malondialdehyde, *in vivo* H_2O_2, and basal and maximum aerobic capacity. *Mech Ageing Dev.* 70: 177–99.
13. Barja G, Cadenas S, Rojas C, López-Torres M, Pérez-Campo R (1994). A decrease of free radical production near critical sites as the main cause of maximum longevity in animals. *Comp Biochem Physiol.* 108B: 501–12.
14. Epstein J, Gershon D (1972). Studies on aging in nematodes. IV. The effect of antioxidants on cellular damage and lifespan. *Mech Ageing Dev.* 1: 257–65.
15. Miquel J, Johnson JE (1975). Effects of various antioxidants and radiation protectants on the life span and lipofuscin of *Drosophila* and C57BL/6J mice. *Gerontologist* 15: 25.
16. Heidrick ML, Hendricks LC, Cook DE (1984). Effect of dietary 2-mercaptoethanol on the life span, immune system, tumour incidence and lipid peroxidation damage in spleen lymphocytes of aging BC3F1 mice. *Mech Ageing Dev.* 27: 341–58.
17. Orr WC, Sohal RS (1994). Extension of life span by overexpression of superoxide dismutase and catalase in *Drosophila melanogaster*. *Science* 263: 1128–30.
18. Kohn RR (1971). Effect of antioxidants on lifespan of C57BL/6J mice. *J Gerontol.* 26: 378–80.
19. Clapp NK, Satterfield LC, Bowles ND (1980). Effects of the antioxidant butylated hydroxytoluene (BHT) on mortality in BALB/c mice. *J Gerontol.* 34: 497–501.
20. Enesco HE, Verdone-Smith C (1980). Alpha-Tocopherol increases lifespan in the rotifer Philovina. *Exp Gerontol.* 15: 335–8.
21. Ledvina M, Hodanova M (1980). The effect of simultaneous administration of tocopherol and sunflower oil on the lifespan of female mice. *Exp Gerontol.* 15: 67–71.
22. Porta EA, Joun NS, Nitta RT (1980). Effects of the type of dietary fat at two levels of vitamin E in Wistar male rats during development and aging. *Mech Ageing Dev.* 13: 1–39.
23. Bozavic V, Enesco HE (1986). Effect of antioxidants on rotifer lifespan and activity. *Age* 9: 41–5.
24. Harris SB, Weindruch R, Smith GS, Mickey MR, Walford RL (1990). Dietary restriction alone and in combination with oral ethoxyquin/2-mercaptoethylamine in mice. *J Gerontol.* 45: B141–7.
25. López-Torres M, Pérez-Campo R, Fernandez A, Barba C, Barja de Quiroga G (1993). Brain glutathione reductase induction increases early survival and decreases lipofuscin accumulation in aging frogs. *J Neurosci Res.* 34: 233–42.
26. López-Torres M, Pérez-Campo R, Rojas C, Cadenas S, Barja de Quiroga G (1993). Simultaneous induction of superoxide dismutase, glutathione reductase, GSH and ascorbate in liver and kidney correlates with survival throughout the life span. *Free Radic Biol Med.* 15: 133–42.
27. Orr WC, Sohal RS (1992). The effects of catalase gene overexpression on life span and resistance to oxidative stress in transgenic *Drosophila melanogaster*. *Arch Biochem Biophys.* 297: 35–41.
28. Seto NOL, Hayashi S, Tener GM (1990). Overexpression of Cu-Zn superoxide dismutase in *Drosophila* does not affect life-span. *Proc Natl Acad Sci USA* 87: 4270–4.
29. Staveley BE, Phillips JP, Hilliker A (1990). Phenotypic consequences of copper-zinc superoxide dismutase overexpression in *Drosophila melanogaster*. *Genome* 33: 867–72.
30. Hang TT, Carlson EJ, Gillespie AM, Shi Y, Epstein CJ (2000). Ubiquitous expression of CuZn superoxide dismutase does not extend life span in mice. *J Gerontol.* 55A: B5–9.
31. Jaarsma D, Haasdijk ED, Grashorn JAC, *et al.* (2000). Human Cu/Zn superoxide dismutase (SOD1) overexpression in mice causes mitochondrial vacuolization, axonal degeneration, and premature motoneuron death and accelerates motoneuron disease in

mice expressing a familial amyotrophic lateral sclerosis mutant SOD1. *Neurobiol Disease.* 7: 623–43.
32. Mockett RJ, Sohal RS, Orr WC (1999). Overexpression of glutathione reductase extends survival in transgenic *Drosophila melanogaster* under hyperoxia but not normoxia. *FASEB J.* 13: 1733–42.
33. Wispe JR, Warner BB, Clarck JC, *et al.* (1992). Human Mn-superoxide dismutase in pulmonary endothelial cells of transgenic mice confers protection from oxygen injury. *J Biol Chem.* 267: 23937–41.
34. Ho YS (1994). Transgenic models for the study of lung biology and disease. *Am J Physiol.* 266: L319–53.
35. Sohal RS (2002). Role of oxidative stress and protein oxidation in the aging process. *Free Radic Biol Med.* 33: 37–44.
36. Reaume AG, Elliot J, Hoffman EK, *et al.* (1996). Motor neurons in Cu/Zn superoxide dismutase-deficient mice develop normally but exhibit enhanced cell death after axonal injury. *Nature Genet.* 13: 43–7.
37. Shefner JM, Reaume AG, Flood DG, *et al.* (1999). Mice lacking cytosolic superoxide dismutase display a distinctive motor axonopathy. *Neurology* 53: 1239–46.
38. Carlsson LM, Jonsson J, Edlund T, Marklund SL (1995). Mice lacking extracellular superoxide dismutase are more sensitive to hyperoxia. *Proc Natl Acad Sci USA* 92: 6264–8.
39. Ho YS, Magnenat JL, Bronson RT, *et al.* (1997). Mice deficient in cellular glutathione peroxidase develop normally and show no increased sensitivity to hyperoxia. *J Biol Chem.* 272: 16644–51.
40. Tsan MF, White JE, Caska B, Epstein CJ, Lee CY (1998). Susceptibility of heterozygous MnSOD gene-knockout mice to oxygen toxicity. *Am J Respir Cell Mol Biol.* 19: 114–20.
41. Ho YS, Gargano M, Cao J (1997). Mice lacking copper/zinc superoxide dismutase show no increased sensitivity to hyperoxia. *Am J Respir Crit Care Med.* 155: A17.
42. Ohlemiller KK, McFadden SL, Ding DL, *et al.* (1999). Targeted deletion of the cytosolic C/Zn-superoxide dismutase gene (Sod1) increases susceptibility to noise-induced hearing loss. *Audiol Neurootol.* 4: 237–46.
43. de Haan JB, Bladier C, Griffiths P, *et al.* (1998). Mice with homologous null mutation for the most abundant glutathione peroxidase, Gpx1, show increased susceptibility to the oxidative stress-inducing agents paraquat and hydrogen peroxide. *J Biol Chem.* 273: 22528–36.
44. Li Y, Huang TT, Carlson EJ, *et al.* (1995). Dilated cardiomyopathy and neonatal lethality in mutant mice lacking manganese superoxide dismutase. *Nature Genet.* 11: 376–81.
45. Melov S, Coskun P, Patel M, *et al.* (1999). Mitochondrial disease in superoxide dismutase 2 mutant mice. *Proc Natl Acad Sci USA* 96: 846–51.
46. Ku HH, Brunk UT, Sohal RS (1993). Relationship between mitochondrial superoxide and hydrogen peroxide production and longevity of mammalian species. *Free Radic Biol Med.* 15: 621–7.
47. Barja G, Cadenas S, Rojas C, Pérez-Campo R, López-Torres M (1994). Low mitochondrial free radical production per unit O_2 consumption can explain the simultaneous presence of high longevity and high aerobic metabolic rate in birds. *Free Radic Res.* 21: 317–28.
48. Calder WA (1985). The comparative biology of longevity and lifetime energetics. *Exp Gerontol.* 20: 161–70.
49. Prinzinger R (1993). Life span in birds and the ageing theory of absolute metabolic scope. *Comp Biochem Physiol.* 105A: 609–15.

50. Sohal RS, Ku HH, Agarwal S, Forster MJ, Lal H (1994). Oxidative damage, mitochondrial oxidant generation and antioxidant defenses during aging and in response to food restriction. *Mech Ageing Dev.* 74: 121–33.
51. Gredilla R, Sanz A, López-Torres M, Barja G (2001). Caloric restriction decreases mitochondrial free radical generation at Complex I and lowers oxidative damage to mitochondrial DNA in the rat heart *FASEB J.* 15: 1589–91.
52. Gredilla R, Barja G, López-Torres M. (2001). Effect of short-term caloric restriction on H_2O_2 production and oxidative DNA damage in rat liver mitochondria and location of the free radical source. *J Bioeng Biomembr.* 33: 279–87.
53. López-Torres M, Gredilla R, Sanz A, Barja G. (2002). Influence of aging and long-term caloric restriction on oxygen radical generation and oxidative DNA damage in rat liver mitochondria. *Free Radic Biol Med.* 32: 882–9.
54. Barja G, Herrero A (2000). Oxidative damage to mitochondrial DNA is inversely related to maximum life span in the heart and brain of mammals. *FASEB J.* 14: 312–18.
55. Herrero A, Barja G (1999). 8-oxodeoxyguanosine levels in heart and brain mitochondrial and nuclear DNA of two mammals and three birds in relation to their different rates of aging. *Aging Clin Exper Res.* 11: 294–300.
56. Richter Ch, Park JW, Ames BN (1988). Normal oxidative damage to mitochondrial and nuclear DNA is extensive. *Proc Natl Acad Sci USA* 85: 6465–7.
57. Barja G (2000). The flux of free radical attack through mitochondrial DNA is related to aging rate. *Aging Clin Exper Res.* 12: 342–55.
58. Cheng KC, Cahill DS, Kasai H, Nishimura S, Loeb LA (1992). 8-Hydroxyguanine, an abundant form of oxidative DNA damage, causes G-T and A-C substitutions. *J Biol Chem.* 267: 166–72.
59. Bielski BHJ, Arudi RL, Sutherland MW (1983). A study of the reactivity of $HO_2/O_2^{\cdot-}$ with unsaturated fatty acids. *J Biol Chem.* 258: 4759–61.
60. Pamplona R, Prat J, Cadenas S, *et al.* (1996). Low fatty acid unsaturation protects against lipid peroxidation in liver mitochondria from longevous species: the pigeon and human case. *Mech Ageing Dev.* 86: 53–66.
61. Gutierrez AM, Reboredo GR, Arcemis CJ, Catalá A (1997). Non-enzymatic lipid peroxidation of microsomes and mitochondria isolated from liver and heart of pigeon and rat. *Int J Biochem Cell Biol.* 32: 73–9.
62. Pamplona R, Portero-Otín M, Requena JR, Thorpe SR, Herrero A, Barja G (1999). A low degree of fatty acid unsaturation leads to lower lipid peroxidation and lipoxidation-derived protein modification in heart mitochondria of the longevous pigeon than in the short-lived rat. *Mech Ageing Dev.* 106: 283–96.
63. Pamplona R, Portero-Otín M, Riba D, *et al.* (1999). Heart fatty acid unsaturation and lipid peroxidation, and aging rate, are lower in the canary and the parakeet than in the mouse. *Aging Clin Exp Res.* 11: 44–9.
64. Pamplona R, Portero-Otín M, Ruiz C, Prat J, Bellmunt MJ, Barja G (1998). Mitochondrial membrane peroxidizability index is inversely related to maximum life span in mammals. *J Lipid Res.* 39: 1989–94.
65. Pamplona R, Portero-Otín M, Ruiz C, Gredilla R, Herrero A, Barja G (1999). Double bond content of phospholipids and lipid peroxidation negatively correlate with maximum longevity in the heart of mammals. *Mech Ageing Dev.* 112: 169–83.
66. Pamplona R, Portero-Otín M, Riba D, *et al.* (2000). Low fatty acid unsaturation: a mechanism for lowered lipoperoxidative modification of tissue proteins in mammalian species with long life span. *J Gerontol.* 55A: B286–91.

67. Couture P, Hulbert AJ (1995). Membrane fatty acid composition of tissues is related to body mass of mammals. *J Membr Biol.* 148: 27–39.
68. Hulbert AJ, Else PL (1999). Membranes as possible pacemakers of metabolism. *J Theor Biol.* 199: 257–74.
69. Hulbert AJ, Else PL (2000). Mechanisms underlying the cost of living in animals. *Ann Rev Physiol.* 62: 207–35.
70. Porter RK, Brand MD (1995). Causes of differences in respiration rate of hepatocytes from mammals of different body mass. *Am J Physiol.* 269: R1213–24.
71. Porter RK, Brand MD (1993). Body mass dependence of H+ leak in mitochondria and its relevance to metabolic rate. *Nature* 362: 628–9.
72. Holmes DJ, Austad SN (1995). The evolution of avian senescence patterns: implications for understanding primary aging processes. *Am Zool.* 35: 307–17.
73. Refsgaard HHF, Tsai L, Stadtman ER (2000). Modifications of proteins by polyunsaturated fatty acid peroxidation products. *Proc Natl Acad Sci USA* 97: 611–16.
74. Marnett LJ (2000). Oxyradicals and DNA damage. *Carcinogenesis* 21: 361–70.
75. Esterbauer H, Schaur RJ, Zollner H (1991). Chemistry and biochemistry of 4-hydroxynonenal, malondialdehyde and related aldehydes. *Free Radic Biol Med.* 11: 81–128.
76. Degenhardt TP, Brinkmann-Frye E, Thorpe SR, Baynes JW (1998). Role of carbonyl stress in aging and age-related diseases. In: O'Brien J, Nursten HE, Crabbe MJC, Ames JM, eds. *The Maillard Reaction in Foods and Medicine.* Cambridge, UK: The Royal Society of Chemistry, pp. 3–10.
77. Cutler RG (1985). Peroxide-producing potential of tissues: inverse correlation with longevity of mammalian species. *Proc Natl Acad Sci USA* 82: 4798–802.
78. Herrero A, Portero-Otín M, Bellmunt MJ, Pamplona R, Barja G (2001). Effect of the degree of fatty acid unsaturation of rat heart mitochondria on their rates of H_2O_2 production and lipid and protein oxidative damage. *Mech Ageing Dev.* 122: 427–43.
79. Gudbjarnason S (1989). Dynamics of n-3 and n-6 fatty acids in phospholipids of heart muscle. *J Intern Med.* 225: 117–28.
80. Hulbert AJ, Rana T, Couture P (2002). The acyl composition of mammalian phospholipids: an allometric analysis. *Comp Biochem Physiol Part B* 132: 515–27.
81. Ivanetich KM, Bradshaw JJ, Ziman MR (1996). Delta-6 desaturase: improved methodology and analysis of the kinetics in a multi-enzyme system. *Biochim Biophys Acta* 1292: 120–32.
82. Portero-Otín M, Bellmunt MJ, Ruiz MC, Barja G, Pamplona R (2001). Correlation of fatty acid unsaturation of the major liver mitochondrial phospholipid classes in mammals to their maximum life span potential. *Lipids* 36: 491–8.
83. Charnock JS, Abeywardena MY, Poletti VM, McLennan PL (1992). Differences in fatty acid composition of various tissues of the marmosset monkey (*Callithrix jacchus*) after different lipid supplemented diets. *Comp Biochem Physiol.* 101A: 387–93.
84. Solà R, La Ville AE, Richard JL, *et al.* (1997). Oleic acid rich diet protects against the oxidative modification of high density lipoprotein. *Free Radic Biol Med.* 22: 1037–45.
85. Brenner RR (1984). Effect of unsaturated fatty acids on membrane structure and enzyme kinetics. *Progr Lipid Res.* 23: 69–96.
86. Sessler A, Ntambi JB (1998). Polyunsaturated fatty acid regulation of gene expression. *J Nutr.* 128: 923–6.
87. Nakamura MT, Cho HP, Clarke SD (2000). Regulation of hepatic delta-6 desaturase expression and its role in the polyunsaturated fatty acid inhibition of fatty acid synthase gene expression in mice. *J Nutr.* 130: 1561–5.

88. Takeuchi Y, Morii H, Tamura M, Hayaishi O, Watanabe Y (1991). A possible mechanism of mitochondrial dysfunction during cerebral ischemia: inhibition of mitochondrial respiratory activity by arachidonic acid. *Arch Biochem Biophys*. 289: 33–8.
89. Hoch FL (1992). Cardiolipins and membrane function. *Biochim Biophys Acta* 1113: 71–133.
90. Troyer D, Fernandes G (1996). Nutrition and apoptosis. *Nutr Res*. 16: 1959–87.
91. Laganiere S, Yu BP (1993). Modulation of membrane phospholipid fatty acid composition by food age and food restriction. *Gerontology* 39: 7–18.
92. Yu BP, Chen JJ, Kang CM, Choe M, Maeng YS, Kristal BS (1996). Mitochondrial aging and lipoperoxidative products. *Ann NY Acad Sci*. 786: 44–56.
93. Pamplona R, Portero-Otín M, Bellmunt MJ, Gredilla R, Barja G (2002). Aging increases Ne-(carboxymethyl)lysine and caloric restriction decreases Ne-(carboxyethyl)lysine and Ne-(malondialdehyde)lysine in rat heart mitochondrial proteins. *Free Radic Res*. 36: 47–54.
94. Choi JH, Kim J, Moon YS, Chung HY, Yu BP (1996). Analysis of lipid composition and hydroxyl radicals in brain membranes of senescent-accelerated mice. *Age* 19: 1–5.
95. Horrobin DF, Manku MS (1987). Genetically low arachidonic acid and high dihomo-gammalinolenic acid levels in Eskimos may contribute to low incidence of coronary heart disease, psoriasis, arthritis and asthma. In: Lands VEM, ed. *Polyunsaturated Fatty Acids and Eicosanoids*. Champaign, Illinois, USA: American Oil Chemists Society, pp. 413–15.

Genomic Instability in Human Premature Aging

Vilhelm A. Bohr and Patricia L. Opresko
Laboratory of Molecular Gerontology, National Institute on Aging, NIH, 5600 Nathan Shock Drive, Baltimore, MD 21224, USA

Human premature aging disorders

A number of premature aging disorders have been described in humans. In patients with these disorders, aging-like symptoms and age-associated diseases appear much earlier than in the average normal individual; thus, the premature aging disorders are useful models for the study of the aging process (Figure 1). Werner syndrome (WS) is the most characterized premature aging disorder. Patients with WS have a large number of signs and symptoms of normal aging at a younger age than normal individuals. However, not all symptoms of WS resemble the normal aging process, and WS is best described as a segmental progeroid disorder. The premature aging disorders include the DNA repair defective disease, xeroderma pigmentosum (XP), which includes seven complementation groups (separate genetic disorders). In this condition, the deficiency of the DNA repair pathway, nucleotide excision, leads to a severely increased incidence of cancer. In Cockayne syndrome, there is also a DNA repair defect, and in addition to the features of premature aging, these individuals have severe neurological deficits. Rothmund-Thomson disease, Hutchinson-Guilford and progeria are other examples of this category of disorders. All of these diseases are very rare conditions in the general population. In several cases, the disorders are associated with a mutation in a single gene, which has by now been identified, cloned and characterized. This means that molecular biochemical experimentation can be done. Complementation assays with transfected mutant cell lines or purified proteins added to extracts from mutant cell lines can be used to study the basis of molecular genetic defects (Figure 1).

T. von Zglinicki (ed.), Aging at the Molecular Level, 65–77.

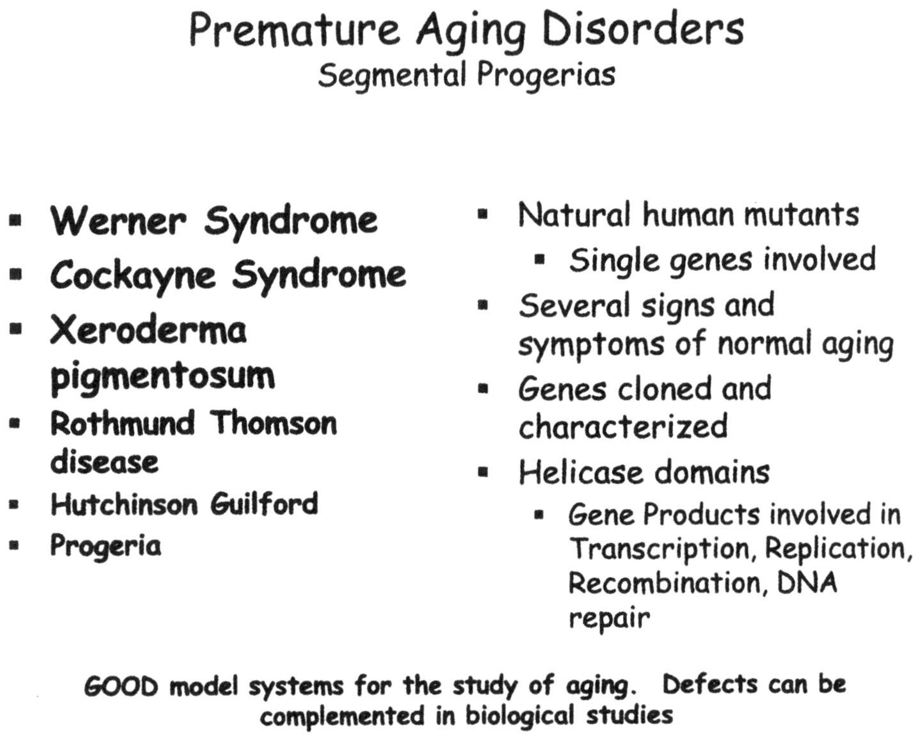

Figure 1. *Important points about premature aging.*

Werner syndrome (WS) and Werner protein (WRN)

Werner Syndrome (WS) is a homozygous recessive disease characterized by early onset of many aspects of normal aging, such as wrinkling of the skin, graying of the hair, cataracts, diabetes, and osteoporosis. Cancers, particularly sarcomas, have been seen in these patients with increased frequency. The symptoms of WS begin to appear around the age of puberty, and most patients die before age 50.

WS is caused by mutations in a gene (*WRN*) belonging to the RecQ family of DNA helicases. The WRN protein (WRN) is a DNA unwinding enzyme (Figure 2). It is a DNA-dependent ATPase and catalyzes strand displacement in a nucleotide-dependent reaction. It also has 3′-5′ exonuclease activity [1, 2]. Evidence *in vitro* indicates that the WRN helicase and exonuclease may act coordinately to remove recombination and repair intermediates [3]. Thus, WRN contains helicase, exonuclease and ATPase activities.

The precise molecular deficiencies involved in the clinical phenotypes of WS are only partially understood. The phenotypical changes in WS cells are shown in Figure 3. The well-documented genomic instability of WS may point to a defect in

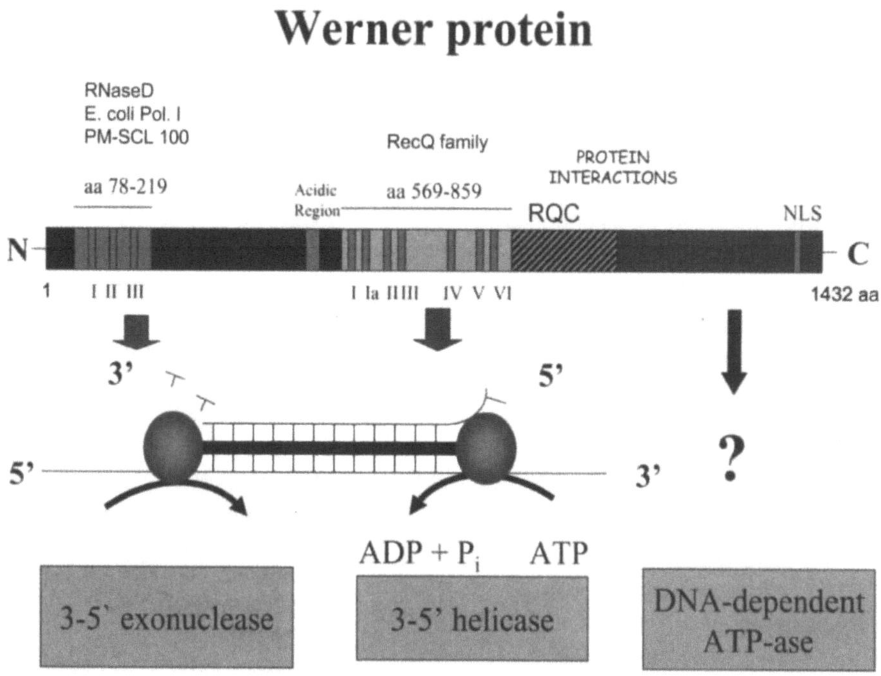

Figure 2. *Functional mapping of the Werner protein.*

replication or recombination. WS cells exhibit premature replicative senescence and delayed S phase progression in some cell lines [4]. A defect in recombination is suggested by the hyper-recombinant phenotype observed in some WS cell lines. WS cell lines are also hypersensitive to drugs that cause DNA interstrand crosslinks [5, 6]. These crosslinks are repaired by recombinational pathways. Moreover, the *WRN* homologs in yeast, *Sgs*1 (*S. cerevisiae*) and *rqh1* (*S. pombe*), as well as the bacterial homolog *recQ*, all suppress illegitimate recombination. WRN, *in vitro,* also has anti-recombinase activity. Thus, there is evidence that WRN functions as an anti-recombinase [for reviews of this material and the following see refs. 1, 7].

Some evidence suggests a role for the WRN in DNA repair. This notion is supported by a report that WS cells are sensitive to the carcinogen 4-nitroquinoline 1-oxide (4-NQO). WS cells are also hypersensitive to topoisomerase I inhibitor camptothecin [8], DNA cross-linking reagents [6] and show mild sensitivity to ionizing radiation [9]. However, WS cells do not exhibit hypersensitivity to UV light or other DNA damaging agents suggesting that a repair deficiency in WS is subtle and perhaps not the primary defect in the disease. A fine structure DNA repair defect in lymphoblast but not fibroblast WS cell lines was reported by this laboratory using the gene-specific and strand-specific repair assays [10].

CELLULAR DEFECTS IN WS

- **Genomic Instability**
  - Chromosomal rearrangements
  - Large spontaneous deletions
- **Telomere maintenance**
  - Shortened telomeres
  - SGS1 acts at telomeres
- **Recombination**
  - Hyper recombination
  - Recombination defects
- **Transcription**
  - RNA polymerase II transcription
    - RNA pol I ?
- **Apoptosis**
  - Attenuated apoptosis
    - p53 mediated
- **Replication**
  - Reduced replicative lifespan
  - Extended S-phase
  - Reduced frequency of initiation sites
  - WRN is part of replication complex
- **DNA Repair (BER, NHEJ)**
  - Hypersensitivity to 4-NQO
  - Hypersensitivity to DNA cross linking agents
  - Reduced repair of psoralen crosslinks (shuttle vector)
  - Hypersensitivity to camptothecin
  - Reduced telomeric repair
  - Reduced transcription coupled repair

Figure 3. *Cellular defects in Werner syndrome.*

In addition to repair or replication defects, evidence from this laboratory suggests that WS cells are defective in transcription [11]. We have used a variety of WS lymphoblast cell lines that carry homozygous mutations in the *WRN* gene to assess the role of *WRN* in transcription. Transcription was measured both *in vivo*, by [^{3}H]-UTP incorporation in the chromatin of permeabilized cells and *in vitro*, by using a plasmid template with a RNA polymerase II specific promoter. The transcription efficiency in different WS cell lines was found to be reduced to 40–60% of that observed in normal cells using both assays [11]. This defect can be complemented by the addition of normal cell extract to the chromatin of WS cells. Furthermore, the addition of purified WRN to the *in vitro* transcription assay stimulates RNA polymerase II driven transcription. These observations suggest that the WRN may act as a general activator of RNA polymerase II transcription.

We and others have demonstrated that WRN helicase can proficiently unwind short DNA duplexes ($\sim$30 bp) in a reaction dependent on nucleoside triphosphate hydrolysis. Unwinding of long DNA duplexes ($>$250 bp) by WRN helicase is dependent on the presence of human Replication Protein A (RPA) [12].

Werner protein, interactions and pathways

In recent years there has been a great deal of interest in the functions of WRN. Biochemical properties and interactions of WRN have been studied. The clarification of these interactions is important in our quest to understand which pathways and processes WRN participates in. A list of reported protein interactions is shown in Figure 4. One of the first established physical and functional interactions of WRN was with RPA [12]. As mentioned, RPA is required for WRN to unwind large DNA duplexes. Since RPA participates in DNA replication and DNA repair, it is likely that WRN may also participate in both of those processes. It was later observed that the tumor suppressor protein p53 also binds to WRN [13], and p53 is, incidentally, known to interact with RPA. Evidence from this laboratory shows that the physical interaction between WRN and p53 is also associated with a functional interaction: p53 inhibits the exonuclease activity of WRN directly [14] and thus may have a

WRN PROTEIN INTERACTIONS

Physical Interactions

RPA
Ku 80/70
p53
human polymerase δ (p50)
PCNA
Topoisomerase I
WHIP
Ubc9
FEN-1
BLM
TRF2
Pol β

Functional Interactions

RPA	*Stimulates WRN helicase*
Ku 80/70	*Stimulates WRN exo*
p53	*Inhibits WRN exonuclease*
Δ polymerase	*Stimulates DNA synthesis by yeast polδ*
FEN-1	*Stimulates FEN-1 nuclease*
BLM	*Inhibits WRN exonuclease*
TRF2	*Stimulates WRN helicase*
Pol β	Stimulated by WRN

Figure 4. *Physical and functional protein interactions of Werner protein.*

function in its regulation. This may imply that these proteins participate in a common pathway. p53 is known to participate in various signal transduction pathways, particularly in cell cycle regulation and apoptosis. Interestingly, WS cells have an attenuated apoptosis process [13], supporting the notion that p53 and WRN both participate in the apoptotic process.

There is evidence for a role of WRN in DNA replication. WS cells have been noted to have a prolonged S-phase [4,] and WRN has been identified as a component of the replication complex in mammalian cells [15]. In addition, recent studies have shown that WRN binds to and stimulates one of the major DNA polymerases, DNA poymerase δ [16]. WRN also interacts physically with another replication protein, proliferating cell nuclear antigen (PCNA), and the functional consequence of this interaction is not yet understood. The previously discussed RPA interaction also supports the notion that WRN functions in replication. WRN also interacts with FEN-1 endonuclease and stimulates the flap cleavage activity of FEN-1 [17], consistent with a possible role in Okazaki fragment processing during lagging strand DNA synthesis. Taken together with the observation that WS cells have prolonged S-phase, it is likely that WRN plays an important role in replication. This could be by resolving complex DNA structures during replication or by affecting the mobility of replication forks after exposure of cells to DNA damage.

As mentioned above, WS cells have increased genomic instability, and this has led to much speculation about whether a DNA repair pathway might be defective in WS. Double strand break repair (DSBR), in general, occurs via two different pathways, one that is homologous (requiring sequence homology) and one that is not, non-homologous end rejoining (NHEJ). WRN interacts physically and functionally with two critical components of the NHEJ pathway, the Ku heterodimer (Ku70, Ku86) [18] and the DNA-dependent protein kinase (DNA-PK) complex [19]. Ku binds very strongly to WRN suggesting that this interaction represents an important pathway in WRN functions. It then strongly stimulates the WRN exonuclease activity [18]. Evidence indicates that the DNA-PK complex (which consists of Ku, the catalytic subunit DNA-PKcs, and DNA) phosphorylates WRN both *in vitro* and *in vivo*, and modulates the WRN exonuclease activity *in vitro* [19]. WS cells are sensitive to agents that form DNA interstrand crosslinks [5, 6], and these adducts in DNA are repaired by DNA recombination.

The biochemical steps of NHEJ specifically involve DNA helicase and exonuclease functions [20], and thus, there is a need in this process for both of the catalytic functions of WRN. So far, *in vitro* studies have not identified a biochemical role for WRN in NHEJ, but there is a good deal of effort on resolution of this question in various laboratories at this time.

The genome instability in WS has been suggested to arise from a failure to resolve alternate DNA structures. Sequences prone to form triple helices are abundant in the human genome. We have tested the ability of WRN helicase to unwind such structures and found that the enzyme is able to unwind a 3′ tailed triple helix DNA substrate in a reaction dependent on nucleoside triphosphate hydrolysis [21]. It is possible that triplex structures are more persistent in WS cells, and this may be a basis for variegated translocation mosaicism seen in WS cells.

During recombinational events such as DNA repair of lesions that include both DNA strands (e.g. crosslinks), intermediate structures are formed in DNA. One such intermediate is the Holliday structure where homologous DNA molecules exchange strands. The 3′ end of the newly synthesized DNA strand invades the daughter duplex structure of the intact DNA molecule. These structures can be generated experimentally *in vitro* and the mechanism of their resolution can be examined. In such an experiment, it appears that the purified WRN enzyme resolves this structure [22]. WRN binds to a single stranded site in the Holliday structure where the ruvA protein also binds. This unwinding is ATP dependent suggesting that it is due to the helicase activity of WRN. Recent studies indicate that WRN and the Bloom syndrome (BLM) helicases efficiently unwind Holliday junctions [23], suggesting a direct role for these enzymes in recombination. A physical and function interaction between WRN and BLM suggests that these enzymes may act in a complex during recombination [24]. Recent evidence from Monnat and co-workers supports a role for WRN in resolving recombination intermediates *in vivo* [9].

Recent studies suggest WRN may also participate in pathways at telomeric ends. WS cells display some defects in telomere metabolism, including increased rates of telomere shortening [25] and deficiencies in repair at telomeres [26]. Furthermore, expression of telomerase in WS cell lines prevents premature replicative senescence [27], and reduces hypersensitivity to 4NQO [28]. Also, the WRN/Ku and WRN/DNA-PK interaction may be important not only for NHEJ, but also for telomere maintenance. The Ku heterodimer and DNA-PKcs localize to telomeres [29], and defects in these enzymes result in accelerated telomere shortening and/or telomere end fusions [30]. Furthermore, Ku interacts with the critical telomere binding protein TRF2. We have recently reported that WRN also interacts with TRF2 both *in vivo* and *in vitro*, and that TRF2 promotes WRN helicase activity *in vitro* [31]. TRF2 also interacts physically and functionally with another member of the RecQ family, the BLM protein [31]. There is recent evidence for WRN and BLM localization with telomeric DNA in some human fibroblast cell lines *in vivo* [32]. In addition, the RecQ yeast homologue, Sgs1, was shown to participate in a telomerase-independent mechanism of telomere lengthening in yeast [32]. Therefore, WRN and BLM may function at telomeres by resolving secondary structure in order to allow access to replication, repair and/or telomere lengthening machinery.

Werner syndrome functions and normal aging

The WRN is a member of the RecQ family of helicases to which the yeast Sgs1 and human Bloom proteins also belong. However, WRN is unusual in that it is the only human member of this family of proteins which contains an exonuclease activity in addition to the helicase function. Until very recently, there has been no information about WRN interactions or about which protein complexes it participates in. In the above, we have discussed various protein interactions involving WRN. Many of these protein interactions are shown in Figure 4. We have outlined some of the major pathways that WRN at this time is considered to participate in, partly based on the known protein interactions (Figure 5). WRN has attracted a lot of interest, and

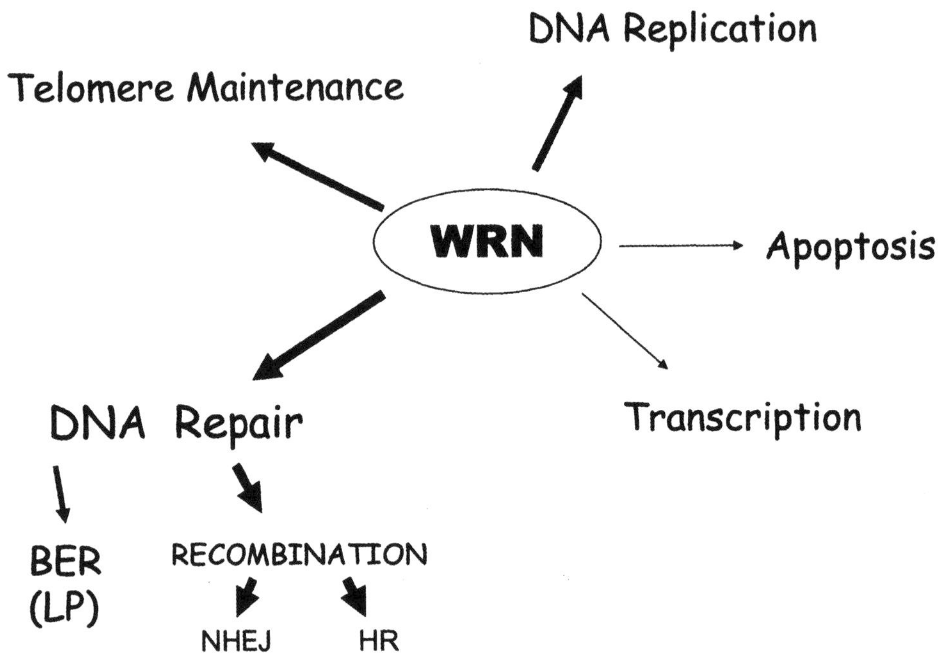

Figure 5. Pathways in which Werner protein is likely to participate.

information is accumulating rapidly about its roles. Soon we should know much more about its interactions and complex formation.

Werner Syndrome is a segmental progeria. This means that the disease is associated with a large number of clinical features that are seen in normal aging at a later stage in life. Whereas WS cannot be equated with the normal aging process, it is of significant interest that a mutation in one gene can lead to so many signs of aging. These include the increased frequency of age associated diseases such as cancers and type II diabetes and they include a large spectrum of phenotypic changes associated with the normal aging process [33]. Thus, insight into the molecular pathways involving the WRN is bound to shed important light on the normal aging process and on why aging is associated with such a high preponderance of disease.

Cockayne's syndrome

Cockayne syndrome (CS) is a rare inherited human genetic disease categorized as a segmental progeria. Affected individuals suffer from postnatal growth failure resulting in cachectic dwarfism, photosensitivity, skeletal abnormalities, mental retardation, progressive neurological degeneration, retinopathy, cataracts, and sensorineural hearing loss [34]. Two complementation groups, CS-A and CS-B, have been identified and the corresponding genes have been cloned. The cellular

DNA REPAIR, TRANSCRIPTION AND CHROMATIN STRUCTURE IN COCKAYNE SYNDROME (CS)

- **DNA Repair**
  - **TCR after UV (NER) defect**
  - **TCR after X-ray and oxidative damage**
  - **Accumulation of base modifications**
  - **Incision defect (BER)**
  - **PCNA relocation defect**
- **Transcription**
  - **Basal transcription defect *in vivo***
  - **Transcription defect *in vitro***
  - **CSB is an elongation factor**
  - **CSB binds to RNA polymerase II**
- **Chromatin Structure**
  - **Lack of nuclear matrix association**
  - **Loose chromatin structure**
  - **Chromatin remodeling**
- **Apoptosis**
  - **Increased apoptosis**

Figure 6. *Cellular characteristics of Cockayne syndrome.*

phenotype of CS includes increased sensitivity to a number of DNA damaging agents including UV radiation, ionizing radiation, and hydrogen peroxide [35] (Figure 6).

Nucleotide excision repair (NER) is a complex process that removes and repairs many types of DNA lesions. NER has two subpathways: (1) the transcription coupled repair (TCR) pathway and (2) the global genome repair (GGR) pathway. The TCR pathway repairs lesions in the transcribed strand of transcriptionally active genes and is dependent on RNA polymerase II. The GGR pathway removes lesions from genes that are transcriptionally active or inactive. CS cells are defective in TCR, but proficient in GGR of UV-induced DNA damage. A characteristic feature of CS cells is that they do not recover the ability to efficiently synthesize RNA after UV damage [35]; this phenotype is consistent with a defect in TCR.

The *CSA* and *CSB* genes have been cloned and their products characterized biochemically. The *CSA* gene product is a 44 kDa protein that belongs to the "WD repeat" family of proteins. Members of this protein family are structural and regulatory proteins, but do not have enzymatic activity. The *CSB* gene product is a 168 kDa protein that belongs to the SWI/SNF family of proteins, which are DNA and RNA helicases with seven conserved sequence motifs. In addition, CSB has an acidic amino acid stretch, a glycine-rich region, and two putative NLS sequences. CSB is a DNA-stimulated ATPase, but is not able to unwind DNA in a conventional strand displacement assay. The human *CSB* cDNA complements the UV sensitivity and the delay in RNA synthesis recovery when transfected into the human CS-B deficient cell line CS1AN.S3.G2 or the hamster CS-B deficient cell line UV61. Gene-specific TCR repair is also recovered when the human *CSB* gene is transfected into UV61 cells [36, 37].

The precise molecular role of CSB is not clear at present. CSB may facilitate repair of active genes by recruiting DNA repair proteins to actively transcribed regions. *In vitro,* CSB exists in a quaternary complex with RNA pol II, DNA and the RNA transcript and ATP hydrolysis is required to form this complex. This quaternary complex recruits another molecular complex which includes the transcription factor IIH (TFIIH) core subunits, p62 and XPB [38]. TFIIH is a complex factor thought to promote local DNA unwinding during transcription initiation by RNA pol II and promoter escape, as well as in NER. Co-immunoprecipitation studies demonstrate that XPG and CSB proteins interact [39]. XPG interacts with multiple components of TFIIH [39] and it is an endonuclease that plays a role in NER. *In vitro,* CSB interacts with CSA and with the NER damage recognition factor XPA [40]. These protein interactions support the hypothesis that CSB participates directly in TCR.

It is possible that CSB indirectly stimulates TCR by facilitating transcription. Members of the SWI/SNF family are involved in regulating transcription, chromatin remodeling, and DNA repair, including such actions as disruption of protein-protein and protein-DNA interactions [reviewed in ref. 41]. The *CSB* gene product could have a similar function. In fact, it is still a matter of debate whether CS is due to a primary defect in transcription or DNA repair. The basal transcription level is lower in human CS-B lymphoblastoid cells and fibroblasts than in normal cells, even without exposure to DNA damaging agents [42]. This transcription defect is complemented by normal cell extracts *in vitro* or by the wild type *CSB* gene *in vivo.* In a reconstituted system, purified CSB protein enhances the rate of transcription by RNA pol II, suggesting that CSB may indirectly stimulate TCR by facilitating the process of transcription [43]. Thus, CSB may be a transcription elongation factor and a repair-coupling factor acting at the site of RNA pol II-blocking lesions, and the CS phenotype may arise from deficiencies in both transcription and DNA repair. The biological function of CSB in these different pathways may be mediated by distinct functional domains of the protein.

As described above, it is well established that the *CSB* phenotype involves a defect in TCR of UV-induced DNA damage. However, evidence also suggests that TCR of oxidative damage may be affected by CSB [44]. Hydrogen peroxide and ionizing radiation induce a large variety of DNA damage including oxidatively damaged bases and single- and double-strand DNA breaks.

It has been shown that primary CS-B cells are slightly more sensitive to γ irradiation; and in addition, they are defective in strand-specific repair of ionizing radiation and hydrogen peroxide-induced thymine glycol. Previously, this laboratory showed that the ability of CS-B cell extracts to carry out incision of 8-oxoguanine lesions *in vitro* is 50% of normal; in contrast, incision of uracil and thymine glycol were normal [45]. 8-Oxoguanine is a base modification caused by oxidative stress from environmental agents and from endogenous metabolic processes, and it is repaired mainly via by BER. Further recent studies from this laboratory show that the CSB protein is involved in BER. We have established stably transfected human cell lines in which functional domains of the *CSB* gene have been mutated [46]. When certain helicase motifs are disrupted, there is a cellular defect in the incision of 8-oxoG containing DNA in cell extracts [46].

Understanding the exact role of CSB is quite challenging. Some of the cellular deficits are listed in Figure 6, also giving clues to the pathways that involve CSB. We are convinced that this important protein participates in not only transcription coupled repair, but also BER, and that it also plays a role in transcription.

Perspectives

We have learned a lot about the molecular function and protein interaction of the premature aging proteins. These proteins appear to participate in a number of DNA related transactions, notably the maintenance of the genomic stability. A major pathway is that of DNA repair, and these proteins appear to be partners in this process, but their role is not essential. It is of increasing interest that WRN and maybe other of these proteins are involved in functions at the telomeric ends, and maybe stabilize the functions there. It is likely that increased understanding of the mechanism of action of these proteins will lead to new insights into the aging process.

References

1. Bohr VA, Cooper M, Orren D, *et al.* (2000). Werner syndrome protein: biochemical properties and functional interactions. *Exp Gerontol.* 35: 695–702.
2. Huang S, Li B, Gray MD, Oshima J, Mian IS, Campisi J (1998). The premature ageing syndrome protein, WRN, is a 3′→5′ exonuclease. *Nat Genet.* 20: 114–16.
3. Opresko PL, Laine JP, Brosh RM, Seidman MM, Bohr VA (2001). Coordinate action of the helicase and 3′ to 5′ exonuclease of Werner syndrome protein. *J Biol Chem.* 276: 44677–87.
4. Poot M, Hoehn H, Runger TM, Martin GM (1992). Impaired S-phase transit of Werner syndrome cells expressed in lymphoblastoid cells. *Exp Cell Res.* 202: 267–73.
5. Bohr VA, Souza PN, Nyaga SG, *et al.* (2001). DNA repair and mutagenesis in Werner syndrome. *Environ Mol Mutagen.* 38: 227–34.
6. Poot M, Yom JS, Whang SH, Kato JT, Gollahon KA, Rabinovitch PS (2001). Werner syndrome cells are sensitive to DNA cross-linking drugs. *FASEB J.* 15: 1224–6.
7. Oshima J (2000). The Werner syndrome protein: an update. *Bioessays* 22: 894–901.
8. Pichierri P, Franchitto A, Moscsso P, Proietti dS, Balajee AS, Palitti F (2000). Werner's syndrome lymphoblastoid cells are hypersensitive to topoisomerase II inhibitors in the G2 phase of the cell cycle. *Mutat Res.* 459: 123–33.
9. Saintigny Y, Makienko K, Swanson C, Emond MC, Monnat RJ (2002). Homologous recombination resolution defect in Werner syndrome. *Mol Cell Biol.* 22: 6971–8.
10. Webb DK, Evans MK, Bohr VA (1996). DNA repair fine structure in Werner's syndrome cell lines. *Exp Cell Res.* 224: 272–8.
11. Balajee AS, Machwe A, May A, *et al.* (1999). The Werner syndrome protein is involved in RNA polymerase II transcription. *Mol Biol Cell.* 10: 2655–68.

12. Brosh RM, Orren DK, Nehlin JO, *et al.* (1999). Functional and physical interaction between WRN helicase and human replication protein A. *J Biol Chem.*, 274: 18341–50.
13. Spillare EA, Robles AI, Wang XW, *et al.* (1999). p53-mediated apoptosis is attenuated in Werner syndrome cells. *Genes Dev.* 13: 1355–60.
14. Brosh RM, Karmakar P, Sommers JA, *et al.* (2001). p53 Modulates the exonuclease activity of Werner syndrome protein. *J Biol Chem.* 276: 35093–102.
15. Lebel M, Spillare EA, Harris CC, and Leder P (1999). The Werner syndrome gene product co-purifies with the DNA replication complex and interacts with PCNA and topoisomerase I. *J Biol Chem.* 274: 37795–9.
16. Kamath-Loeb AS, Johansson E, Burgers PM, Loeb LA (2000). Functional interaction between the Werner Syndrome protein and DNA polymerase delta. *Proc Natl Acad Sci USA* 97: 4603–8.
17. Brosh RM, von Kobbe C, Sommers JA, *et al.* (2001). Werner syndrome protein interacts with human flap endonuclease 1 and stimulates its cleavage activity. *EMBO J.* 20: 5791–801.
18. Cooper MP, Machwe A, Orren DK, Brosh RM, Ramsden D, Bohr VA (2000). Ku complex interacts with and stimulates the Werner protein. *Genes Dev.* 14: 907–12.
19. Karmakar P, Piotrowski J, Brosh RM, *et al.* (2002). Werner protein is a target of DNA-dependent protein kinase *in vivo* and *in vitro*, and its catalytic activities are regulated by phosphorylation. *J Biol Chem.* 277: 18291–302.
20. Featherstone C, Jackson SP (1999). DNA-dependent protein kinase gets a break: its role in repairing DNA and maintaining genomic integrity. *Br J Cancer* 80(Suppl. 1): 14–19.
21. Brosh RM, Majumdar A, Desai S, Hickson ID, Bohr VA, Seidman MM (2001). Unwinding of a DNA triple helix by the Werner and Bloom syndrome helicases. *J Biol Chem.* 276: 3024–30.
22. Constantinou A, Tarsounas M, Karow JK, *et al.* (2000). The Werner's syndrome protein (WRN) translocates Holliday junctions *in vitro* and co-localises with RPA upon replication arrest. *EMBO Rep.* 1, 80–84.
23. Mohaghegh P, Karow JK, Brosh RM, Bohr VA, Hickson ID (2001). The Bloom's and Werner's syndrome proteins are DNA structure-specific helicases. *Nucl Acids Res.* 29: 2843–9.
24. von Kobbe C, Karmakar P, Dawut L, *et al.* (2002). Colocalization, physical, and functional interaction between Werner and Bloom syndrome proteins. *J Biol Chem.* 277: 22035–44.
25. Schulz VP, Zakian VA, Ogburn CE, *et al.* (1996). Accelerated loss of telomeric repeats may not explain accelerated replicative decline of Werner syndrome cells. *Hum Genet.* 97: 750–4.
26. Kruk PA, Rampino NJ, Bohr VA (1995). DNA damage and repair in telomeres: relation to aging. *Proc Natl Acad Sci USA* 92: 258–62.
27. Wyllie FS, Jones CJ, Skinner JW, *et al.* (2000). Telomerase prevents the accelerated cell ageing of Werner syndrome fibroblasts. *Nat Genet.* 24: 16–17.
28. Hisama FM, Chen YH, Meyn MS, Oshima J, Weissman SM (2000). WRN or telomerase constructs reverse 4-nitroquinoline 1-oxide sensitivity in transformed Werner syndrome fibroblasts. *Cancer Res.* 60: 2372–6.
29. Hsu HL, Gilley D, Blackburn EH, Chen DJ (1999). Ku is associated with the telomere in mammals. *Proc Natl Acad Sci USA* 96: 12454–8.
30. Bailey SM, Meyne J, Chen DJ, *et al.* (1999). DNA double-strand break repair proteins are required to cap the ends of mammalian chromosomes. *Proc Natl Acad Sci USA* 96: 14899–904.

31. Opresko PL, von Kobbe C, Laine JP, Harrigan J, Hickson ID, Bohr VA (2002). Telomere-binding protein TRF2 binds to and stimulates the Werner and Bloom syndrome helicases. *J Biol Chem.* 277: 41110–9.
32. Johnson FB, Marciniak RA, McVey M, Stewart SA, Hahn WC, Guarente L (2001). The Saccharomyces cerevisiae WRN homolog Sgs1p participates in telomere maintenance in cells lacking telomerase. *EMBO J.* 20: 905–13.
33. Martin GM (1997). The pathobiology of the Werner syndrome. *FASEB J.* 11: A1449.
34. Friedberg EC (1996). Cockayne syndrome – a primary defect in DNA repair, transcription, both or neither? *Bioessays* 18: 731–8.
35. Balajee AS and Bohr VA (2000). Genomic heterogeneity of nucleotide excision repair. *Gene* 250: 15–30.
36. Brosh RM, Balajee AS, Selzer RR, Sunesen M, De Santis LP, Bohr VA (1999). The ATPase domain but not the acidic region of Cockayne syndrome group B gene product is essential for DNA repair. *Mol Biol Cell* 10: 3583–94.
37. Orren DK, Dianov GL, and Bohr VA (1996). The human CSB (ERCC6) gene corrects the transcription-coupled repair defect in the CHO cell mutant UV61. *Nucl Acids Res.* 24: 3317–22.
38. Tantin D (1998). RNa polymerase II elongation complexes containing the Cockayne syndrome group B protein interact with a molecular complex conating the transcription factor IIH components xeroderma pigmentosum B and p62. *J Biol Chem.* 273: 27794–9.
39. Iyer N, Reagan MS, Wu KJ, Canagarajah B, Friedberg EC (1996). Interactions involving the human RNA polymerase II transcription/nucleotide excision repair complex TFIIH, the nucleotide excision repair protein XPG, and Cockayne syndrome group B (CSB) protein. *Biochemistry* 35: 2157–67.
40. Henning K, Li L, Legerski R, *et al.* (1995). The Cockayne syndrome complementation group A gene encodes a WD-repeat protein which interacts with CSB protein and a subunit of the RNA pol II transcription factor IIH. *Cell* 82: 555–66.
41. Pazini MJ, Kadonaga JT (1997). SWI2/SNF2 and related proteins: ATP-driven motors that disrupt protein-DNA interactions? *Cell* 88: 737–40.
42. Balajee AS, May A, Dianov GL, Friedberg EC, and Bohr VA (1997). Reduced RNA polymerase II transcription in intact and permeabilized Cockayne syndrome group B cells. *Proc Natl Acad Sci USA* 94: 4306–11.
43. Selby CP, Drapkin R, Reinberg D, Sancar A (1997). RNA polymerase II stalled at a thymine dimer: footprint and effect of excision repair. *Nucl Acid Res.* 25: 787–93.
44. Le Page F, Klungland A, Barnes DE, Sarasin A, Boiteux S (2000). Transcription coupled repair of 8-oxoguanine in murine cells: the oggl protein is required for repair in nontranscribed sequences but not in transcribed sequences. *Proc Natl Acad Sci USA* 97: 8397–402.
45. Dianov G, Bischoff C, Sunesen M, Bohr VA (1999). Repair of 8-oxoguanine in DNA is deficient inCockayne syndrome group B cells. *Nucl Acids Res.* 27: 1365–8.
46. Tuo J, Muftuoglu M, Chen C, *et al.* (2001). The Cockayne Syndrome group B gene product is involved in general genome base excision repair of 8-hydroxyguanine in DNA. *J Biol Chem.* 276: 45772–9.

Oxidative Damage, Somatic Mutations and Cellular Aging

Rita A. Busuttil[1], Miguel Rubio[2], Martijn E.T. Dollé[1], Judith Campisi[4] and Jan Vijg[1,3]

[1]Sam and Ann Barshop Center for Longevity and Aging Studies, University of Texas Health Science Center, San Antonio, TX 78245, USA; [2]Life Sciences Division, Lawrence Berkeley National Laboratory, 1 Cyclotron Road, Berkeley, CA 94720, USA; [3]Geriatric Research Education and Clinical Center, South Texas Veterans Health Care System, San Antonio, TX 78229, USA; [4]Buck Institute for Age Research, 8001 Redwood Blvd, Novato, CA 94945, USA

Introduction

Many theories have been proposed to explain the aging process. One of the more widely accepted of these is the free radical theory of aging. First presented in 1954, it postulates that aging is caused by the accumulation of damage to the macromolecules of somatic cells by free radicals [1].

Free radicals are molecules that have been modified by the removal of a single electron so as to yield a single unpaired electron in their outer orbit. This configuration is unstable and as a result these molecules have a tendency to capture electrons from neighboring molecules in an attempt to restore their stable, resting state. By the very process of extracting an electron from a nearby molecule, the free radical damages its neighbors, and often sets off a series of chain reactions as molecules obtain electrons from one another.

The formation of free radicals has been attributed to a variety of causes. In humans and animals the main cause of free radical formation is oxidative phosphorylation. Other causes of free radical formation include radiation, metabolism of toxic compounds and an excess of transition metals such as iron and copper, which in the presence of superoxide anions and hydrogen peroxide may be converted into highly toxic hydroxyl radicals via the Fenton reaction [2].

Cells possess a variety of defense mechanisms that help protect against free radicals. These defense mechanisms include antioxidant enzymes. One group of antioxidant enzymes is known as the superoxide dismutases or SODs, which can be

T. von Zglinicki (ed.), Aging at the Molecular Level, 79–90.

found in three isoforms in mammals. CuZn-SOD (SOD1) is located in the cytosol, Mn-SOD (SOD2) is found in the mitochondria and Fe-SOD (SOD3) is found in the extracellular space. SODs reduce superoxide anions to hydrogen peroxide, which is then reduced to water by either catalase (found primarily in the peroxisomes) or glutathione peroxidase (Gpx) (located in the mitochondria and cytosol [3]). In addition to these antioxidant defense enzymes several non-enzymatic free radical scavengers help cells inactivate free radicals [4], which are an inevitable consequence of oxidative stress [4].

Oxidative stress is the harmful condition that occurs when the delicate balance between free radical production and their reduction by antioxidants is disturbed. Despite their potential for harming biological molecules, free radicals – at low concentrations – also act as second messengers for certain physiological processes, such as cell proliferation signals. Thus, cells must and do tolerate low levels of free radicals. Nonetheless, damage from free radicals can accumulate, and oxidative stress results in deleteriously high free radical levels. As well as being a potentially major cause of aging, free radicals can also contribute to many diseases and causes of death that are associated with advanced age, including neurodegenerative disorders, such as Alzheimer's disease [1].

One major type of somatic damage inflicted by free radicals is damage to DNA. DNA damage can result in mutations, which in turn can lead to cancer and possibly other aging-related degenerative phenotypes. In this chapter we will discuss the possible contribution of mutation accumulation to aging processes in cells both *in vivo* and in culture.

Aging *in vivo* and in culture

Aging can be defined in practical terms as a series of time-related processes occurring in the adult individual that ultimately brings life to a close. However, as proposed by Hayflick and Moorhead, the basic mechanisms responsible for the aging and death of an organism can also take place in individual cells. These investigators found that human diploid fibroblast cells have only a limited capacity to divide in culture, which has since been widely interpreted as aging at the cellular level [5]. The phenomenon is often referred to as replicative senescence. Evidence indicates that the primary mechanism by which replicative senescence occurs in most human cells is telomere shortening.

Telomeres are the repetitive DNA sequence [(TTAGGG)n (in humans and other vertebrates)] and associated proteins located at the ends of linear chromosomes [6, 7]. During normal DNA replication between 50 and 200 bases are left unreplicated at the 3′ end of the telomere. Consequently, each daughter molecule is shorter than the parental strand [5]. It is believed that when telomeres shorten beyond a certain point (to approximately 4–6 kb in human cells [7]), the cell will stop dividing [8]. Some cells (such as many human cancer cells) are able to bypass telomere shortening, and hence replicative senescence, and are said to be replicatively immortal.

The most common mechanism by which cells become replicatively immortal is by expressing an enzyme known as telomerase. In most normal somatic cells the expression of telomerase, a ribonucleoprotein capable of adding TTAGGG repeats to chromosome ends, is suppressed. Hence, telomere repeat loss cannot be repaired or reversed. Critically short telomeres can ultimately lead to chromosomal aberrations, which is probably why normal cells respond to short telomeres by arresting proliferation. The artificial reconstitution of telomerase in normal cells can prevent replicative senescence and appears to essentially immortalize many normal human cells [9]. This strongly suggests that telomerase is a major and perhaps the most important genome stability system in actively proliferating human cells.

Senescent cells arrest growth with a G1 DNA content and are unable to replicate [10] nor can they be stimulated to divide by the addition of any physiological mitogens [7]. This irreversible arrest of proliferation is accompanied by changes in cellular function. Indeed, it is conceivable that cell cycle arrest during passaging in culture is part of a more general deteriorative process which might be similar to aging *in vivo*.

Independent of telomere shortening, cells can still adopt a senescent phenotype when challenged with stimuli that have the potential to damage the genome (or cause neoplastic transformation by other means). In mouse cells, senescence takes place independently of telomere shortening. Telomeres of rodent cells are much longer than those of human cells and in some cases these cells also express telomerase. It has been suggested that the senescent phenotype observed in mouse cells may in fact be due to stressful culture conditions [11]. Indeed, normal cells undergo a senescence arrest when exposed to a variety of factors, including but not limited to clastogens such as ionizing radiation.

The phenomenon of telomere shortening is dependent on oxidative stress. It has been observed that altered antioxidant defense levels greatly affect the rate of telomere shortening in culture. Von Zglinicki and his colleagues found that culturing human WI-38 fibroblasts under high oxygen tension (40% O_2) causes telomeres to shorten prematurely and consequently the replicative life span of the cells is significantly reduced [12]. Studies have also shown that in cells with high antioxidant defense capabilities telomere shortening proceeds at a slower rate [13] than in cells with low antioxidant defenses [14]. This suggests that telomere length may be determined by the ratio of oxidative stress and antioxidant defense levels in the cell [10]. Saretzki *et al.* [10] suggested that antioxidant defenses might determine telomere length *in vivo* as well as in culture.

We propose that, since oxidative stress leads to DNA damage, which in turn may result in mutation accumulation, genomic instability could be a causal factor in aging, including the senescence response in cultured primary cells. Other endpoints of oxidative stress, that could contribute to aging, are cell death and inhibition of transcription (Figure 1).

Monitoring mutation accumulation during aging *in vivo* and *in vitro*

To monitor the accumulation of mutations in cell populations, especially in different organs and tissues *in vivo*, is not trivial. Initially cytogenetic techniques were used to demonstrate the presence of large chromosomal mutations in metaphases of

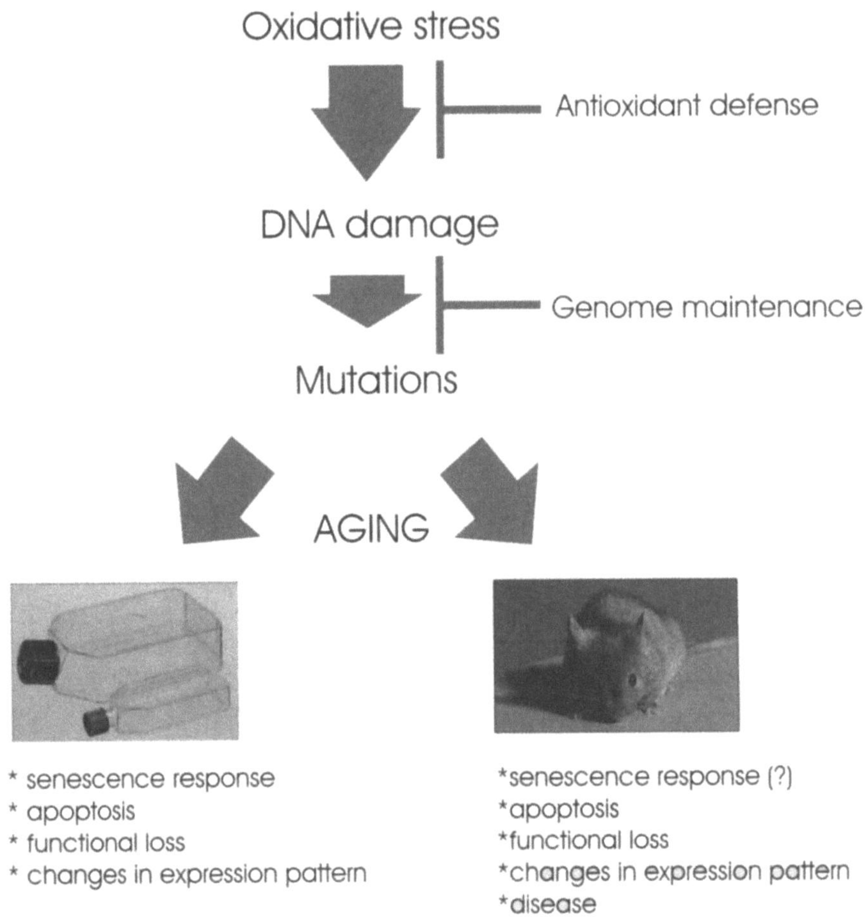

Figure 1*. Schematic representation of the hypothetical pathway that may lead to aging both in vitro and in vivo. DNA damage, continuously induced through oxidative stress, ultimately generates mutations. Mutations may lead to aging, both in vitro and in vivo.*

peripheral blood lymphocytes and other proliferating cells. Curtis and co-workers [15] were among the first to demonstrate an increase in chromosomal mutations with aging in mouse liver, using hepatocyte metaphases after partial hepatectomy. Tests were later developed utilizing HPRT and other endogenous selectable target genes [see ref. 16 for a review], allowing for the determination of multiple types of mutations, but mainly in peripheral blood lymphocytes. A major disadvantage of such systems is that only actively proliferating cells can be successfully used. Given that the large majority of cells in adult humans and animals undergo cell division only rarely, such assays are a poor reflection of the *in vivo* situation.

With the development of transgenic mouse models harboring chromosomally integrated bacterial reporter genes as part of bacteriophage lambda or plasmid vectors that can be recovered from chromosomal DNA, it became possible to directly test the hypothesis that somatic mutations in a neutral gene accumulate with age in different organs and tissues [17, 18]. It has now been demonstrated that the mutation frequency at most reporter loci is the same and not different on average from endogenous autosomal loci such as MHC [19, 20] and APRT [21]. Since reporter loci are neutral and not expressed in the mouse, they reflect levels of genomic instability more faithfully than endogenously expressed genes, which may suffer from selection bias [22].

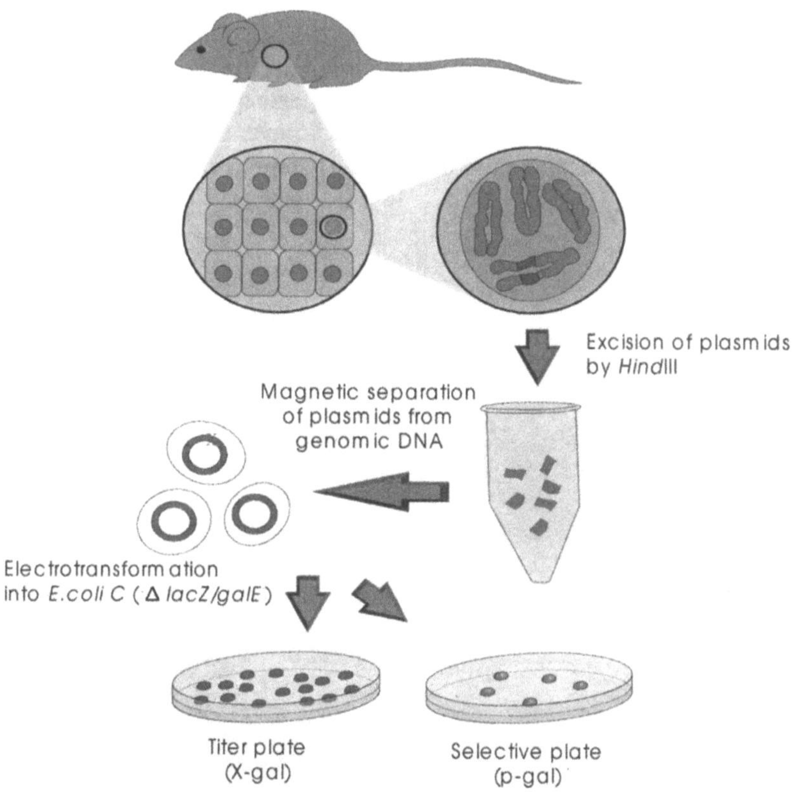

Figure 2*. Schematic of the LacZ-plasmid model. In this system, plasmids are rescued by excision of genomic DNA with HindIII, followed by their separation from the mouse genomic DNA using magnetic beads, precoated with a lacI repressor protein [24]. The plasmids are then ligated and transferred to Escherichia coli C (ΔlacZ, galE⁻) using electrotransformation. A small amount of transformants are plated on X-gal to determine the total number of plasmids rescued. The remainder is plated on the lactose-analogue p-gal, to select only the cells harboring a mutant lacZ. The mutation frequency is the ratio of the colonies on the selective plate versus colonies on the titer plate (times the dilution factor).*

Initially, mutant *lacZ* genes were detected as colorless plaques or colonies among blue, wild type ones. Later a positive selection system based on an *E.coli* host with an inactivated *galE* gene was developed [23]. Upon receiving a wild type *lacZ* gene and grown in the presence of the lactose analogue phenyl-β-D-galactoside (p-gal), such cells produce UDP-galactose, which is highly toxic when it cannot be converted to UDP-glucose. Thus, on medium containing p-gal, only those cells harboring a mutant *lacZ* gene can give rise to a colony (Figure 2). This plasmid-based system is able to detect point mutations, small deletions and insertions, as well as large rearrangements involving the mouse genome.

Mutation accumulation in aging cells and tissues

Using the plasmid-based system Dollé *et al.* have demonstrated that mutant frequencies at the lacZ transgene increase with age in the liver, from about 4×10^{-5} in young adults to about 15×10^{-5} in animals of around 30 months of age. An age-related increase was also found in small intestine, spleen and heart, whilst there was virtually no age-dependent difference in mutant frequency in the brain and testis [24, 25] (Figure 3).

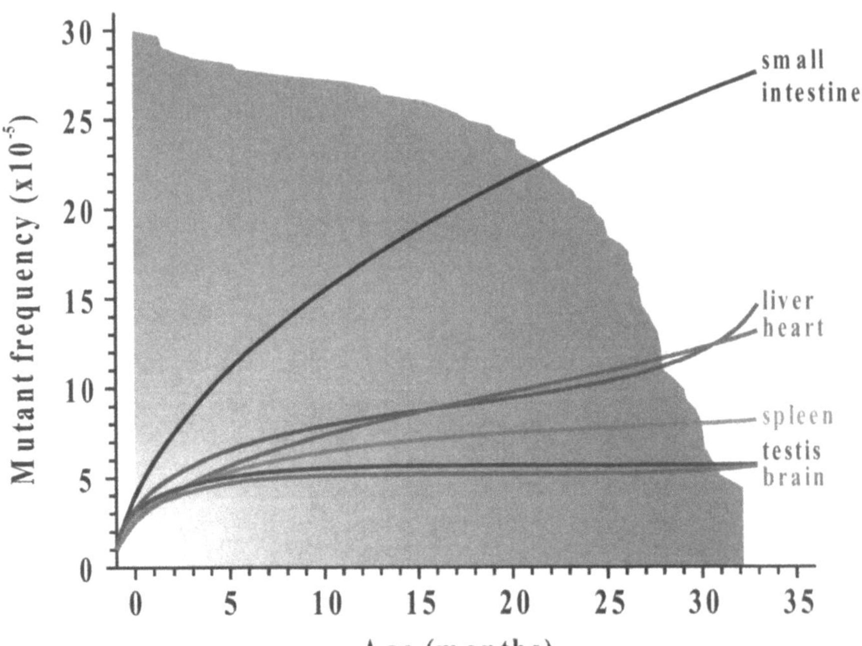

__Figure 3__. Spontaneous lacZ mutant frequencies of various tissues in transgenic line 60 mice. The gray fading area represents the survival curve of the lacZ-plasmid transgenic mice which is not different from normal control mice of the same strain.

Dollé *et al.* also observed striking organ specificity with respect to the mutational spectra of the old animals. In the small intestine and brain, the vast majority of mutations that accumulated were point mutations (and the small increase in the brain was almost totally due to point mutations). In the liver and heart, however, large deletion mutations were a prominent part of the spectrum [24, 25].

Whilst the origin of the age-accumulated mutations, as well as the organ-specificity of mutation rate and type, is unclear at present, it is tempting to speculate that they are related in some way to oxidative stress. Indeed, when fibroblasts were isolated from a 14-day old embryo and brought into culture the mutation frequency increased 3-fold (Figure 4). It is reasonable to assume that this increase is caused by the high oxygen tension (typically 20%) associated with standard culture conditions. What is the rationale of comparing this with the aging process *in vivo*?

Aging of cells in culture is obviously different from the aging process in an organism, in the sense that there is no longer an aging phenotype involving different organs and tissues. Nevertheless, it is entirely possible that the harsh environment of the culture conditions, including high oxygen tension, would have similar effects as the aging process *in vivo*. Therefore, oxidative stress in primary cell cultures could have the same molecular consequences as oxidative stress in the constituent cells of an animal (Figure 4).

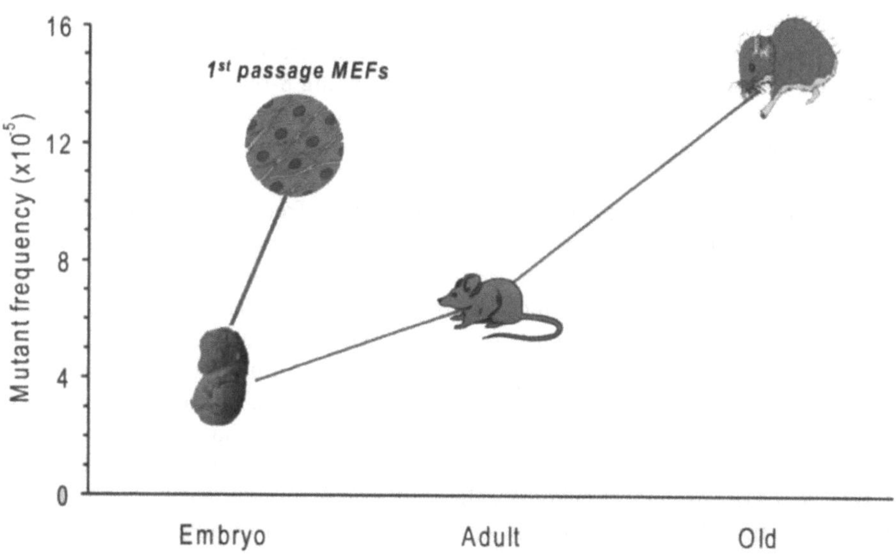

Figure 4. *Conceptual visualization of the possible parallel between aging in vivo and aging in vitro. The average increase in mutation frequency of mouse tissues with aging is compared with the increase in mutation frequency of MEFs placed immediately into standard culture conditions (Submitted for publication).*

Shay and Wright [26] have proposed that standard tissue culture conditions, lacking most extracellular matrix molecules, containing 20% O_2 and levels of serum and other additives, may induce an oxidative stress level ultimately leading to damage accumulation in mouse cells that will eventually result in their growth arrest [26]. Interestingly, Shay and Wright referred to the original work of Hart and Setlow [27] suggesting that cells from short-lived species such as rodents have a lower DNA repair capacity than cells from longer-lived species such as humans. This inferior genome maintenance capacity could also explain why mouse primary cells undergo replicative senescence so much earlier than human cells. Hence, the maintenance and repair capacities of longer-lived species may have evolved in such a way so as to allow these species to deal with damage and other cellular problems which are known to accumulate with age.

In this respect, it is interesting that certain mutagenic agents, such as ionizing radiation, mitomycin C and free radicals, readily induce senescence in primary cells. It is tempting to speculate that such agents are all capable of inducing the senescence response through their main common endpoint: DNA mutations, including genome rearrangements.

It has been demonstrated that simply culturing human fibroblasts in 2–3% oxygen enables them to undergo 20–50% more population doublings [28–32] when compared to those cells cultivated under standard culture conditions. Parrinello *et al.* [28a] have shown that this is even more true in mouse embryonic fibroblasts (see also Figure 5a). Evidence has been obtained that high oxygen levels are responsible for oxidative damage and cellular senescence in human cells by using 8-oxoguanine (oxo^8Gua) as a marker [28]. Oxo^8Gua is a lesion formed during an attack of guanine residues in DNA by hydroxyl radicals and its level was demonstrated to be much higher in cells cultured at 20% O_2.

Since there was an immediate increase in mutation frequency when MEFs were placed at 20% O_2 (Figure 4) we were interested to determine the long-term effects of culturing cells at 20% O_2 on mutation frequency. As can be seen in Figure 5 (Busuttil *et al.*, submitted) we observed an increase in mutation frequency during *in vitro* life span of cells cultured under standard conditions, which was associated with a period of slower growth. This may be similar to *in vivo* situation due to cellular senescence. As cells or animals age, oxidative stress increases and as a consequence so do the number of mutations. Those cells maintained at 3% O_2 maintained a constant mutation frequency (Figure 5). These results suggest that a high mutation load may adversely affect the growth of primary cells.

In order to test if mutations per se can cause a growth disadvantage we treated MEFs with UV-radiation (at population doubling 3.6), continued to culture them at 3% O_2 and determined their mutation frequency at a series of population doublings after the treatment (Figure 6). After seven population doublings, when we would expect that the amount of DNA damage is greatly reduced as a consequence of dilution and DNA repair, a significant number of mutations continued to persist. After continued passaging, the mutation frequency gradually decreased to almost normal levels at 17 population doublings. This strongly indicates selective elimina-

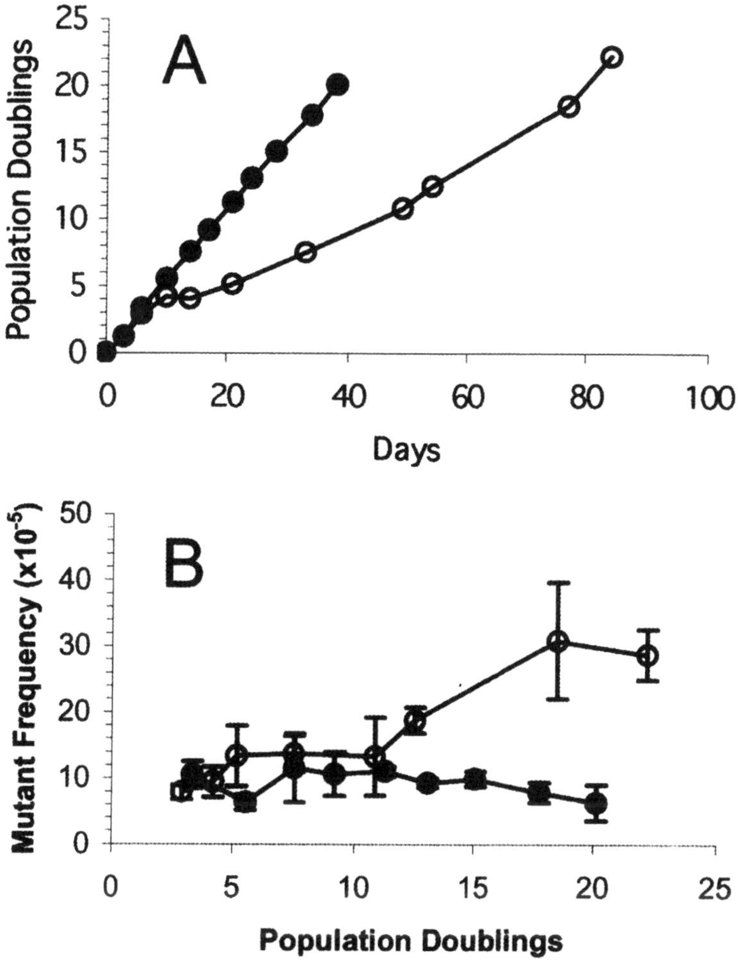

Figure 5. Cumulative growth curve (A) and mutant frequency (B) of C57/Bl6 lacZ-plasmid mouse embryonic fibroblasts cultured under 3% (●) or 20% O_2 (○) (Submitted for publication).

tion of cells with a high mutation frequency, suggesting that random mutations in the genome are associated with a growth disadvantage, as has been shown for *E.coli* [33].

Discussion and future prospects

In this paper we have argued that the accumulation of mutations in the DNA of the genome, including large rearrangements, could be a major causal factor in cell senescence, both in culture and *in vivo*, primarily as a consequence of oxidative stress. This would be in keeping with the well-documented induction of the senescence response in primary cultured cells by mutagenic agents, especially those that lead to

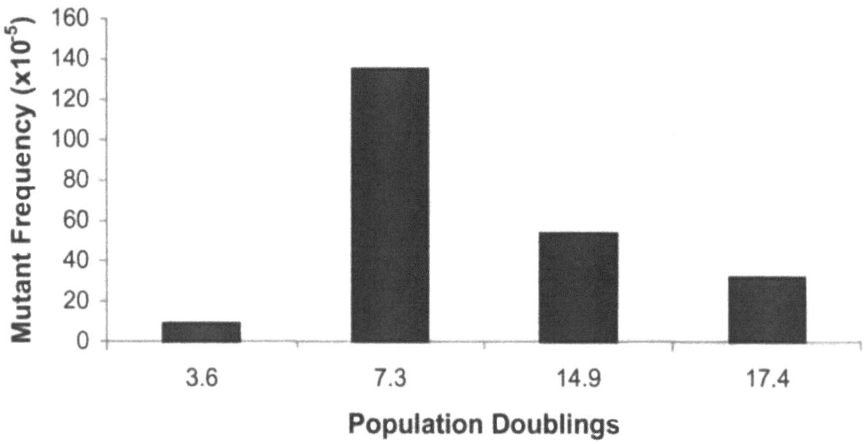

Figure 6. *Declining number of LacZ mutations induced by 5 J/m^2 in MEFs at PD 3.6 (Submitted for publication).*

genome rearrangements, such as mitomycin C and ionizing radiation, or oxidative stress. In fact, there is now convincing evidence that telomere instability, which can be exacerbated by oxidative stress, is the single most important cause of replicative senescence in human primary cells.

The examples provided in this paper show that, in the mouse, mutations at a lacZ reporter locus increase with age in a tissue-specific manner. Similarly, almost immediately upon their establishment in culture, fibroblasts isolated from mouse embryos show a significant increase in mutation frequency. This is likely to be due to the high conditions of oxidative stress associated with culture conditions. Evidence was then also provided that an increased mutation load is associated with growth inhibition and that low mutation frequencies offer a selective advantage to cultured primary cells. Apart from other, non-mutagenic factors, it is conceivable that increased mutation loads could induce a senescence response, possibly through adverse effects on normal patterns of gene expression.

To further investigate the possible relationship between genomic instability and cellular aging, mutation accumulation can be studied in genetically modified cells and animals. For example, over-expression of genes involved in antioxidant defense in cultured primary cells would be expected to decrease mutation accumulation and postpone senescence. Similarly, mice with over-expressed antioxidant genes, or kept under conditions associated with reduced oxidative stress, such as caloric restriction, would be expected to have a lower rate of mutation accumulation and live longer.

Acknowledgments

These studies were supported by NIA grant PO1 AG 17242.

References

1. Harman D (1956). Aging: a theory based on free radical and radiation chemistry. *J Gerontol.* 11: 298–300.
2. Gutteridge JM, Quinlan GJ, Kovacic P (1998). Phagomimetic action of antimicrobial agents. *Free Radic Res.* 28: 1–14.
3. Wallace DC, Melov S (1998). Radicals r'aging. *Nat Genet.* 19: 105–6.
4. Wickens AP (2001). Ageing and the free radical theory. *Respir Physiol.* 128: 379–91.
5. Hayflick L, Moorhead PS (1961). The serial cultivation of human diploid cell strains. *Exp Cell Res.* 25: 585–621.
6. Meyne J, Ratliff RL, Moyzis RK (1989). Conservation of the human telomere sequence (TTAGGG)n among vertebrates. *Proc Natl Acad Sci USA* 86: 7049–53.
7. Campisi J (2001). Cellular senescence as a tumor-suppressor mechanism. *Trends Cell Biol.* 11: S27–31.
8. Sozou PD, Kirkwood TB (2001). A stochastic model of cell replicative senescence based on telomere shortening, oxidative stress, and somatic mutations in nuclear and mitochondrial DNA. *J Theor Biol.* 213: 573–86.
9. Bodnar AG, Ouellette M, Frolkis M, *et al.* (1998). Extension of life-span by introduction of telomerase into normal human cells. *Science* 279: 349–52.
10. Saretzki G, Von Zglinicki T (2002). Replicative aging, telomeres, and oxidative stress. *Ann NY Acad Sci.* 959: 24–9.
11. Sherr CJ, DePinho RA (2000). Cellular senescence: mitotic clock or culture shock? *Cell* 102: 407–10.
12. von Zglinicki T, Saretzki G, Docke W, Lotze C (1995). Mild hyperoxia shortens telomeres and inhibits proliferation of fibroblasts: a model for senescence? *Exp Cell Res.* 220: 186–93.
13. Lorenz M, Saretzki G, Sitte N, Metzkow S, von Zglinicki T (2001). BJ fibroblasts display high antioxidant capacity and slow telomere shortening independent of hTERT transfection. *Free Rad Biol Med.* 31: 824–31.
14. von Zglinicki T, Serra V, Lorenz M, *et al.* (2000). Short telomeres in patients with vascular dementia: an indicator of low antioxidative capacity and a possible risk factor? *Lab Invest.* 80: 1739–47.
15. Curtis H, Crowley C (1963). Chromosome aberrations in liver cells in relation to the somatic theory of aging. *Radiat Res.* 19: 337–344.
16. Vijg J (2000). Somatic mutations and aging: a re-evaluation. *Mutat Res.* 447: 117–35.
17. Gossen JA, de Leeuw WJ, Tan CH, *et al.* (1989). Efficient rescue of integrated shuttle vectors from transgenic mice: a model for studying mutations *in vivo*. *Proc Natl Acad Sci USA* 86: 7971–5.
18. Boerrigter ME, Dollé ME, Martus HJ, Gossen JA, Vijg J (1995). Plasmid-based transgenic mouse model for studying *in vivo* mutations. *Nature* 377: 657–9.
19. Dempsey JL, Odagiri Y, Morley AA (1993). *In vivo* mutations at the H-2 locus in mouse lymphocytes. *Mutat Res* 285: 45–51.

20. Shao C, Deng L, Henegariu O, *et al.* (1999). Mitotic recombination produces the majority of recessive fibroblast variants in heterozygous mice. *Proc Natl Acad Sci USA* 96: 9230–5.
21. Grist SA, McCarron M, Kutlaca A, Turner DR, Morley AA (1992). *In vivo* human somatic mutation: frequency and spectrum with age. *Mutat Res.* 266: 189–96.
22. da Cruz AD, Curry J, Curado MP, Glickman BW (1996). Monitoring hprt mutant frequency over time in T-lymphocytes of people accidentally exposed to high doses of ionizing radiation. *Environ Mol Mutagen.* 27: 165–75.
23. Gossen JA, Vijg J (1993). A selective system for lacZ- phage using a galactose-sensitive *E. coli* host. *Biotechniques* 14: 326, 330.
24. Dollé ME, Giese H, Hopkins CL, Martus HJ, Hausdorff JM, Vijg J (1997). Rapid accumulation of genome rearrangements in liver but not in brain of old mice. *Nat Genet.* 17: 431–4.
25. Dollé ME, Snyder WK, Gossen JA, Lohman PH, Vijg J (2000). Distinct spectra of somatic mutations accumulated with age in mouse heart and small intestine. *Proc Natl Acad Sci USA* 97: 8403–8.
26. Shay JW, Wright WE (2001). Aging. When do telomeres matter? *Science* 291: 839–40.
27. Hart RW, Setlow RB (1974). Correlation between deoxyribonucleic acid excision-repair and life-span in a number of mammalian species. *Proc Natl Acad Sci USA* 71: 2169–73.
28. Chen Q, Fischer A, Reagan JD, Yan LJ, Ames BN (1995). Oxidative DNA damage and senescence of human diploid fibroblast cells. *Proc Natl Acad Sci USA* 92: 4337–41.
28a. Parrinello S, Samper E, Krtolica A, Goldstein J, Melov S, Campisi J (2003). Oxygen sensitivity severely limits the replicative lifespan of murine fibroblasts. *Nat Cell Biol.* 5(8): 741–7.
29. Balin AK, Fisher AJ, Carter DM (1984). Oxygen modulates growth of human cells at physiologic partial pressures. *J Exp Med.* 160: 152–66.
30. Saito H, Hammond AT, Moses RE (1995). The effect of low oxygen tension on the *in vitro*-replicative life span of human diploid fibroblast cells and their transformed derivatives. *Exp Cell Res.* 217: 272–9.
31. Packer L, Fuehr K (1977). Low oxygen concentration extends the lifespan of cultured human diploid cells. *Nature* 267: 423–5.
32. Shigenaga MK, Ames BN (1991). Assays for 8-hydroxy-2′-deoxyguanosine: a biomarker of *in vivo* oxidative DNA damage. *Free Rad Biol Med.* 10: 211–6.
33. Elena SF, Lenski RE (1997). Test of synergistic interactions among deleterious mutations in bacteria. *Nature* 390: 395–8.

Mitochondria and Aging

Martin Barron and Doug Turnbull
Department of Neurology, The Medical School, University of Newcastle, Newcastle upon Tyne, NE2 4EH, United Kingdom

Introduction

Animal cells rely on oxidative phosphorylation to supply the chemical energy necessary for life. In this process electrons are transferred, via a series of membrane-bound enzymes (complexes I-IV) and mobile electron carriers (ubiquinone and cytochrome *c*) to molecular oxygen resulting in the formation of water [1]. The chemical energy involved in these electron transfers is stored as a transmembrane chemical gradient established by the translocation of protons across the membrane by complexes I, III and IV [2, 3]. This stored energy is then coupled to the formation of ATP by the controlled flow of protons down their chemical gradient within complex V, the ATP synthase.

The use of molecular oxygen as an "electron sink" accelerates the process of oxidative phosphorylation. However, oxygen can also interact at an earlier stage by accepting electrons from ubisemiquinone, a reduced form of the mobile electron carrier ubiquinone. This results in the formation of the superoxide radical and the subsequent formation of many other damaging reactive oxygen species (ROS) [4].

Mitochondria

Mitochondria are organelles found in the cytosol of eukaryotic cells. They subserve a variety of functions including intermediary metabolism (citric acid cycle, β-oxidation), urea cycle, calcium storage, haem biosynthesis and, most importantly, ATP production by the process of oxidative phosphorylation as outlined above. Mitochondria appear as circular or oval profiles of 0.5–2.0 μm diameter or cylinders up to 5.0 μm in length by electron microscopy [5]. Their size and number is variable according to cell type. Cells with high metabolic activity, and therefore high ATP requirements, generally contain thousands of mitochondria [5]. Although the above

T. von Zglinicki (ed.), Aging at the Molecular Level, 91–106.

description is the most common, it is increasingly recognized that *in vivo* mitochondria are highly dynamic structures exhibiting budding, fusion and fission [6]. Therefore it may be inaccurate to speak of a unitary mitochondrion but rather to think of a mitochondrial syncitium.

Mitochondria are composed of two membranes, the inner membrane enclosing a central compartment, the matrix, and an outer membrane which creates a compartment between it and the inner membrane called the inter-membrane space. The inner membrane is the site of oxidative phosphorylation and shows a number of specializations for this purpose. The protein content is high (ca. 70%) due to the presence of the oxidative enzyme complexes mentioned above; the membrane contains a highly hydrophobic lipid, cardiolipin, which provides high transmembrane electrochemical insulation; finally, the surface area of the inner membrane is maximized due to extensive folding forming structures called cristae. Within the matrix, as well as the enzymes responsible for other mitochondrial processes mentioned above, are the components of the mitochondrial genetic machinery; mitochondrial DNA (mtDNA), RNA and ribosomes [5].

Mitochondrial DNA

The mitochondrion contains the only extranuclear source of DNA in animal cells (mtDNA) [7]. MtDNA is a closed, circular, double-stranded molecule of 16 569 bp [8, 9]. It encodes 37 genes including 2 ribosomal RNA's (12S & 16S), 22 tRNAs and 13 polypeptides which are essential components of the respiratory chain (Table 1). Several copies of the mitochondrial genome are present in the matrix [10].

The human mitochondrial genome shows extreme economy in its organization, containing little non-coding DNA and no introns (Figure 1). MtDNA is thought to be particularly vulnerable to oxidative attack since it is located in the mitochondrial matrix close to the electron transport chain. Furthermore, mtDNA is not protected by histones [11, 12] unlike nuclear DNA and its repair processes are thought to be less efficient [13]. As a consequence, the mtDNA mutation rate is estimated to be 10 times greater than that for nuclear DNA [14].

Over the last decade, defects of the mitochondrial genome have been increasingly recognized as important causes of disease and much of what we now understand of mitochondrial genetics is based upon the study of these patients [15]. The molecular defect in patients is either a rearrangement of mtDNA (deletion or duplication) or a point mutation (involving either an RNA or protein-encoding gene). These mutations can affect all copies of the mitochondrial genome within a cell – termed homoplasmy or there may be a mixture of mutated and wild-type genomes in the same cell – heteroplasmy [16]. The vast majority of mtDNA mutations described are recessive and in the presence of heteroplasmy high levels of mutated mtDNA with low levels of wild-type mtDNA have to be present within an individual cell before there is a biochemical defect that compromises ATP production.

Table 1. Distribution of nuclear and mitochondrial encoded respiratory chain subunits

Complex	Total subunits	Nuclear encoded	Mitochondrial encoded
I: NADH ubiquinone oxidoreductase	41	34	7 (ND1,2,3,4,4L,5,6)
II: Succinate ubiquinone oxidoreductase	4	4	0
III: Ubiquinol cytochrome c oxidoreductase	11	10	1 (Cytochrome b)
IV: Cytochrome c oxidase	13	10	3 (COX 1,2,3)
V: ATP Synthase	14	12	2 (ATPase 6,8)

Mitochondrial theory of aging

The free radical theory of aging contends that aging is due to the random accumulation of damage in cells and tissues due to a life-times exposure to endogenously generated free radicals and reactive oxygen species [17]. The mitochondrial theory of aging [18, 19] developed from this when it was appreciated that the greatest intracellular source of free radicals and reactive oxygen species was the mitochondrion [20]. While this initial study claimed that 1–2% of cellular oxygen was converted to superoxide by the mitochondria a more accurate, recent study placed the level at 0.1% [21]. Nevertheless, over a lifetime, this rate of production is likely to be significant.

Evidence for the mitochondrial theory of aging

The results of experiments using transgenic *Drosophila melanogaster,* in which superoxide dismutase-1 (SOD-1) and catalase were overexpressed, repeatedly show an increase in the mean and maximum lifespan [22–25]. Both SOD-1 and catalase are found in mitochondria as well as the cytosol. A *Caenorhabditis elegans* strain with a loss of function mutation in the Clk-1 gene shows an increased lifespan [26, 27]. This gene is homologous to the yeast and rat CAT5/CoQ7 gene concerned with the synthesis of Coenzyme Q (CoQ, ubiquinone) [28, 29]. CoQ is implicated as a principal effector in the mitochondrial theory since it is thought to be the prime site for the generation of ROS in the electron transport chain.

Mitochondrial composition and structure have been shown to alter during aging. Decreases in mitochondrial size and the mitochondria/myofibril ratio have been observed in rat myocardium during aging while a decrease in mitochondrial number accompanied by an increase in their size has also been observed [30, 31]. The lipid content of mitochondria, especially cardiolipin, has been shown to decline markedly during aging [32–34].

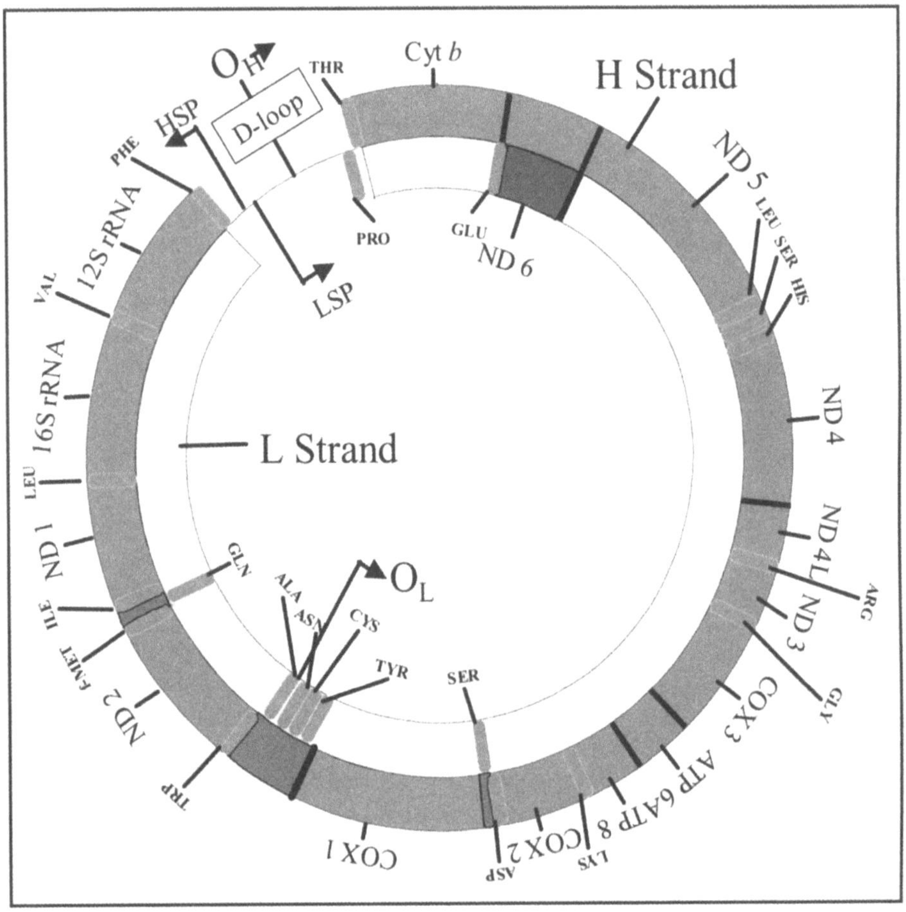

Figure 1. *Organization of the human mitochondrial genome. HSP and LSP are the heavy and light strand promoters while O_H and O_L are the origins of heavy and light strand replication respectively. ND, COX, ATP and Cyt b represent NADH ubiquinone oxidoreductase, cytochrome c oxidase, F_1F_0 ATP synthase and apocytochrome b of ubiquinol: cytochrome c oxidoreductase. The small, three-lettered abbreviations represent the mitochondrial transfer RNA genes for the corresponding amino acid: 16S and 12S represent the mitochondrial ribosomal RNA genes.*

The rate of ROS production by mitochondria is said to increase with age [35–37]. This has not been demonstrated directly and may largely be the result of an age-related deficiency of antioxidant defenses [38–41]. Recent reports using cells where 8-oxoguanine glycosylase had been targeted to mitochondria, suggests that the normal repair-rate of oxidative damage in mtDNA may be inadequate in the face of severe oxidative insult [42, 43]. The cells in these studies showed increased repair of menadione induced oxidative damage to mtDNA and extended survival following such oxidative insults.

More indirect evidence for mitochondrial involvement in aging comes from the phenomenon of caloric restriction. In rodents, severely restricting dietary intake has been shown to delay the onset of most age-related diseases, diminish physiological deficits and extend the average and maximum species lifespan by 30–40% [44]. Caloric restriction has widespread physiological effects including alterations in the rate of intermediary metabolism, increased resistance to infection, improved gluco-regulation but, of importance here, it enhances the ability of the organism to withstand stress, in particular oxidative stress [45].

Caloric restriction has been shown to inhibit oxidative damage to lipids [46], proteins [36] and nucleic acids [47]. This resistance to oxidative stress may be largely due to augmented expression of antioxidant defenses such as copper/zinc and manganese superoxide dismutases [48]. Furthermore, there may be increased expression of complex IV of the respiratory chain during caloric restriction [48] which may have the effect of hastening electron flow through the respiratory chain and thereby limit the opportunity for premature reduction of molecular oxygen.

Further evidence in support of a mitochondrial role in the aging process comes from recent work in *C. elegans* [49, 50]. Two recent papers have used RNA interference (RNAi) for gene inactivation in this organism. In a systematic RNAi screen of 5690 *C. elegans* genes for longevity, genes important for mitochondrial function stood out as a principal group affecting longevity [49]. In one example a mutation in a nuclear gene was found that significantly impaired mitochondrial function with lower ATP production and oxygen utilization, but enhanced resistance to heat shock and hydrogen peroxide. In a different study, the activity of the electron transport chain and ATP synthase were lowered by RNAi [50]. These perturbations reduced body size and behavioral rates and extended adult life-span. Restoring messenger RNA to near-normal levels during adulthood did not correct either the biochemical defect or the phenotype. It seems that the developing animal contains a regulatory system that monitors mitochondrial activity early in life and establishes rates of respiration and aging that persists into adulthood. These studies also highlight the major differences between species, since defects of mitochondrial oxidations in man lead to severe disease and often death [51, 52]. Certainly, lowering of ATP synthesis to the degree seen in *C. elegans* would not be compatible with longevity in man.

Age-related changes in respiratory chain activity

Age-related changes in oxidative phosphorylation have been reported [53–57]. These biochemical studies predominantly used skeletal muscle, but mitochondrial activity in this tissue is heavily dependent on physical demand. Therefore, these studies may simply reflect reduced physical activity in the elderly rather than a genuine age-related decline [58].

Cytochrome *c* oxidase deficient cells in aging

COX deficient muscle fibres are a characteristic of mitochondrial diseases in general [59], e.g., they are widespread in Kearns–Sayre syndrome and myoclonic epilepsy with ragged red fibres (MERFF) and demonstrate a biochemical deficit in mitochondrial function. Often, affected muscle fibres show accumulations of structurally abnormal mitochondria ("ragged red fibres"). Such cells are seen during aging albeit to a lesser degree [58, 60, 61].

The development of a dual histochemical assay [62] has allowed us to demonstrate COX deficient cells in a wide range of aging human tissues [63, 64]. Succinate dehydrogenase (SDH) is encoded exclusively by nuclear genes and therefore it is independent of mtDNA. Thus, when other components of the respiratory chain are absent because of mtDNA damage, SDH activity should still be present. In the dual COX/SDH assay therefore, COX deficient cells do not accumulate the brown reaction product during histochemistry for COX but appear blue following incubation in SDH medium.

The studies of Müller-Hocker relied solely on the histochemical method for COX [60, 61, 65]. In this, COX deficient cells do not accumulate reaction product and appear white. In patients with mitochondrial disease detection of COX deficient muscle fibres is uncomplicated because their numbers are high. In aging tissues however, the incidence of COX deficient cells is generally low ($<1\%$) [66], making non-reactive cells difficult to detect. The dual COX/SDH assay, by demonstrating SDH activity retained in COX deficient cells, facilitates detection. Additionally, the assay allows the distinction to be made between viable and necrotic cells because both COX and SDH activity would be lost from the latter.

Using this technique, we have shown age-related increases of COX deficient human hippocampal neurones and choroid plexus cells [67]. In a recent study we have shown an age-related increase in the numbers of COX deficient cones in human retina [68]. The distribution of such cones showed topographical differences, being highest in the macular region (up to 1.5% in the fovea) and lowest in the peripheral retina (up to 0.25%). This is of interest because the macula is the area of the human retina most frequently affected by age-related pathology.

Mitochondrial DNA in aging

A relationship between inherited mtDNA variants and aging has been proposed [69] although the findings are discrepant between populations [70]. These are complex studies however, and are dependent on large populations. These studies have also revealed the paradoxical finding that the same polymorphisms in mtDNA can predispose both to longevity and to complex diseases [71].

An age-related increase in the level of acquired mtDNA mutation has been frequently reported. Rearrangements, including deletions and partial duplications, of mtDNA are the most common alterations observed in human skeletal muscle, heart, brain, skin, retina, ovary and sperm [55, 68, 72–87]. A large number of different mtDNA deletions have been shown in human skeletal muscle using long-extension PCR [88]. Point mutations of human mtDNA also accumulate with age [89]; the nt8344 (MERFF), nt3243 (MELAS) mutations show an age-related rise [87, 90–93], although this has not been found by all workers [94].

Evidence from patients with mitochondrial diseases has shown that even deletions of the mitochondrial genome are functionally recessive. Thus patients rarely develop symptoms or even a biochemical defect unless the level of mutated mtDNA is high (>60%) within a tissue. Since the total amount of mtDNA within a cell will be limited, the increased level of mutated mtDNA will result in lower amounts of wild-type mtDNA and this presumably is the cause of the reduced oxidative phosphorylation within individual cells. The situation in aging is that the total amount of an individual mutated mtDNA within a tissue is very low and therefore unlikely to induce a biochemical defect unless there is focal accumulation of mutations within cells. In skeletal muscle cytochrome *c* oxidase (COX) negative fibres were found to contain high levels of a variety of different deletions by *in situ* hybridization or PCR [64, 95], while similar expansions of individual deletions have been shown in cardiomyocytes using long-extension PCR [96]. These observations suggest that if mtDNA does have a role in human aging then clonal expansion of individual mtDNA mutations within individual cells is an important phenomenon.

Clonal expansion of mitochondrial DNA mutations

A number of theories seek to explain why clonal expansion of mutant mtDNA occurs. One of the earliest suggested that mutated mtDNA had a replicative advantage allowing it to out-compete the wild type genome [97–100]. This is plausible for deleted mtDNA molecules because they are smaller and therefore would complete a round of replication more rapidly than a wild type molecule. Point mutations present a problem however because they do not alter the size of the genome. Replicative advantage involving point mutations is therefore unlikely unless such mutations confer an enhanced engagement with the replication machinery [101].

Using mathematical modelling it has been suggested that there is no requirement for a selective process for clonal expansion [102, 103]. In this model, using conservative estimates of mitochondrial mutation, replication and fission rates, the majority of mutations are shown to arise and disappear over the maximum human lifespan of 120 years. However, occasionally, with no applied selective pressure, some mutations show an enormous amplification. Thus, clonal expansion is sufficiently explained by random drift of the mtDNA population within cells. However, this explanation is modelled on observations in aging human skeletal muscle and thus may not be pertinent to other tissues, although a similar model, applied to epithelial cells, has produced similar results [104]. Furthermore, the mtDNA mutation rate is not necessarily equal in all tissues but may vary according to cell type (e.g., post-mitotic versus proliferative) and/or metabolic activity. Evidence suggests that the somatic mutation rate for mtDNA may be several hundred-fold greater than the nuclear mutation rate (cf. 10 times the nuclear mutation rate usually quoted) [105]. Thus, in certain tissues the frequency of COX deficient cells may be higher than in skeletal muscle, a situation which we have observed in aging human extraocular muscle (ca. 7%), human choroid plexus (ca. 7%) and in human and murine ciliary epithelium where COX deficient cells can reach levels of 30 and 50% respectively (unpublished observations).

Animal models of relevance to the mitochondrial theory of aging

Recently, animal models attempting to link mitochondrial oxidative stress to cellular damage have been produced. The adenine nucleotide translocators are integral membrane proteins from the inner mitochondrial membrane which catalyze the exchange of extramitochondrial ADP for ATP generated in the matrix. The ANT-1 knockout mouse lacks expression of the adenine nucleotide translocator type-1 usually present at high levels in heart and skeletal muscle and at lower levels in brain, kidney and eye [106]. The homozygote knockout shows increased serum lactate, mitochondrial myopathy including "ragged red fibres" and a hypertrophic cardiomyopathy. Since ADP cannot enter, the mitochondria of these animals are unable to diminish the proton gradient generated by the electron transport chain. Under such conditions electron flow is inhibited and the local concentration of molecular oxygen increases (state 4 respiration). ROS production is therefore increased as the transfer of electrons from ubisemiquinone to oxygen is favored. Hydrogen peroxide levels are increased 8-fold in skeletal muscle of these animals and 3-fold in the heart. Glutathione peroxidase expression is increased 3-fold in both heart and muscle whereas SOD-2 is increased 6-fold in skeletal muscle but not at all in heart. Very little change in the levels of mtDNA deletions was detected by long-extension PCR in skeletal muscle but the levels seen in 16–20-month old mouse hearts were equivalent to those seen in normal 32-month old animals. The low levels of deletions in skeletal muscle may be due to adequate upregulation of SOD-2 in this tissue although the development of myopathy is puzzling.

A SOD-2 knockout mouse has also been developed [107, 108]. The homozygous state is lethal; the animals die at around 8 days of age from dilated cardiomyopathy with hepatic involvement. Analysis of heart, skeletal muscle and brain showed that the primary target of superoxide in these animals was the FeS centres in complexes I, II and aconitase. Supply of a synthetic SOD mimic (manganese-5,10,15,20-tetrakis-(-4-benzoic acid)-porphyrin) resulted in an extension of lifespan to around 16 days. These animals developed a fatal spongiform encephalopathy probably because the SOD mimic could not cross the blood/brain barrier.

A heterozygote SOD-2 knockout showed a 50% reduction in SOD-2 activity [109]. The decreased SOD-2 activity correlated with diminished activity of FeS centre proteins (aconitase and complex I), increases in the levels of carbonylated protein and 8-hydroxydeoxyguanosine in mtDNA [109]. Age-related changes were also seen in this mouse [110]. Three age groups of SOD-2 deficient mice were investigated (5 months; 10–14 months; 20–25 months): all age groups showed a decreased mitochondrial membrane potential ($\Delta\Psi$), decreased state 3 (the oxidative phosphorylation rate under conditions of high matrix ADP concentration where electron flow through the respiratory chain and reduction of molecular oxygen by complex IV are the highest) and increased state 4 respiration rate. An upregulation of respiratory chain complexes was also detected and perhaps represents an attempt to diminish oxidative stress by facilitating electron flow. An increased level of lipid peroxidation was seen with age as was the level of apoptotic hepatocytes. Interestingly, levels of peroxidation decreased dramatically in the aged SOD-2 deficient mouse hepatocytes while they increased in normal age-matched controls. The authors suggest this is because premature accumulation of lipid peroxides in SOD-2 deficient hepatocytes leads to their rapid loss by apoptosis, leaving a population of cells with reduced levels of oxidative damage.

Age-related deficits in tissue function

Despite the presence of COX deficient cells within a tissue it is difficult to appreciate how the low numbers of such cells (ignoring the recent observations above) could lead to the profound decreases in function associated with aging. A possible explanation is cell loss by apoptosis.

There is evidence that mitochondria are major regulators of apoptosis through the opening of the mitochondrial permeability transition pore (mtPTP) [111]. The mtPTP is thought to be composed of a large number of components connecting the mitochondrial outer and inner membranes. These include the inner membrane ANT, the outer membrane voltage-dependent channel, Bax, Bcl-2 and cyclophilin D [112-114]. Opening of the PTP is associated with excessive Ca^{2+} uptake, decreased $\Delta\Psi$, increased ADP/ATP ratios and oxidative stress [114, 115].

Cells respond to oxidative stress in a broad spectrum of ways including mitotic proliferation, transient adaptation, permanent growth arrest, apoptosis or necrosis [116]. The dosage and type of oxidant leads to different cell death mechanisms; low doses of oxidants are associated with apoptosis whereas high dose/potency oxidants lead to necrosis. Intermediate modes of cell death occur between these extremes in a dose-dependent manner [117]. ATP supply is thought to be the major determinant regulating the selection of cell death mechanism. Following the opening of the PTP the $\Delta\Psi$ collapses as protons flow back into the mitochondrial matrix. Collapse of the $\Delta\Psi$ reverses the action of the ATP synthase so cellular supplies of ATP are rapidly hydrolyzed. If ATP hydolysis proceeds unchecked, insufficient ATP is available to complete the apoptotic program and necrosis ensues [118, 119]. Low doses of oxidants are thought to produce transient opening of the PTP such that the $\Delta\Psi$ is maintained at a level consistent with ATP production or conservation [118, 119].

During aging it is likely that oxidant generation will be relatively low but persistent. Persistent oxidant generation may lead to mtDNA damage and consequent dysfunction. As the oxidative stress increases, affected cells become increasingly vulnerable to apoptosis. Occasionally, a COX deficient cell may be produced because its mutant mtDNA population rapidly expands. Such cells would generate little ATP and therefore fail to execute apoptosis while retaining enough to remain viable. Thus, COX deficient cells may be a visible sign of a more aggressive underlying process affecting cells with milder mitochondrial dysfunction.

Conclusion

The available evidence suggests that mitochondrial dysfunction plays a role in aging. However, the role of mitochondria and mtDNA abnormalities in human aging remains contentious [120]. Interesting theories are constantly being proposed for phenomena such as clonal expansion which will drive future research in the area. One of the major questions to be addressed is the extent to which mitochondrial abnormalities contribute to the decline in function seen in aging within specific tissues. It is clear that the mitochondrial theory is unable to explain all of the phenomena which constitute aging. To this end more integrated theories are being developed which include mitochondrial dysfunction as a major contributor to aging in association with other phenomena such as telomere attrition, endogenous oxidative stress, DNA damage and repair and apoptosis.

Acknowledgments

We are very grateful to Professor Tom Kirkwood, Professor Bob Lightowlers and Dr Patrick Chinnery for very helpful discussions on the role of mitochondria in aging, and to the Wellcome Trust for continued support for our work.

References

1. Chance B (1965). Reaction of oxygen with the respiratory chain in cells and tissues. *J Gen Physiol.* 49(Suppl.): 163–95.
2. Chance B, Mela L (1966). A hydrogen ion concentration gradient in a mitochondrial membrane. *Nature* 212: 369–72.
3. Mitchell P, Moyle J (1967). Chemiosmotic hypothesis of oxidative phosphorylation. *Nature* 213: 137–9.
4. Boveris A, Cadenas E, Stoppani AO (1976). Role of ubiquinone in the mitochondrial generation of hydrogen peroxide. *Biochem J.* 156: 435–44
5. Carroll M (1989). *Organelles.* New York: Guilford Press.
6. Turnbull DM, Lightowlers RN (2001). Might mammalian mitochondria merge? *Nature Genet.* 7: 895–96.
7. Nass MM (1966). The circularity of mitochondrial DNA. *Proc Natl Acad Sci USA* 56: 1215–22.
8. Anderson S, Bankier AT, Barrell BG, *et al.* (1981) Sequence and organization of the human mitochondrial genome. *Nature* 290: 457–65.
9. Andrews RM, Kubacka I, Chinnery PF, Lightowlers RN, Turnbull DM, Howell N (1999) Reanalysis and revision of the Cambridge reference sequence for human mitochondrial DNA. *Nature Genet.* 23: 147.
10. Bogenhagen D, Clayton DA (1974). The number of mitochondrial deoxyribonucleic acid genomes in mouse L and human HeLa cells. Quantitative isolation of mitochondrial deoxyribonucleic acid. *J Biol Chem.* 249: 7991–5.
11. Caron F, Jacq C, Rouviere-Yaniv J (1979). Characterization of a histone-like protein extracted from yeast mitochondria. *Proc Natl Acad Sci USA* 76: 4265–9.
12. DeFrancesco L, Attardi G (1981). In situ photochemical crosslinking of HeLa cell mitochondrial DNA by a psoralen derivative reveals a protected region near the origin of replication. *Nucleic Acids Res.* 9: 6017–30.

13. Howell N (1999). Human mitochondrial diseases: answering questions and questioning answers. *Int Rev Cytol.* 186: 49–116.
14. Brown WM, George M, Jr., Wilson AC (1979). Rapid evolution of animal mitochondrial DNA. *Proc Natl Acad Sci USA* 76: 1967–71.
15. Chinnery PF, Howell N, Andrews RA, Turnbull DM (1999). Clinical mitochondrial genetics. *J Med Genet.*36: 425–36.
16. Lightowlers RN, Chinnery PF, Turnbull DM, Howell N (1997). Mammalian mitochondrial genetics: heredity, heteroplasmy and disease. *Trends Genet.* 13: 450–5.
17. Harman D (1956). Aging: a theory based on free radical and radiation chemistry. *J Gerontol.* 11: 298–300.
18. Harman D (1972). The biologic clock: the mitochondria? *J Am Geriatr Soc.* 20: 145–7.
19. Miquel J, Economos AC, Fleming J, Johnson JE Jr (1980). Mitochondrial role in cell aging. *Exp Gerontol.* 15: 575–91.
20. Chance B, Sies H, Boveris A (1979). Hydroperoxide metabolism in mammalian organs. *Physiol Rev.* 59: 527–605.
21. Hansford RG, Hogue BA, Mildaziene V (1997). Dependence of H_2O_2 formation by rat heart mitochondria on substrate availability and donor age. *J Bioenerg Biomembr* 29: 89–95.
22. Orr WC, Sohal RS (1994). Extension of life-span by overexpression of superoxide dismutase and catalase in *Drosophila melanogaster. Science* 263: 1128–30.
23. Sohal RS, Agarwal A, Agarwal S, Orr WC (1995). Simultaneous overexpression of copper- and zinc-containing superoxide dismutase and catalase retards age-related oxidative damage and increases metabolic potential in *Drosophila melanogaster. J Biol Chem.* 270: 15671–4.
24. Parkes TL, Elia AJ, Dickinson D, Hilliker AJ, Phillips JP, Boulianne GL (1998). Extension of *Drosophila* lifespan by overexpression of human SOD1 in motorneurons. *Nature Genet.* 19: 171–4.
25. Sun J, Tower J (1999). FLP recombinase-mediated induction of Cu/Zn-superoxide dismutase transgene expression can extend the life span of adult *Drosophila melanogaster* flies. *Mol Cell Biol.* 19: 216–28.
26. Wong A, Boutis P, Hekimi S (1995). Mutations in the clk-1 gene of *Caenorhabditis elegans* affect developmental and behavioral timing. *Genetics* 139: 1247–59.
27. Lakowski B, Hekimi S (1996). Determination of life-span in *Caenorhabditis elegans* by four clock genes. *Science* 272: 1010–13.
28. Jonassen T, Marbois BN, Kim L, *et al.* (1996). Isolation and sequencing of the rat Coq7 gene and the mapping of mouse Coq7 to chromosome 7. *Arch Biochem Biophys.* 330: 285–9.
29. Marbois BN, Clarke CF (1996). The COQ7 gene encodes a protein in *Saccharomyces cerevisiae* necessary for ubiquinone biosynthesis. *J Biol Chem.* 271: 2995–3004.
30. Frenzel H, Feimann J (1984). Age-dependent structural changes in the myocardium of rats. A quantitative light- and electron-microscopic study on the right and left chamber wall. *Mech Ageing Dev.* 27: 29–41.
31. Ledda M, Martinelli C, Pannese E (2001). Quantitative changes in mitochondria of spinal ganglion neurons in aged rabbits. *Brain Res Bull.* 54: 455–9.
32. Paradies G, Ruggiero FM (1990). Age-related changes in the activity of the pyruvate carrier and in the lipid composition in rat-heart mitochondria. *Biochim Biophys Acta* 1016: 207–12.

33. Ruggiero FM, Cafagna F, Petruzzella V, Gadaleta MN, Quagliariello E (1992). Lipid composition in synaptic and nonsynaptic mitochondria from rat brains and effect of aging. *J Neurochem.* 59: 487–91.
34. Shigenaga MK, Hagen TM, Ames BN (1994). Oxidative damage and mitochondrial decay in aging. *Proc Natl Acad Sci USA* 91: 10771–8.
35. Sohal RS, Sohal BH (1991). Hydrogen peroxide release by mitochondria increases during aging. *Mech Ageing Dev.* 57: 187–202.
36. Sohal RS, Ku HH, Agarwal S, Forster MJ, Lal H (1994). Oxidative damage, mitochondrial oxidant generation and antioxidant defenses during aging and in response to food restriction in the mouse. *Mech Ageing Dev.* 74: 121–33.
37. Souza-Pinto NC, Croteau DL, Hudson EK, Hansford RG, Bohr VA (1999). Age-associated increase in 8-oxo-deoxyguanosine glycosylase/AP lyase activity in rat mitochondria. *Nucleic Acids Res.* 27: 1935–42.
38. Sohal RS, Arnold L, Orr WC (1990). Effect of age on superoxide dismutase, catalase, glutathione reductase, inorganic peroxides, TBA-reactive material, GSH/GSSG, NADPH/NADP+ and NADH/NAD+ in *Drosophila melanogaster. Mech Ageing Dev.* 56: 223–35.
39. Semsei I, Rao G, Richardson A (1991). Expression of superoxide dismutase and catalase in rat brain as a function of age. *Mech Ageing Dev.* 58: 13–19.
40. Tian L, Cai Q, Wei H (1998). Alterations of antioxidant enzymes and oxidative damage to macromolecules in different organs of rats during aging. *Free Radic Biol Med.* 24: 1477–84.
41. Lu CY, Lee HC, Fahn HJ, Wei YH (1999). Oxidative damage elicited by imbalance of free radical scavenging enzymes is associated with large-scale mtDNA deletions in aging human skin. *Mutat Res.* 423: 11–21.
42. Dobson AW, Xu Y, Kelley MR, LeDoux SP, Wilson GL (2000). Enhanced mitochondrial DNA repair and cellular survival after oxidative stress by targeting the human 8-oxoguanine glycosylase repair enzyme to mitochondria. *J Biol Chem.* 275: 37518–23.
43. Rachek LI, Grishko VI, Musiyenko SI, Kelley MR, LeDoux SP, Wilson GL (2002). Conditional targeting of the DNA repair enzyme hOGG 1 into mitochondria. *J Biol Chem.* 277: 44932–7.
44. Masoro E, J. (2000). Caloric restriction and aging: an update. *Exp Gerontol.* 35: 299
45. Weindruch R, Keenan KP, Caney JM, *et al.* (2001). Caloric restriction mimetics: metabolic interventions. *J Gerontol Series A: Biol Sci.* 56: 20–33.
46. Lass A, Sohal BH, Weindruch R, Forster MJ, Sohal RS (1998). Caloric restriction prevents age-associated accrual of oxidative damage to mouse skeletal muscle mitochondria. *Free Radic Biol Med.* 25: 1089–97.
47. Kaneko T, Tahara S, Matsuo M (1997). Retarding effect of dietary restriction on the accumulation of 8-hydroxy-2′-deoxyguanosine in organs of Fischer 344 rats during aging. *Free Radic Biol Med.* 23: 76–81.
48. Sreekumar R, Unnikrishanan J, Fu A, *et al.* (2002). Effects of caloric restriction on mitochondrial function and gene transcripts in rat muscle. *Am J Physiol Endocrinol Metab.* 238: E38–43.
49. Lee SS, Lee RYN, Fraser AG, Kamath RS, Ahringer J, Ruvkin G (2003). A systematic RNAi screen identifies a critical role for mitochondria in *C. elegans* longevity. *Nature Genet.* 33: 40–8.
50. Dillin A, Hsu A-L, Arantes-Oliveira N, *et al.* (2002). Rates of behaviour and aging specified by mitochondrial function during developement. *Science* 298: 2398–401.

51. Leonard JV, Schapira AH (2000). Mitochondrial respiratory chain disorders I: mitochondrial DNA defects. *Lancet* 355: 299–304.
52. Leonard JV, Schapira AVH (2000). Mitochondrial respiratory chain disorders II: neurodegenerative disorders and nuclear gene defects. *Lancet* 355: 389–94.
53. Trounce I, Byrne E, Marzuki S (1989). Decline in skeletal muscle mitochondrial respiratory chain function: possible factor in aging [see comments]. *Lancet* 1: 637–9
54. Yen TC, Chen YS, King KL, Yeh SH, Wei YH (1989). Liver mitochondrial respiratory functions decline with age. *Biochem Biophys Res Commun.* 165: 944–1003.
55. Cooper JM, Mann VM, Schapira AH (1992). Analyses of mitochondrial respiratory chain function and mitochondrial DNA deletion in human skeletal muscle: effect of ageing. *J Neurol Sci.* 113: 91–8.
56. Bowling AC, Mutisya EM, Walker LC, Price DL, Cork LC, Beal MF (1993). Age-dependent impairment of mitochondrial function in primate brain. *J Neurochem.* 60: 1964–7.
57. Boffoli D, Scacco SC, Vergari R, Solarino G, Santacroce G, Papa S (1994). Decline with age of the respiratory chain activity in human skeletal muscle. *Biochim Biophys Acta* 1226: 73–82.
58. Brierley EJ, Johnson MA, James OF, Turnbull DM (1996). Effects of physical activity and age on mitochondrial function. *Q J Med.* 89: 251–8.
59. Chinnery PF, Turnbull DM (1997). Clinical features, investigation, and management of patients with defects of mitochondrial DNA [editorial]. *J Neurol Neurosurg Psychiatr.* 63: 559–63.
60. Muller-Hocker J (1989). Cytochrome-c-oxidase deficient cardiomyocytes in the human heart – an age-related phenomenon. A histochemical ultracytochemical study. *Am J Pathol.* 134: 1167–73.
61. Muller-Hocker J (1990). Cytochrome c oxidase deficient fibres in the limb muscle and diaphragm of man without muscular disease: an age-related alteration. *J Neurol Sci.* 100: 14–21.
62. Sciacco M, Bonilla E (1996). Cytochemistry and immunocytochemistry of mitochondria in tissue sections. *Methods Enzymol.* 264: 509–21.
63. Old SL, Johnson MA (1989). Methods of microphotometric assay of succinate dehydrogenase and cytochrome c oxidase activities for use on human skeletal muscle. *Histochem J.* 21: 545–55.
64. Brierley EJ, Johnson MA, Lightowlers RN, James OF, Turnbull DM (1998). Role of mitochondrial DNA mutations in human aging: implications for the central nervous system and muscle. *Ann Neurol.* 43: 217–23.
65. Muller-Hocker J, Schneiderbanger K, Stefani FH, Kadenbach B (1992). Progressive loss of cytochrome c oxidase in the human extraocular muscles in ageing – a cytochemical-immunohistochemical study. *Mutat Res.* 275: 115–24.
66. Cottrell DA, Blakely EL, Johnson MA, Borthwick GM, Ince PI, Turnbull DM (2001). Mitochondrial DNA mutations in disease and ageing. *Novartis Found Symp.* 235: 234–43; discussion 243–6.
67. Cottrell DA, Blakely EL, Johnson MA, Ince PG, Borthwick GM, Turnbull DM (2001). Cytochrome c oxidase deficient cells accumulate in the hippocampus and choroid plexus with age. *Neurobiol Aging* 22: 265–72.
68. Barron MJ, Johnson MA, Andrews RM, *et al.* (2001). Mitochondrial abnormalities in ageing macular photoreceptors. *Invest Ophthalmol Vis Sci.* 42: 3016–22.
69. Tanaka M, Gong J-S, Zhang J, Yoneda M, Yagi K (1998). Mitochondrial genotype associated with longevity. *Lancet* 351: 185–6.

70. Ross OA, McCormack R, Curran MD, *et al.* (2001). Mitochondrial DNA polymorphism: its role in longevity of the Irish population. *Exp Gerontol.* 36: 1161–78.
71. Rose G, Passarino G, Carrieri G, *et al.* (2001). Paradoxes in longevity: sequence analysis of mtDNA haplogroup J in centenarians. *Eur J Hum Genet.* 9: 701–7.
72. Cortopassi GA, Arnheim N (1990). Detection of a specific mitochondrial DNA deletion in tissues of older humans. *Nucleic Acids Res.* 18: 6927–33.
73. Ikebe S, Tanaka M, Ohno K, *et al.* (1990). Increase of deleted mitochondrial DNA in the striatum in Parkinson's disease and senescence. *Biochem Biophys Res Commun.* 170: 1044–8.
74. Linnane AW, Baumer A, Maxwell RJ, Preston H, Zhang CF, Marzuki S (1990). Mitochondrial gene mutation: the ageing process and degenerative diseases. *Biochem Int.* 22: 1067–76.
75. Corral-Debrinski M, Horton T, Lott MT, Shoffner JM, Beal MF, Wallace DC (1992). Mitochondrial DNA deletions in human brain: regional variability and increase with advanced age. *Nature Genet.* 2: 324–29.
76. Corral-Debrinski M, Shoffner JM, Lott MT, Wallace DC (1992). Association of mitochondrial DNA damage with aging and coronary atherosclerotic heart disease. *Mutat Res.* 275: 169–80.
77. Cortopassi GA, Shibata D, Soong NW, Arnheim N (1992). A pattern of accumulation of a somatic deletion of mitochondrial DNA in aging human tissues. *Proc Natl Acad Sci USA* 89: 7370–4.
78. Simonetti S, Chen X, DiMauro S, Schon EA (1992). Accumulation of deletions in human mitochondrial DNA during normal aging: analysis by quantitative PCR. *Biochim Biophys Acta* 1180: 113–22.
79. DiDonato S, Zeviani M, Giovannini P, *et al.* (1993). Respiratory chain and mitochondrial DNA in muscle and brain in Parkinson's disease patients. *Neurology* 43: 2262–8.
80. Kitagawa T, Suganuma N, Nawa A, *et al.* (1993). Rapid accumulation of deleted mitochondrial deoxyribonucleic acid in postmenopausal ovaries. *Biol Reprod.* 49: 730–6.
81. Lee HC, Pang CY, Hsu HS, Wei YH (1994). Differential accumulations of 4,977 bp deletion in mitochondrial DNA of various tissues in human ageing. *Biochim Biophys Acta* 1226: 37–43.
82. Lezza AM, Boffoli D, Scacco S, Cantatore P, Gadaleta MN (1994). Correlation between mitochondrial DNA 4977-bp deletion and respiratory chain enzyme activities in aging human skeletal muscles. *Biochem Biophys Res Commun.* 205: 772–9.
83. Pang CY, Lee HC, Yang JH, Wei YH (1994). Human skin mitochondrial DNA deletions associated with light exposure. *Arch Biochem Biophys.* 312: 534–8.
84. Yang JH, Lee HC, Lin KJ, Wei YH (1994). A specific 4977-bp deletion of mitochondrial DNA in human ageing skin. *Arch Dermatol Res.* 286: 386–90.
85. Kao S, Chao HT, Wei YH (1995). Mitochondrial deoxyribonucleic acid 4977-bp deletion is associated with diminished fertility and motility of human sperm. *Biol Reprod.* 52: 729–36.
86. Barreau E, Brossas JY, Courtois Y, Treton JA (1996). Accumulation of mitochondrial DNA deletions in human retina during aging. *Invest Ophthalmol Vis Sci.* 37: 384–91.
87. Fayet G, Jansson M, Sternberg D, *et al.* (2002). Ageing muscle: clonal expansions of mitochondrial DNA point mutations and deletions cause focal impairment of mitochondrial function. *Neuromusc Disord.* 12: 484–93.
88. Melov S, Shoffner JM, Kaufman A, Wallace DC (1995). Marked increase in the number and variety of mitochondrial DNA rearrangements in aging human skeletal muscle

[published erratum appears in *Nucleic Acids Res.* (1995) 23: 4938]. *Nucleic Acids Res.* 23: 4122–6.
89. Nekhaeva E, Bodyak ND, Kraytsberg Y, *et al.* (2002). Clonally expanded mtDNA point mutations are abundant in individual cells of human tissues. *Proc Natl Acad Sci USA* 99: 5521–6.
90. Lin MT, Simon DK, Ahn CH, Kim LM, Beal MF (2002). High aggregate burden of somatic mtDNA point mutations in aging and Alzheimer's disease brain. *Hum Mol Genet.* 11: 133–45.
91. Munscher C, Rieger T, Muller-Hocker J, Kadenbach B (1993). The point mutation of mitochondrial DNA characteristic for MERRF disease is found also in healthy people of different ages. *FEBS Lett.* 317: 27–30.
92. Zhang C, Linnane AW, Nagley P (1993). Occurrence of a particular base substitution (3243 A to G) in mitochondrial DNA of tissues of ageing humans. *Biochem Biophys Res Commun.* 195: 1104–10.
93. Michikawa Y, Mazzucchelli F, Bresolin N, Scarlato G, Attardi G (1999). Aging-dependent large accumulation of point mutations in the human mtDNA control region for replication. *Science* 286: 774–9.
94. Murdock DG, Christacos NC, Wallace DC (2000). The age-related accumulation of a mitochondrial DNA control region mutation in muscle, but not brain, detected by a sensitive PNA-directed PCR clamping based method. *Nucleic Acids Res.* 28: 4350–5.
95. Muller-Hocker J, Seibel P, Schneiderbanger K, Kadenbach B (1993). Different in situ hybridization patterns of mitochondrial DNA in cytochrome c oxidase-deficient extraocular muscle fibres in the elderly. *Virchows Arch A Pathol Anat Histopathol.* 422: 7–15.
96. Khrapko K, Bodyak N, Thilly WG, *et al.* (1999). Cell-by-cell scanning of whole mitochondrial genomes in aged human heart reveals a significant fraction of myocytes with clonally expanded deletions. *Nucleic Acids Res.* 27: 2434–41.
97. Yoneda M, Chomyn A, Martinuzzi A, Hurko O, Attardi G (1992). Marked replicative advantage of human mtDNA carrying a point mutation that causes the MELAS encephalomyopathy. *Proc Natl Acad Sci USA* 89: 11164–8.
98. Attardi G, Yoneda M, Chomyn A (1995). Complementation and segregation behavior of disease-causing mitochondrial DNA mutations in cellular model systems. *Biochim Biophys Acta* 1271: 241–8.
99. Hofhaus G, Gattermann N (1999). Mitochondria harbouring mutant mtDNA – a cuckoo in the nest? *Biol Chem.* 380: 871–7.
100. Diaz F, Bayona-Bafaluy MP, Rana M, Mora M, Hao H, Moraes CT (2002). Human mitochondrial DNA with large deletions repopulates organelles faster than full-length genomes under relaxed copy number control. *Nucleic Acids Res.* 30: 4626–33.
101. Takeda K, Takahashi S, Onishi A, Hanada H, Imai H (2000). Replicative advantage and tissue-specific segregation of RR mitochondrial DNA between C57BL/6 and RR heteroplasmic mice. *Genetics* 155: 777–83.
102. Chinnery PF, Samuels DC (1999). Relaxed replication of mtDNA: a model with implications for the expression of disease. *Am J Hum Genet.* 64: 1158–65.
103. Elson JL, Samuels DC, Turnbull DM, Chinnery PF (2001). Random intracellular drift explains the clonal expansion of mitochondrial DNA mutations with age. *Am J Hum Genet.* 68: 802–6.
104. Coller HA, Khrapko K, Bodyak ND, Nekhaeva E, Herrero-Jimenez P, Thilly WG (2001). High frequency of homoplasmic mitochondrial DNA mutations in human tumors can be explained without selection. *Nature Genet.* 28: 147–50.

105. Khrapko K, Coller HA, Andre PC, Li XC, Hanekamp JS, Thilly WG (1997). Mitochondrial mutational spectra in human cells and tissues. *Proc Natl Acad Sci USA* 94: 13798–803.
106. Graham BH, Waymire KG, Cottrell B, Trounce IA, MacGregor GR, Wallace DC (1997). A mouse model for mitochondrial myopathy and cardiomyopathy resulting from a deficiency in the heart/muscle isoform of the adenine nucleotide translocator. *Nature Genet.* 16: 226–34.
107. Li Y, Huang TT, Carlson EJ, *et al.* (1995). Dilated cardiomyopathy and neonatal lethality in mutant mice lacking manganese superoxide dismutase. *Nature Genet.* 11: 376–81.
108. Melov S, Coskun P, Patel M, *et al.* (1999). Mitochondrial disease in superoxide dismutase 2 mutant mice. *Proc Natl Acad Sci USA* 96: 846–51.
109. Williams MD, Van Remmen H, Conrad CC, Huang TT, Epstein CJ, Richardson A (1998). Increased oxidative damage is correlated to altered mitochondrial function in heterozygous manganese superoxide dismutase knockout mice. *J Biol Chem.* 273: 28510–15.
110. Kokoszka JE, Coskun P, Esposito LA, Wallace DC (2001). Increased mitochondrial oxidative stress in the Sod2 (+/–) mouse results in the age-related decline of mitochondrial function culminating in increased apoptosis. *Proc Natl Acad Sci USA* 98: 2278–83.
111. Susin SA, Zamzami N, Kroemer G (1998). Mitochondria as regulators of apoptosis: doubt no more. *Biochim Biophys Acta* 1366: 151–65.
112. Zoratti M, Szabo I (1995). The mitochondrial permeability transition. *Biochim Biophys Acta* 1241: 139–76.
113. Petit PX, Susin SA, Zamzami N, Mignotte B, Kroemer G (1996). Mitochondria and programmed cell death: back to the future. *FEBS Lett.* 396: 7–13.
114. Green DR, Reed JC (1998). Mitochondria and apoptosis. *Science* 281: 1309–12.
115. Liu X, Kim CN, Yang J, Jemmerson R, Wang X (1996). Induction of apoptotic program in cell-free extracts: requirement for dATP and cytochrome c. *Cell* 86: 147–57.
116. Davies KJ (1999). The broad spectrum of responses to oxidants in proliferating cells: a new paradigm for oxidative stress. *IUBMB Life* 48: 41–7.
117. Cai J, Jones DP (1999). Mitochondrial redox signaling during apoptosis. *J Bioenerg Biomembr.* 31: 327–34.
118. Lemasters JJ, Qian T, Bradham CA, *et al.* (1999). Mitochondrial dysfunction in the pathogenesis of necrotic and apoptotic cell death. *J Bioenerg Biomembr.* 31: 305–19.
119. Pedersen PL (1999). Mitochondrial events in the life and death of animal cells: a brief overview. *J Bioenerg Biomembr.* 31: 291–304.
120. Lightowlers RN, Jacobs HT, Kajander OA (1999). Mitochondrial DNA – all things bad? *Trends Genet.* 15: 91–3.

Biological Clocks in the Aging Cell

Petra Boukamp
DKFZ (German Cancer Research Center), Division of Genetics of Skin Carcinogenesis, Heidelberg, Germany

Introduction

Since the pioneering studies of Hayflick and Moorhead [1] it is well established that normal human fibroblasts have a limited division potential in culture. As the culture reaches a certain number of population doublings, the growth potential gradually declines. The number of non-dividing cells increases and the culture enters the ultimate state of terminal and irreversible arrest termed replicative senescence [2]. This replicative senescence is not restricted to fibroblasts but similarly seen in a number of normal cell types such as endothelial cells, T-lymphocytes, adrenocortical, smooth muscle, glial, and pancreatic β-cells. The number of divisions that cells undergo before they stop dividing, the "Hayflick limit," inversely correlates with donor-age suggesting that cellular senescence is not a mere tissue culture phenomenon – due to lack of correct signals – but may indeed be a general life-restricting mechanism which is controlled by an endogenous clock [reviewed in ref. 3].

Cellular senescence is a continuous process which includes changes in cell morphology, i.e., the cells enlarge and take on a flattened and irregular cell shape [4]. Similarly, the protein expression pattern changes and finally the cells enter an irreversible G0/G1 arrest [5]. It was shown that the final growth arrest correlates with de-phosphorylation of the retinoblastoma gene pRb, increased expression of the cell cycle inhibitors p21 [6] and p16 [7] as well as P53 up-regulation [reviewed in ref. 8], all required to induce cell cycle blockade. Most importantly, senescent cells do not immediately die. Under appropriate conditions, they can stay viable for many months if not years [9] and are resistant to apoptosis [10, 11]. During this time they are metabolically active and express a number of cytokines, matrix proteins and other gene products such as the senescence-related SA-β galactosidase at pH 6.0 [12], i.e., they show a pattern of gene expression highly specific for this final stage. Thus, senescence represents a specific program that is only terminated by cell death after a long non-proliferative but otherwise active "phase."

T. von Zglinicki (ed.), Aging at the Molecular Level, 107–119.

However, it should also be emphasized that senescence is a population phenomenon and a phenomenon defined for cultured cells. It is, therefore, not surprising that it is still a matter of debate whether senescence occurs in the organism *in situ*. Similarly, senescence can not be studied at the single cell level. Furthermore, early experiments revealed a high variation of the number of maximal population doublings from culture to culture as well as variations in the life-span of individual clones from the same mass culture. Even the doubling potential between two cells arising from a single mitotic event varied by as many as eight population doublings [13]. Nevertheless, the sequence of changes, reproducibly detected when investigating long-term cultures from the same or different fibroblast strain, demonstrate that cellular senescence is a genetically programmed process.

First evidence for the genetic basis came from cell fusion experiments. It was already shown in the early 1980s that fusion between immortal and normal cells yielded hybrids with a finite life span. In addition, fusion between young and old cells yielded hybrids with a remaining life span of the old cells [14]. All this suggested that (i) the senescent phenotype is dominant and (ii) the clock of the old cells is fixed and dominant over the clock of the young cells. In a first approach to determine the "clock," i.e., to determine genes responsible for senescence, fusion of more than 40 immortal cell lines were performed. This established as few as 4 different complementation groups for indefinite division [reviewed in ref. 15]. Interestingly, six of seven fibroblasts lines immortalized by the introduction of SV-40, belonged to the same complementation group and seven individually derived T- and B-cell lines were assigned to a common complementation group. Although the complementation groups in general did not directly correlate with cell type, tumor type, or expression of specific oncogenes, these findings nevertheless suggested a common senescence pathway for cells immortalized by similar mechanisms, e.g., SV-40.

The experimental approach was further improved by introducing single human chromosomes instead of fusing complete cells. Using the method of micro cell-mediated chromosome transfer [for review see ref. 16] a number of different chromosomes were shown to cause senescence. Certain chromosomes seemed to be representative for a specific complementation group, e.g., chromosome 4 for complementation group B, chromosome 1q for complementation group C, and chromosome 7 for complementation group D [all reviewed in ref. 17]. On the other hand, a number of chromosomes were determined by micro cell-mediated chromosome transfer which induced senescence in only individual immortal cell lines. These included chromosomes 2, 3p, 6q, 11, 18, and X. Since these chromosomes did not full fill the genetic criteria of being common "senescence inducers" it was proposed that the observed loss of proliferation was not a true genetic effect but a dosage effect. Alternatively, it was not excluded that other complementation groups may exist or that these cells may contain recessive changes in more than one senescence-related gene or pathway.

So far, all these studies only allowed to identify one gene which was mapped to chromosome 4 and termed MORF4, mortality factor on human chromosome 4 [18]. Although transfection of MORF4 into immortal cells of complementation group B resulted in a senescence-like growth arrest, the function of its gene product and its causal relationship to senescence still needs to be established.

What is the role of oxidative damage?

For a long time, one of the most favored hypothesis of cellular aging was the hypothesis of oxidative damage, i.e., the concept that accumulation of damage may be responsible for limiting the life span of cells. Indeed, it was shown for normal human lung fibroblasts that during the last 15 population doublings, the level of reactive oxygen species increased two- to three-fold and by providing anti-oxidants in the medium the proliferation could be extended for about 20 pd [19]. This and the many more examples present in the literature are in agreement with the interpretation that the anti-oxidant defense decreases with "age" (as does the immune system in humans), thus allowing an increased damage accumulation and a final damage-dependent growth arrest. However, it is difficult to reconcile the idea of an endogenous clock, i.e., "cells knowing their own age," with oxidative damage as a principal cause of senescence. Mostly, oxidative damage is understood as a stochastic phenomenon, which seems not compatible with the relatively fixed number of population doublings until a cell culture becomes senescent and the fact that senescence is seen as a genetically programmed process [20].

On the other hand, neither generation of reactive oxygen species nor cellular responses to damage are simply stochastic. It was discussed that most superoxide radicals are generated by cytochrome P566 and ubiquinone in the inner membrane of the mitochondria [21] and that therefore oxygen radical production is not a random process, but precisely site specific [22]. As discussed below and elsewhere in this volume, oxidative stress is instrumental for telomere shortening, one of the most probable clock mechanisms identified so far. Thus, the current vision of the free radical theory of aging is far from being understood.

DNA methylation as mitotic clock

Progressive loss of DNA methylation was proposed as a candidate clock mechanism [23, 24]. In human cells only a minor portion of the CpG di-nucleotides is unmethylated while 60% to 90% are methylated at the carbon 5 position. Although the role of DNA methylation in gene regulation was long debated it is now well established that DNA methylation of CpG islands correlates with transcriptional silencing and that methylation is, therefore, responsible for at least some transcriptional regulation [for review see 25]. It is, however, still unsolved whether DNA methylation interferes with transcription directly or rather marks parts of the DNA that are in a transcriptionally incompetent state. As suggested from recent work, gene silencing by methylation is associated with the formation of condensed chromatin that is enriched for hypo-acetylated histones (see below) [26, 27]. Furthermore, it was shown that methylated genes were initially transcriptionally active but after a few hours were packed into less accessible chromatin structures [28].

Irrespective of the final mechanism, Wilson and Jones [23] already showed in the early 1980s that DNA methylation decreased in aging normal cells but not in immortal ones. On the contrary, regional hyper-methylation is found in tumor cells. Thus, successive hypo-methylation could result in a successive activation of genes required to restrict the life span of a cell and to allow gene expression patterns to change towards those characteristic of aged cells. In this context, it is of interest to note that the cell cycle inhibitor p16 needs to be inactivated for cells to overcome senescence. In many immortal and tumor cells this is obtained by methylation of the p16 promoter [29]. Similar to p16 also p21 is up-regulated in aging cells and it was recently shown that the level of p21 was increased by inhibition of methyl-transferases [30, 31]. This may suggest that, in addition to inducing changes in the methylation status, methyl-transferases can directly inhibit gene transcription. Methyl-transferases contain a repressive domain that can interact with repressive complexes. Since these repressive complexes include histone deacetylases, a close interaction between methylation status and chromatin structure as determined by histone deacetylases seems likely [summarized in ref. 20].

A role for methylation in cellular senescence is further strengthened by the findings that DNA methyl-transferase is a down-stream effector of cellular transformation [32]. It was reported that in Balb/c 3T3 cells transformed with SV-40, the T antigen caused up-regulation of the DNA methyl-transferase and global genomic DNA methylation. Furthermore, since treatment of these cells with anti-sense oligonucleotides to methyl-transferase prior to introduction of SV-40 prevented transformation, the data suggest that transformation requires silencing of certain genes which are necessary for cellular senescence. Thus, it is tempting to speculate that the process of continuous de-methylation might at least in part account for the subsequent changes in gene expression observed to occur during the process of cellular aging and final growth arrest. In addition, it would allow a cell-type specific pattern of de-methylation and with that would explain the different changes seen in the different cell types during the aging process.

Acetylation/deacetylation and chromatin structure

Similar to methylation/demethylation, also acetylation/deacetylation of histones in the nucleosomes gained a lot of attention lately and may account for a clock-like mechanism. Post-translational acetylation of specific lysine residues in the amino-terminal ends of the highly conserved core histones H2A, H2B, H3, and H4 by histone acetyltransferases is believed to neutralize their positive charge. This in turn leads to a more open DNA conformation and thus allows access of transcription factors to activate target genes. Deacetylation, on the contrary, causes depletion of the acetyl-groups which restores the positive charge and thereby causes condensation of the nucleosome structure. Thus, hyper-acetylated histones are linked to transcriptionally active loci whereas hypo-acetylated histones are linked to transcriptionally silenced loci [33].

The role for histone deacetylation in aging came from a number of studies using the budding yeast *S. cerevisiae* as a model system. It was shown that transcriptional silencing of genes located near the ends of the chromosomes, the telomeres, was correlated with aging. The genes involved in silencing include the SIR proteins (silence information regulator). SIR 2-4, as a complex, are recruited to the nucleus. Here SIR2 represses transcription by deacetylating the histones and SIR3 while SIR4 binds to the hypo-acetylated histones and compacts the chromatin [34, 35].

It was recently shown that histone de-acetylation inhibitors can activate telomerase in otherwise telomerase-negative cells. Own experiments show that in skin keratinocytes which regenerate continuously *in vivo* but only show a defined life-time *in vitro*, the histone acetylation pattern changed after transferring the cells into culture [36]. Interestingly, these *in vivo* telomerase-positive cells also rapidly lost their ability to express telomerase activity. However, telomerase was reactivated in culture when treating the cells with the histone deacetylation inhibitor FR901228. It will now be of interest to determine whether continuous treatment with FR901228 will prolong the life-span of the human skin keratinocytes, i.e., whether continuous inhibition of histone deacetylation will prevent expression of genes required for cellular aging.

Along the same lines it was recently shown that the human SIR2 regulates the p53 gene by binding to and de-acetylating it [37, 38]. Over-expression of SIR2 prevented p53 activation, while SIR2 inactivation caused enhanced p53 activity. Since it was also demonstrated that over-expression of SIR2 resulted in extended life-span while inhibition of SIR2 caused a decrease, changes in the pattern of histone acetylation/deacetylation, if not being the internal clock itself, may well contribute to the process of replicative senescence of cells in culture as well as to longevity or aging of organisms [reviewed in ref. 39].

Telomeres as clock

The presently most favored hypothesis for a counting mechanism is the "telomere hypothesis of cellular aging" [40]. The ends of the chromosomes are capped by very specific DNA structures, the telomeres. In contrast to the coding sequences of the chromosome the vertebrate telomeric DNA is built up by only three bases forming many thousand repeats of the hexa-nucleotide TTAGGG. In agreement with the lack of one base, the telomeres do not provide specific genetic information but represent a "source of expandable DNA" [41]. This telomeric DNA was found to shorten progressively during *in vitro* propagation of normal human fibroblasts [42]. From this it was proposed that telomere attrition might be the mitotic clock, the counting mechanism that restricts the number of possible DNA replications before critically short telomeres would signal the cell to stop dividing, i.e., to senesce [43].

The reason for telomere shortening is likely to be twofold. On the one hand, DNA replication faces an "end-replication problem" [44]. DNA polymerase can not replicate linear DNA to the outermost end. During lagging strand synthesis RNA primers are required for replication of the Okazaki fragments. When the primers are replaced, standard polymerases are unable to fill the space of the outermost one, thus leaving a gap with each round of replication. This phenomenon, primarily termed the "theory of marginotomy" [45], should result in a rather linear rate of telomere loss during active proliferation of the cells. However, when investigating individual cells, the telomere lengths vary tremendously between different chromosomes. Although there may be indications that genetic factors determine the primary telomere lengths of each individual chromosome [46], firm evidence is still missing that in cultured cells the level of telomere shortening is equal on each chromosome and that the relationship in telomere length between different chromosomes is maintained throughout the entire growth period. This would, however, be required if one specific chromosome would always reach the critical length first. It also would suggest that senescence should occur simultaneously in all cells of a given population. Although all of this does not argue against a principal role of replication-dependent telomere shortening as the mitotic clock it clearly shows that other mechanisms have to cooperate in order to explain (i) the inability to find a specific chromosome that marks cellular senescence as well as (ii) the observed heterogeneity in replicative senescence seen in parallel cultures of the same parental cells [13].

One such mechanism could be oxidative stress. Accelerated telomere shortening combined with a decrease in replicative life span was found in cultured cells grown under stress conditions such as chronic hypoxia, or hydrogen peroxide. On the contrary, growth in the presence of free radical scavengers prolonged their life span and this was correlated with a slow down in telomere erosion [for review see ref. 47]. It was shown that UV-A radiation can induce telomeric double strand breaks which cause shortening of the telomeres even in non-dividing cells [48, 49]. Furthermore, primary evidence exists that also in humans *in vivo* accelerated telomere shortening is linked to disease states in which oxidative stress plays a causal role [50–52]. All this suggests that oxidative damage is a potent candidate to cooperate with replication-dependent telomere loss in order to generate the pattern of telomere attrition typically observed in the different studies.

Thus, irrespective of the factors that finally contribute to telomere erosion, the major function of telomeres is to provide sufficient DNA that is dispensable and thereby allows many rounds of replication without loosing chromosomal integrity. Germ line, immortal and tumor cells express the enzyme complex telomerase that is able to compensate for this replication-dependent telomere loss by addition of telomeric sequences *de novo*. Telomerase is, therefore, able to stabilise telomere length in these cells. In many normal somatic tissues telomerase is turned off during the later stages of embryonic development. Even in regenerative tissues which remain telomerase-positive [for references see ref. 53], activity seems too low to counteract the proliferation-dependent telomere loss. Thus, all tissues analyzed in more detail so far show an inverse correlation between donor age and telomere length [for review see ref. 54]. Similarly, .cultured normal cells show a continuous telomere loss suggesting that this telomere erosion may be the internal clock, the counting mechanism for cellular aging.

Further evidence for this hypothesis was provided by showing that forced expression of telomerase in otherwise negative cells could overcome the cellular clock and caused an elongated life span in many cell types [55]. Direct evidence for the counting mechanism was also provided by an experiment by Steinert *et al.* [56]. They could show that a "cre"-excisable telomerase, when introduced into telomerase-negative fibroblasts, caused a telomere elongation of 2.5 kb within 7 pd and excision of telomerase after these 7 pd by the cre-recombinase led to a continuous telomere loss thereafter. However, these cells showed a life time extension of 50 divisions which directly reflected the gain of telomere length when considering a 50 bp loss with each round of replication due to the end replication problem.

Thus, although a lot argues for the "telomere hypothesis of cellular aging," the question remains whether telomere shortening is the causal event (the clock work) for aging or just a marker (the hand of the clock) of an as yet unidentified mechanism. As mentioned above, treatment of the keratinocytes with the histone deacetylation inhibitor FR109228 reactivated telomerase activity and it is now of interest to determine whether this or other events can reset the "telomere clock."

Fooling the clock

It is no longer questioned that telomere length stabilization is required for indefinite growth because otherwise telomeres would shorten in a proliferation-dependent manner and finally induce a p53-dependent damage pathway. Thus it is not surprising that mechanisms exist to eliminate the clock. One such mechanism is the above mentioned up-regulation of the ribonucleoprotein complex telomerase. Telomerase is able to add telomeric sequences *de novo* and thereby to counteract the proliferation-dependent telomere loss. Presently, it still needs to be elucidated whether telomerase up-regulation can indeed reset the clock, i.e., whether it either (i) protects the present status, (ii) rejuvenates the cells, or (iii) allows expression of new programs. We tried to address this question by comparative expression array analysis of young and old mortal cells versus young and old hTERT immortalized cells. Preliminary data suggest that long-term growth (about 60 pd) after induction of telomerase is correlated with the expression of a large number of genes not conspicuous before (Boukamp and Hergenhahn, unpublished observations). Thus, we cannot exclude that telomerase-dependent immortalization does more than only stopping or resetting the aging clock.

In addition to telomerase there is also evidence for an alternative mechanism of telomere lengthening (ALT) [for review see ref. 57]. Some cells without telomerase activity are able to maintain or increase their telomere length by recombination events [58]. In contrast to telomerase-stabilized telomeres, the ALT telomeres are characterized by a great heterogeneity in size – ranging from hardly detectable to abnormally long telomeres – within one cell and the presence of specific PML (promyelocytic leukemia) bodies which contain extra-chromosomal DNA, telomere-binding proteins and proteins involved in DNA replication and recombination. Although the ALT mechanism is only found in a restricted number of cell lines and tumors there is suggestive evidence that in telomerase knock out mice which have gone through several generations and suffer from telomere erosion ALT is activated [59, 60]. This demonstrates that different mechanisms exist to bypass senescence and with that to overrule the endogenous clock.

Other counting mechanisms?

Early fusion experiments between immortal and normal human fibroblasts have shown that all hybrids exhibited a limited doubling potential [reviewed in ref. 61]. In agreement with these studies, fusion between immortal skin keratinocytes [62] and fibroblasts caused hybrids which senesced after 6 to 8 passages. However, all hybrids analyzed had maintained their full telomerase activity (Boukamp, unpublished data). Furthermore, we have previously shown that the same HaCaT cells senesced when a normal copy of chromosome 3 was introduced. By using deletion chromosomes, the senescence locus was further assigned to 3p12 to p14 [63]. In most other cell lines, however, telomerase was inhibited and telomere shortening was proposed as a potential mechanism for senescence [64]. Thus, it is tempting to speculate that in analogy to the fact that a minor fraction of cells is able to escape senescence by ALT, a small fraction of cells also may be able to enter senescence by a non-telomere mechanism. Thus, identification and functional characterization of the putative senescence gene on chromosome 3p12 to p14 will help to elucidate this alternative mechanism.

Do different cell types use the same "mitotic clock"?

All the above experimental evidences account well for fibroblasts and some other cell types. However, what is causing senescence in epidermal keratinocytes? It is well established from labelling experiments that epidermal keratinocytes *in vivo* only replicate 4 to 5 times before they stop dividing and enter the differentiation pathway [for review see ref. 65]. There is also no evidence to believe that this is different in culture and it is remarkable that cells from the epidermis, a tissue that proliferates continuously throughout life time of a human being, show a significantly shorter life span *in vitro* than fibroblasts which only proliferate rarely (upon wounding or to replace dead cells). While the maximal life span of a culture is likely dependent on the number of stem cells brought to and maintained in culture, the counting mechanism for the 4 to 5 pd per keratinocyte is still unclear. So far neither the genes are known for defining the number of potential divisions nor are the genes known that cause transition from proliferation to differentiation. Since the number of pds is the same in young and aged people, telomere erosion is unlikely to be responsible. Thus, we have to await that this very active field of research will provide an answer in near future and will elucidate whether similarities exist between the "mitotic clock" of fibroblasts and proliferation-restricting counting mechanism in keratinocytes.

Soluble factors (hormones) as internal clocks?

This raises one more question. Senescence or aging, as we presently see it, is a population phenomenon. Thus, we can not exclude that juxtacrine or paracrine acting factors are playing an important role. Own observations with epithelial cells indicate that immortalization with HPV16 or SV-40 caused an elongated life span also of those cells which were not positive for the T-antigen and which were finally eliminated (crisis). This suggests that the "immortal cells" can pass survival signals also to non-immortal neighbouring cells and allowing them to go far beyond the stage of senescence (also called mortality state 1 = M1 [66]), i.e., to potentially reach the mortality stage 2 (M2) a state of cell crisis characterized by critically short telomeres. If factors or cell-cell interaction mechanisms exist that allow life extension, it is reasonable to hypothesise that there is also a time-restricting interaction, i.e., that senescence of a few cells is transmitted to the entire population. While these factors are still hypothetical, there is evidence for the human being that hormones such as melatonin may be involved in the aging process *in situ*.

The pineal gland as a central aging clock?

It is an "old" hypothesis that hormonal variations occur with aging and it was in 1991 that Finch introduced his "hypothalamic clock." Other possibilities involved the pineal gland and its secretion product melatonin [for references see ref. 67]. Melatonin is secreted by the pineal glands at night time in humans and mammals thereby generating the circadian rhythms in plasma and urine melatonin levels. The absolute concentration reaches its peak at the age between 2 and 5 and thereafter steadily declines. It has been proposed that the decline in melatonin secretion at the end of the first decade of life which depends on the time of sexual maturity, the pubertal stage, rather than the individual's age *per se*, is directly correlated with aging. Plasma levels decline markedly in most people by the sixth or seventh decade of life. Thus it was suggested that "the pineal could serve as an extremely accurate physicochemical timing device that objectively records the biological age" [67]. In other words the pineal is the "aging clock" system and melatonin the actor which, because of circulating in the body, allows communication with every cell in the organism. So far, however, proof for this hypothesis is still missing.

Circadian clocks

Along the same line, we can not ignore the fact that a cellular machinery of proteins exists that regulate expression of a circadian rhythm. Since also this requires a defined counting mechanism, it would be tempting to speculate that both processes – circadian clock and aging clock – are related. First hints for such a hypothesis came from a recent report showing that two circadian clock genes, *clock* and *period1* are expressed in human skin, cultured human keratinocytes, dermal fibroblasts and melanocytes [68]. Although the function of these genes in the skin still needs to be investigated the fact that both genes are expressed in normal human cells need to make us alert and open-minded to the fact that the clock mechanism of cellular senescence may not yet be fully understood.

Conclusions

Although telomere erosion is presently favored as the dominant "mitotic clock," senescence is a very complex process and one would not be surprised to see that senescence is a combination of several mechanisms or at least that several mechanisms contribute to senescence. It also remains to be elucidated whether telomere erosion is the active counting mechanism (the clockwork with the driving force, the balance spring) or is only a visible sign (the hand of the clock) that allows to register the changes. Taken together, from all evidences provided above, it is tempting to speculate that counting could be a passive process performed by telomere shortening. However, a timed program in gene expression occurs and this is likely to be induced by genetic as well as epigenetic changes. Thus, the mechanism of cellular senescence, which still needs to be defined in detail, may well turn out to be a result of a multi-mechanistically and fine tuned process with a regulator, the "mitotic clock," which is far from being understood.

Acknowledgment

This work was supported by grants from the DFG (Bo1246/4-3) and EU (TACIT).

References

1. Hayflick L, Moorhead PS (1961). The serial cultivation of human diploid cell strains. *Exp Cell Res.* 25: 585–621.
2. Smith RJ, Hayflick L (1974). Variation in the life-span of clones derived from human diploid cell strains. *J Cell Biol.* 62: 48–53.
3. Hayflick L (1979). Progress in cytogerontology. *Mech Ageing Dev.* 9: 393–408.
4. Bayreuther KH, Rodemann HP, Hommel R, Dittmann K, Abietz M, Francz PI (1988). Human skin fibroblasts in vitro differentiate along a terminal cell lineage. *Proc Natl Acad Sci USA* 85: 5112–6.
5. Sherwood SW, Rush D, Ellsworth JL, Schimke RT (1988). Defining cellular senescence in IRM-90 cells: a flow cytometric analysis. *Proc Natl Acad Sci USA* 85: 9086–90.
6. Johnson M, Dimitrov D, Vojta PJ, *et al.* (1994). Evidence for a p53-independent pathway for upregulation of SDI1/CIP1/WAF1/p21 RNA in human cells. *Mol Carcinog.* 11: 59–64.
7. Hara E, Smith R, Parry D, Tahara H, Stone S, Peters G (1996). Regulation of p16CDKN2 expression and its implications for cell immortalization and senescence. *Mol Cell Biol.* 16: 859–67.
8. Donehower LA (2002). Does p53 affect organismal aging? *J Cell Physiol.* 192: 23–33.
9. Matsumura T, Pfendt EA, Hayflick L (1979). DNA synthesis in the human diploid cell strain WI-38 during *in vitro* aging: an autoradiography study. *J Gerontol.* 34: 323–7.
10. Smith JR, Pereira-Smith OM (1996). Replicative senescence: implications for *in vivo* aging and tumor suppression. *Science* 273: 63–7.
11. Campisi J (1996). Replicative senescence: an old lives' tail? *Cell* 84: 497–500.
12. Dimri GP, Lee X, Basile G, *et al.* (1995). A biomarker that identifies senescent human cells in culture and in aging skin *in vivo. Proc Natl Acad Sci USA* 92: 9363–7.
13. Smith JR, Whitney RG (1980). Intraclonal variation in proliferative potential of human diploid fibroblasts: stochastic mechanism for cellular aging. *Science* 207: 82–4.
14. Pereira-Smith OM, Smith JR (1983). Evidence for the recessive nature of cellular immortality. *Science* 221: 964–6.

15. Tominaga K, Olgun A, Smith JR, Pereira-Smith OM (2002). Genetics of cellular senescence. *Mech Ageing Dev.* 123: 927–36 (Review).
16. Stanbridge EJ (1992). Functional evidence for human tumour suppressor genes: chromosome and molecular genetic studies. *Cancer Surv.*12: 5–24 (Review).
17. Ran Q, Pereira-Smith OM (2000). Genetic approaches to the study of replicative senescence. *Exp Gerontol.* 35: 7–13 (Review).
18. Bertram MJ, Berube NG, Hang-Swanson X, *et al.* (1999). Identification of a gene that reverses the immortal phenotype of a subset of cells and is a member of a novel family of transcription factor-like genes. *Mol Cell Biol.* 19: 1479–85.
19. Atamna H, Parler-Martinez A, Ames BN (2000). N-t-butyl hydroxylamine, a hydrolysis product of alpha-phenyl-N-t-butyl nitrone, is more potent in delaying senescence in human lung fibroblasts. *J Biol Chem.* 275: 6741–8.
20. Young J, Smith JR (2000). Epigenetic aspects of cellular senescence. *Exp Gerontol.* 35: 23–32.
21. Nohl H, Jordan W (1986). The mitochondrial site of superoxide formation. *Biochem Biophys Res Commun.* 138: 533–9.
22. Fleming JE, Bensch KG (1991). Oxidative stress as a causal factor in differentiation and aging: a unifying hypothesis. *Exp Gerontol.* 26: 511–17.
23. Wilson VL, Jones PA (1983). DNA methylation decreases in aging but not in immortal cells. *Science* 220: 1055–7.
24. Hoal-van Helden EG, van Helden PD (1989). Age-related methylation changes in DNA may reflect the proliferative potential of organs. *Mutat Res.* 219: 263–6.
25. Bannister AJ, Schneider R, Kouzarides T (2002). Histone methylation: dynamic or static? *Cell* 109: 801–6 (Review).
26. Eden S, Hashimshony T, Keshet I, Cedar H, Thorne AW (1998). DNA methylation models histone acetylation. *Nature* 394: 842.
27. Jones PA, Laird PW (1999). Cancer epigenetics comes of age. *Nature Genet.* 21: 163–7 (Review).
28. Kass SU, Landsberger N, Wolffe AP (1997). DNA methylation directs a time-dependent repression of transcription initiation. *Curr Biol.* 7: 157–65.
29. McGregor F, Muntoni A, Fleming J, *et al.* (2002). Molecular changes associated with oral dysplasia progression and acquisition of immortality: potential for its reversal by 5-azacytidine. *Cancer Res.* 62: 4757–66.
30. Fournel M, Sapieha P, Beaulieu N, Besterman JM, MacLeod AR (1999). Down-regulation of human DNA-(cytosine-5) methyltransferase induces cell cycle regulators p16(ink4A) and p21(WAF/Cip1) by distinct mechanisms. *J Biol Chem.* 274: 24250–6.
31. Milutinovic S, Knox JD, Szyf M (2000). DNA methyltransferase inhibition induces the transcription of the tumor suppressor p21(WAF1/CIP1/sdi1). *J Biol Chem.* 275: 6353-6359.
32. Slack A, Cervoni N, Pinard M, Szyf M (1999). DNA methyltransferase is a downstream effector of cellular transformation triggered by simian virus 40 large T antigen. *J Biol Chem.* 274: 10105–12.
33. Strahl BD, Allis CD (2000). The language of covalent histone modifications. *Nature* 403: 41–5.
34. Galy V, Olivo-Marin JC, Scherthan H, Doye V, Rascalou N, Nehrbass U (2000). Nuclear pore complexes in the organization of silent telomeric chromatin. *Nature* 403: 108–12.
35. Defossez P, Lin S, McNabb DS (2001). Sound silencing: the Sir2 protein and cellular senescence. *Biol Essays* 23: 327–32.

36. Greulich Bode K, Lindenmaier H, Figueroa R, *et al.* Telomerase and telomere length regulation in normal human skin keratinocytes *in vitro* and *in vivo*. Submitted.
37. Vaziri H, Dessain SK, Ng Eaton E, *et al.* (2001). hSIR2(SIRT1) functions as an NAD-dependent p53 deacetylase. *Cell* 107: 149–59.
38. Luo J, Nikolaev AY, Imai S, *et al.* (2001). Negative control of p53 by Sir2alpha promotes cell survival under stress. *Cell* 107: 137–48.
39. Chang KT, Min K-T (2002). Regulation of lifespan by histone deacetylase. *Ageing Res Rev.* 1: 313–26.
40. Shay JW, Wright WE (2001). Ageing and cancer: the telomere and telomerase connection. *Novartis Found Symp.* 235: 116–25; discussion 125–9; 146–9 (Review).
41. Wright WE, Shay JW (2001). Cellular senescence as a tumor-protection mechanism: the essential role of counting. *Curr Opin Genet Dev.* 11: 98–103 (Review).
42. Harley CB, Futcher AB, Greider CW (1990). Telomeres shorten during ageing of human fibroblasts. *Nature* 345: 458–60.
43. Martens UM, Chavez EA, Poon SS, Schmoor C, Lansdorp PM (2000). Accumulation of short telomeres in human fibroblasts prior to replicative senescence. *Exp Cell Res.* 256: 291–99.
44. Watson, JD (1972). Origin of concatemeric T7 DNA. *Nat New Biol.* 239: 197–201.
45. Olovnikov AM (1972). The immune response and the process of marginotomy in lymphoid cells. *Vestn Akad Med Nauk SSSR* 27: 85–7 (Review).
46. Slagboom PE, Droog S, Boomsma DI (1994). Genetic determination of telomere size in humans: a twin study of three age groups. *Am J Hum Genet.* 55: 876–82.
47. von Zglinicki (2002). Oxidative stress shortens telomeres. *Trends Biochem Sci.* 27: 339–44.
48. Bar-Or D, Thomas GW, Rael LT, Lau EP, Winkler JV (2001). Asp-Ala-His-Lys (DAHK) inhibits copper-induced oxidative DNA double strand breaks and telomere shortening. *Biochem Biophys Res Commun.* 282: 356–60.
49. Oikawa S, Tada-Oikawa S, Kawanishi S (2001). Site-specific DNA damage at the GGG sequence by UVA involves acceleration of telomere shortening. *Biochemistry* 40: 4763–8.
50. von Zglinicki T, Serra V, Lorenz M, *et al.*(2000). Short telomeres in patients with vascular dementia: an indicator of low antioxidative capacity and a possible risk factor? *Lab Invest.* 80: 1739–47.
51. Samani NJ, Boultby R, Butler R, Thompson JR, Goodall AH (2001). Telomere shortening in atherosclerosis. *Lancet* 358: 472–3.
52. Brummendorf TH, Maciejewski JP, Mak J, Young NS, Lansdorp PM (2001). Telomere length in leukocyte subpopulations of patients with aplastic anemia. *Blood* 97: 895–900.
53. Bachor C, Bachor OA, Boukamp P (1999). Telomerase is active in normal gastrointestinal mucosa and not up-regulated in precancerous lesions. *J Cancer Res Clin Oncol.* 125: 453–60.
54. Granger MP, Wright WE, Shay JW (2002). Telomerase in cancer and aging. *Crit Rev Oncol Hematol.* 41: 29–40 (Review).
55. Li H, Liu JP (2002). Signaling on telomerase: a master switch in cell aging and immortalization. *Biogerontology* 3: 107–16.
56. Steinert S, Shay JW, Wright WE (2000). Transient expression of human telomerase extends the life span of normal human fibroblasts. *Biochem Biophys Res Commun.* 273: 1095–8.
57. Henson JD, Neumann AA, Yeager TR, Reddel RR (2002). Alternative lengthening of telomeres in mammalian cells. *Oncogene* 21: 598–610 (Review).
58. Neumann AA, Reddel RR (2002). Telomere maintenance and cancer – look, no telomerase. *Nat Rev Cancer* 2: 879–84.

59. Hande MP, Samper E, Lansdorp P, Blasco MA (1999). Telomere length dynamics and chromosomal instability in cells derived from telomerase null mice. *J Cell Biol.* 144: 589–601.
60. Herrera E, Martinez-A C, Blasco MA (2000). Impaired germinal center reaction in mice with short telomeres. *EMBO J.* 19: 472–81.
61. Bérubé NG, Smith JR, Pereira-Smith OM (1998). The genetics of cellular senescence. *Am J Hum Genet.* 62: 1015–19.
62. Boukamp P, Petrussevska RT, Breitkreutz D, Hornung J, Markham A, Fusenig NE (1988). Normal keratinization in a spontaneously immortalized aneuploid human keratinocyte cell line. *J Cell Biol.* 106: 761–71.
63. Fusenig NE, Boukamp P (1998). Multiple stages and genetic alterations in immortalization, malignant transformation, and tumor progression of human skin keratinocytes. *Mol Carcinog.* 23: 144–58.
64. Cuthbert AP, Bond J, Trott DA, *et al.* (1999). Telomerase repressor sequences on chromosome 3 and induction of permanent growth arrest in human breast cancer cells. *J Natl Cancer Inst.* 91: 37–45.
65. Fuchs E, Raghavan S (2002). Getting under the skin of epidermal morphogenesis. *Nat Rev Genet.* 3: 199–209 (Review).
66. Wright WE, Pereira-Smith OM, Shay JW (1989). Reversible cellular senescence: implications for immortalization of normal human diploid fibroblasts. *Mol Cell Biol.* 9: 3088–92.
67. Kloeden PE, Rossler R, Rossler OE (1993). Timekeeping in genetically programmed aging. *Exp Gerontol.* 28: 109–18 (Review).
68. Zanello SB, Jackson DM, Holick MF (2000). Expression of the circadian clock genes clock and period1 in human skin. *J Invest Dermatol.* 115: 757–60.

Telomeric Damage in Aging

Thomas von Zglinicki
Henry Wellcome Laboratory for Biogerontology, Newcastle University, General Hospital, Newcastle, NE4 6BE, UK

Introduction

In recent years, a broad consensus has emerged that aging is not brought about by a genetically fixed programme, which is executed step by step, but rather the result of a limited ability to keep with stress and the resultant damage in all its various forms. Evolutionary biology provided a strong argument by showing that with few exceptions any "programmed senescence" would (i) hardly contribute to limiting lifespan in the wild (because death in the wild is nearly exclusively due to external causes, not to aging) and (ii) even if it were beneficial for the species, would hardly increase reproductive fitness of the individual and thus could not be selected for in evolution. On the other hand, both these arguments suggest a trade-off that limits the resources put into somatic maintenance and thus, sustainability of the soma [1]. There is a wealth of data demonstrating an important role for molecular damage and for cellular defence and repair mechanisms in aging, which is reviewed elsewhere in this volume. However, there are also strong data indicating that "clocks" are operating on a molecular scale, especially in cellular aging processes (reviewed in the Chapter 8, this volume). These clocks count biological time, for instance in numbers of cell divisions, and after a reproducible number of divisions trigger signalling pathways that block cellular growth. Telomeres are the most clearly established of these molecular clocks. Now, if cellular replicative senescence is programmed by a running clock it seems to follow that the evolutionary arguments that govern organismic aging do not apply to cellular senescence. Thus, either cellular senescence is fundamentally different from aging, or aging can be a programmed process after all.

The present chapter will review evidence that reconciles this seemingly contradictory alternatives with respect to telomeres.

T. von Zglinicki (ed.), Aging at the Molecular Level, 121–129.

Replicative senescence and the senescent phenotype

Cellular senescence has been defined by Hayflick [2] as the ultimate and irreversible loss of replicative capacity occurring in primary somatic cell culture after a reproducible number (under constant culture conditions) of population doublings (PD). This number is now often termed the "Hayflick limit." There are good arguments to see this process as a cellular model for, or rather a cellular form of, aging. These arguments include a general correlation between donor age or donor species lifespan and Hayflick limit, shorter Hayflick limits for cultures from donors with premature aging syndromes and a higher fraction of cells displaying markers of a senescent phenotype in tissues from older donors [3, 4].

In 1990 it was shown that as human primary fibroblasts are cultured towards the end of their replicative lifespan, telomere length gradually decreased [5]. When it was demonstrated that the activity of telomerase, the enzyme that counteracts telomere shortening by re-elongating telomeres, was strong in immortal cells but essentially absent from human somatic cells and that human primary fibroblasts could be immortalized by transfection of hTERT, the catalytic subunit of human telomerase [6], telomere shortening was firmly established as a biological clock for cellular senescence. It is clear now that it is not the length of the telomeres themselves that triggers a senescent arrest but rather some property that has been described as uncapping [7] or dysfunctional telomeres [8]. It has been shown that telomeres form terminal loops *in vivo* and *in vitro*, which are stabilized by a number of telomere-binding proteins, notably TRF2 [9]. It is probable that this loop provides telomere capping function and that its opening and the resultant unscheduled exposure of the terminal single-stranded telomeric overhang triggers the signal transduction pathway towards senescence [10, 11]. Telomere shortening would destabilize the loop and thus increase the probability of uncapping. Accordingly, overexpression of TRF2, which stabilises the telomeric loop, shifts the telomere length threshold towards shorter telomeres [12], while disruption of TRF2 function by expression of a dominant negative modification of the protein leads to arrest or apoptosis depending on the cell type [13]. Thus, cells from different donors might enter senescence with widely different telomere lengths [14], possibly in relation to their TRF2 activity.

Still, in any given cell culture, telomere shortening down to a threshold length seems to be the best known predictor of senescence. As telomere length varies considerably between chromosomes and chromosome arms, the one or few shortest telomere(s) are expected to play the most important role [15], although other data indicate that average telomere length has a stronger predictive value than the length of any single telomere [16].

Telomere shortening is widely attributed to the so-called "end-replication problem", which, as shown long ago [17, 18], arises from two basic properties of all known DNA polymerases: they work only in 5′ to 3′ direction, and a short primer is needed as a starting point. That means, the lagging strand is replicated in short Okazaki fragments, each starting from its own RNA primer. At the very distal end of the lagging strand, primer removal cannot be followed by fill-in synthesis, resulting in a shortening of the end of the lagging strand by at least the size of the RNA primer (8 to 12 nucleotides). Whether this effect alone can explain the observed telomere shortening rates under standard cell culture conditions (up to 200 base pairs per PD) is not clear. Telomeres end in a single-stranded overhang of the 3′ (G-rich) strand of about 100–300 nucleotides in length. It is not clear whether all [19] or only half [20] of the telomeres actually have this overhang and to what extent processing of the telomeric ends contributes to telomere shortening [21, 22]. Additional factors governing the stress-dependency of telomere shortening will be dealt with below.

The telomere-driven check point is by far not the only one that, if activated, induces a senescence-like growth arrest. Human epithelial cells, for instance, encounter a telomere-independent, p16-dependent growth arrest that might be triggered in response to suboptimal culture conditions [23]. Overexpression of oncogenes, like activated RAS or RAF, induces a senescence-like arrest in primary human or mouse cells [24–26]. Modification of chromatin towards a more decondensed structure, for instance by inhibitors of histone deacetylases, induces a senescent phenotype [27]. Last but not least, DNA-damaging stress does induce cellular growth arrest. Stresses include different types of radiation [28], drugs generating DNA double strand breaks [29] and different means to generate oxidative stress like increase of the ambient oxygen tension [30, 31] and treatment with hydrogen peroxide [32, 33] or butylhydroperoxide [34]. Growth arrest can be induced rather immediately by acute intense stress [35, 36] or more slowly under chronic or semi-chronic treatment regimens [30, 31, 34], thus suggesting stress-induced growth arrest being a premature form of senescence [37].

Mechanistically and phenotypically, telomere-driven senescence and stress-induced arrest are very similar. Telomere dysfunction induces growth arrest via activation of ATM/ATR [13, 38], and, eventually, p53 and (in human but not mouse fibroblasts) p16 [39]. While there are differences in p53 phosphorylation patterns between senescent and IR- or UV-irradiated cells [40], patterns between senescent and oxidatively stressed cells are more similar [40, 41]. Serine 15 is targeted by ATR/ATM, and it is found similarly phosphorylated under all examined conditions. Likewise, phenotypic markers of senescence, from cellular morphology over expression of senescence-associated β-galactosidase activity [42] to expression of senescence-associated genes at the mRNA [33, 43] or protein level [44], have shown to be significantly overlapping if not identical between senescent and stressed cells.

So it appears that essentially the same phenotype is triggered via hardly distinguishable pathway(s) either by random stress and damage or by a pre-programmed cell division counter. It has been suggested that the term "senescence" should be reserved for only those arrests which occur as a result of counting (which in practical terms most often equates telomere-driven arrest although there might be other potential clocks as well, see Chapter 8, this volume), while stress-induced, premature arrests should not be regarded as senescence at all, but rather as cell culture artefacts [45–47]. This distinction were useful, if the assumption that telomeres were pre-programmed "honest" counters of cell divisions would hold. However, as shown below, this is not the case. On the contrary, telomere shortening is largely stress-dependent.

Telomere shortening and DNA damage

Well-established human fibroblast lines like MRC-5 or WI-38 proliferate for about 50 PD in mass culture and shorten their telomeres by about 50 to 100 base pairs per PD under standard conditions. By increasing the ambient oxygen partial pressure from 21 to 40% mitochondrial respiration is activated and more reactive oxygen species (ROS) are generated as shown for instance by increased levels of intracellular peroxides [48] or protein carbonyls [49] and faster accumulation of lipofuscin [49, 50]. At the same time, the replicative lifespan decreases to very few PD, and the rate of telomere shortening increases four to tenfold [31], so that cells are growth-inhibited with the same telomere length as under standard culture conditions [14, 31]. This result has now been confirmed by a number of independent groups in different cell types [for review see ref. 51].

How can oxidative stress modify the rate of telomere shortening? In G0-arrested fibroblasts under oxidative stress, telomeres accumulate single-strand breaks at levels about one order of magnitude higher than that found elsewhere in the genome [10, 22, 52] because telomeric single-strand break repair is of low efficiency [53]. During DNA replication, damaged sites stall polymerases transiently. At the telomere, this might lead to premature termination of replication and to larger amounts of telomere shortening as would be produced by the end replication problem alone [54]. In fact, single-strand breaks that accumulate during G0 arrest are quantitatively transferred into telomere loss within 1–2 PD following release of cells from the arrest [52].

Having shown that stress accelerates telomere shortening, the next question to ask was whether stress and stress-induced damage would play any role for telomere shortening under standard cell culture conditions. First, we obtained correlative evidence. In Fanconi Anemia fibroblasts, steady-state levels of oxidative stress are higher and telomeres shorten faster [55]. Comparing fibroblast strains from normal, healthy donors, we found significant differences in antioxidant defence capacity and Hayflick limit [14]. Again, telomere shortening rate under standard conditions (as well as under hyperoxia) showed a significant inverse correlation to antioxidant defence [48]. Interestingly, the major antioxidant enzyme responsible for these strain-to-strain differences was identified as extracellular superoxide dismutase (EC-SOD) rather than Cu/Zn-SOD or Mn-SOD [56, 57]. In these experiments, fibroblast strains with the best antioxidant defence were found to shorten their telomeres by only 5–20 base pairs per PD. This is very close to the lower limit for the end-replication problem as set by the size of the RNA primer. One can conclude that the DNA replication machinery has the ability to position the last primer at the very end of the copied strand, so that typical telomere shortening rates under standard culture conditions of 50–100 base pairs per PD would to a large extent be caused by stress-dependent damage with a comparatively minor contribution from the end-replication problem [22].

Intervention experiments confirmed this conclusion. The spin trap α-phenyl-t-butyl-nitrone is a potent free radical scavenger that prolongs the replicative lifespan of human fibroblasts under standard culture conditions by 20–30% [58, 59], decreased intracellular peroxide levels and slows down telomere shortening significantly [22]. A stable derivative of vitamin C, ascorbic acid phosphoric ester magnesium salt, decreased the level of oxidative stress, extended replicative lifespan and slowed down telomere shortening in human vascular endothelial cells [60] and fibroblasts [61]. Ubiquinol can be targeted to mitochondria and acts there as a potent antioxidant [62]. It prolongs the replicative lifespan of human fibroblasts and minimizes their telomere shortening [63]. Finally, overexpression of the antioxidant enzyme, EC-SOD, in human fibroblasts increases their superoxide dismutase activity, antioxidant defence capacity and replicative lifespan and decreases their telomere shortening rate [57]. Thus, ROS-mediated damage of telomeres is a major course of telomere shortening even under "normal" conditions *in vitro* [22, 51].

Replicative senescence is a stress response

The reviewed data confirm that telomere shortening is to a large extent governed by externally and internally generated stress and the cell's ability to cope with it. In other words, telomeres are not pre-programmed clocks but rather sentinels for genomic damage and mutation risk [64]. The specifically low repair efficiency in telomeres [53, 65] ensures preferential damage accumulation in a genomic region that is non-coding but, via accelerated shortening, able to elicit stable growth arrest. This sentinel function of telomeres limits the potentially dangerous consequences of DNA damage in all coding sequences without the need of actually having to screen the whole genome.

The data also confirm that the postulated distinction between telomere-driven "real" replicative senescence and stress-induced "senescent phenotype" is a highly arbitrary one. Under most conceivable conditions, telomere-driven senescence is at least partially stress-induced, and the pace of the telomeric "clock" is highly variable. There is no reason to assume a fundamental difference between the *in-vivo* and the *in-vitro* situation: While conditions in the body are surely more favourable for cell survival than those under atmospheric oxygen tension *in vitro*, they are there as well defined by a compromise between formation of and defence against ROS and other toxic by-products. Thus, senescence is one of the generalized cellular responses to molecular stress, similar to, but less grave than apoptosis [51].

If this is so, then replicative senescence does not only occur in tissues as organisms including man age. It must also have significant consequences for aging of organisms. Replicative senescence appears to be a pleiotrophic trait [66], on one hand limiting the risk of uncontrolled tumour growth, while on the other hand generating a "bystander" effect which is pro-inflammatory [67] and pro-cancerogenic [68] and thus, "pro-aging" [69]. Finally, stress-dependent telomere shortening, being a cause for cellular senescence including, possibly, immunosenescence [70], might not only be a contributor to the aging process. It appears also to be a biomarker of aging. Recent research shows inverse correlations between telomere length in human lymphocytes and incidence of age-related degenerative diseases like stroke-dependent (vascular) dementia [48], atherosklerosis [71] and even multi-cause human mortality [72]. A tempting hypothesis is that these correlations reflect a common underlying cause for age-related morbidity and mortality and telomere length: oxidative damage.

References

1. Kirkwood TBL, Austad SN (2000). Why do we age? *Nature* 408: 233–8.
2. Hayflick L, Moorehead PS (1961). The serial cultivation of human diploid cell strains. *Exp Cell Res.* 25: 585–621.
3. Campisi J (2001). From cells to organisms: what can we learn about aging from cells in culture? *Exp Gerontol.* 36: 607–18.
4. Faragher RG, Kipling D (1998). How might replicative senescence contribute to human ageing? *Bioessays* 20: 985–91.
5. Harley CB, Futcher AB, Greider CW (1990). Telomeres shorten during ageing of human fibroblasts. *Nature* 345: 458–60.
6. Bodnar AG, Ouellette M, Frolkis M, *et al.* (1998). Extension of life-span by introduction of telomerase into normal human cells. *Science* 279: 349–52.
7. Blackburn EH (2000). Telomere states and cell fates. *Nature* 408: 53–6.
8. Chin L, Artandi SE, Shen Q, *et al.* (1999). p53 deficiency rescues the adverse effects of telomere loss and cooperates with telomere dysfunction to accelerate carcinogenesis. *Cell* 97: 527–38.
9. Griffith JD, Comeau L, Rosenfield S, *et al.* (1999). Mammalian telomeres end in a large duplex loop. *Cell* 97: 503–14
10. Saretzki G, Sitte N, Merkel U, Wurm RE, von Zglinicki T (1999). Telomere shortening triggers a p53-dependent cell cycle arrest via accumulation of G-rich single stranded DNA fragments. *Oncogene* 18: 5148–58.
11. von Zglinicki T (2001). Telomeres and replicative senescence: is it only length that counts? *Cancer Lett.* 168: 111–16.
12. Karlseder J, Smogorzewska A, de Lange T (2002). Senescence induced by altered telomere state, not telomere loss. *Science* 295: 2446–50.
13. Karlseder J, Broccoli D, Dai Y, Hardy S, de Lange T (1999). p53- and ATM-dependent apoptosis induced by telomeres lacking TRF2. *Science* 283: 1321–5.
14. Serra V, von Zglinicki T (2002). Human fibroblasts *in vitro* senesce with a donor-specific telomere length. *FEBS Lett.* 516: 71–4.
15. Hemann MT, Strong MA, Hao LY, Greider C (2001). The shortest telomere, not average telomere length, is critical for cell viability and chromosome stability. *Cell* 107: 67–77.
16. Martens UM, Chavez EA, Poon SS, Schmoor C, Lansdorp PM (2000). Accumulation of short telomeres in human fibroblasts prior to replicative senescence. *Exp Cell Res.* 256: 291–9.
17. Olovnikov AM (1973). A theory of marginotomy: the incomplete copying of template margin in enzymic synthesis of polynucleotides and biological significance of the problem. *J Theor Biol.* 41: 181–90.
18. Watson JD (1972). Origin of concatemeric DNA. *Nature New Biol.* 239: 197–201.
19. Makarov VL, Hirose Y, Langmore JP (1997). Long G tails at both ends of human chromosomes suggest a C strand degradation mechanism for telomere shortening. *Cell* 88: 657–66.
20. Wright WE, Tesmer VM, Huffman KE, Levene SD, Shay JW (1997). Normal human chromosomes have long G-rich telomeric overhangs at one end. *Genes Dev.* 11: 2801-9.
21. Huffmann KE, Levene SD, Tesmer VM, Shay JW, Wright WE (2000). Telomere shortening is proportional to the size of the G-rich telomeric overhang. *J Biol Chem.* 275: 19719–22.
22. von Zglinicki T, Pilger R, Sitte N (2000). Accumulation of single-strand breaks is the major cause of telomere shortening in human fibroblasts. *Free Radic Biol Med.* 28: 64–74.

23. Stampfer MR, Yaswen P (2003). Human epithelial cell immortalisation as a step in carcinogenesis. *Cancer Lett.* 194: 199–208.
24. Dimri GP, Martinez JL, Jacobs JJ, *et al.* (2002). The Bmi-1 oncogene induces telomerase activity and immortalises human mammary epithelial cells. *Cancer Res.* 62: 4736–45.
25. Ferbeyre G, de Stanchina E, Lin AW, *et al.* (2002). Oncogenic ras and p53 cooperate to induce cellular senescence. *Mol Cell Biol.* 22: 3497–508.
26. Lin AW, Barradas M, Stone JC, van Aelst L, Serrano M, Lowe SW (1998). Premature senescence involving p53 and p16 is activated in response to constitutive MEK/MAPK mitogenic signalling. *Genes Dev.* 12: 3008–19.
27. Ogryzko VV, Hirai TH, Russanova VR, Barbie DA, Howard BH (1996). Human fibroblast commitment to a senescence-like state in response to histone deacetylase inhibitors is cell cycle dependent. *Mol Cell Biol.* 16: 5210–218.
28. Herskind C, Rodemann HP (2000). Spontaneous and radiation-induced differentiationof fibroblasts. *Exp Gerontol.* 35: 747–55.
29. Robles SJ, Adami GR (1998). Agents that cause DNA double strand breaks lead to p16INK4a enrichment and the premature senescence of normal fibroblasts. *Oncogene* 16: 1113–23.
30. Balin AK, Goodman DB, Rasmussen H, Cristofalo VJ (1977). The effect of oxygen and vitamin E on the lifespan of human diploid cells *in vitro. J Cell Biol.*74: 58–67.
31. von Zglinicki T, Saretzki G, Docke W, Lotze C (1995). Mild hyperoxia shortens telomeres and inhibits proliferation of fibroblasts: a model for senescence? *Exp Cell Res.* 220: 186–93.
32. Chen Q, Ames BN (1994). Senescence-like growth arrest induced by hydrogen peroxide in human diploid fibroblast F65 cells. *Proc Natl Acad Sci USA* 91: 4130–4
33. Frippiat C, Chen QM, Zdanov S, Magalhaes JP, Remacle J, Toussaint O (2001). Subcytotoxic H_2O_2 stress triggers a release of transforming growth factor-beta 1, which induces biomarkers of cellular senescence of human diploid fibroblasts. *J Biol Chem.* 276: 2531–7.
34. Dumont P, Burton M, Chen QM, *et al.* (2000). Induction of replicative senescence biomarkers by sublethal oxidative stresses in normal human fibroblast. *Free Rad Biol Med.* 28: 361–73.
35. Chen QM, Prowse KR, Tu VC, Purdom S, Linskens MH (2001). Uncoupling the senescent phenotype from telomere shortening in hydrogen peroxide-treated fibroblasts. *Exp Cell Res.* 265: 294–303.
36. Gorbunova V, Seluanov A, Pereira-Smith OM (2002). Expression of human telomerase (hTERT) does not prevent stress-induced senescence in normal human fibroblasts but protects the cells from stress-induced apoptosis and necrosis. *J Biol Chem.* 277: 38540–9.
37. Toussaint O, Medrano EE, von Zglinicki T (2000). Cellular and molecular mechanisms of stress-induced premature senescence (SIPS) of human diploid fibroblasts and melanocytes. *Exp Gerontol.* 35: 927–45.
38. Rouse J, Jackson SP (2002). Interfaces between the detection, signaling, and repair of DNA damage. *Science* 297: 547–51.
39. Smogorzewska A, De Lange T (2002). Different telomere damage signaling pathways in human and mouse cells. *EMBO J.* 21: 4338–48.
40. Webley K, Bond JA, Jones CJ, *et al.* (2000). Posttranslational modifications of p53 in replicative senescence overlapping but distinct from those induced by DNA damage. *Mol Cell Biol.* 20: 2803–8.
41. Xie S, Wang Q, Wu H, *et al.* (2001). Reactive oxygen species induced-phosphorylation of p53 on serine-20 is mediated in part by Plk3. *J Biol Chem.* 276: 36194–9.

42. Dimri GP, Lee X, Basile G, *et al.* (1995). A biomarker that identifies senescent human cells in culture and in aging skin *in vivo*. *Proc Natl Acad Sci USA* 92: 9363–7.
43. Saretzki G, Feng J, von Zglinicki T, Villeponteau B (1998). Similar gene expression pattern in senescent and hyperoxic-treated fibroblasts. *J Gerontol.* 53A: B438–42.
44. Dierick JF, Kalume DE, Wenders F, *et al.* (2002). Identification of 30 protein species involved in replicative senescence and stress-induced premature senescence. *FEBS Lett.* 531: 499–504.
45. Sherr CJ, DePinho RA (2000). Cellular senescence: mitotic clock or culture shock? *Cell* 102: 407–10.
46. Wright WE, Shay JW (2001). Cellular senescence as a tumor-protection mechanism: the essential role of counting. *Curr Opin Genet Dev.* 11: 98–103.
47. Wright WE, Shay JW (2002). Historical claims and current interpretations of replicative aging. *Nature Biotechnol.* 20: 682–8.
48. von Zglinicki T, Serra V, Lorenz M, *et al.* (2000). Short telomeres in patients with vascular dementia: an indicator of low antioxidative capacity and a possible risk factor? *Lab Invest.* 80: 1739–47.
49. Sitte N, Merker K, Von Zglinicki T, Davies KJ, Grune T (2000). Protein oxidation and degradation during cellular senescence of human BJ fibroblasts: part II – aging of nondividing cells. *FASEB J.* 14: 2503–10.
50. Sitte N, Merker K, Grune T, von Zglinicki T (2001). Lipofuscin accumulation in proliferating fibroblasts *in vitro*: an indicator of oxidative stress. *Exp Gerontol.* 36: 475–86.
51. von Zglinicki T (2002). Oxidative stress shortens telomeres. *Trends Biochem Sci.* 27: 339–44.
52. Sitte N, Saretzki G, von Zglinicki T (1998). Accelerated telomere shortening in fibroblasts after extended periods of confluency. *Free Rad Biol Med.* 24: 885–93.
53. Petersen S, Saretzki G, von Zglinicki T (1998). Preferential accumulation of single-stranded regions in telomeres of human fibroblasts. *Exp Cell Res.* 239: 152–60.
54. von Zglinicki T (2000). Role of oxidative stress in telomere length regulation and replicative senescence. *Ann NY Acad Sci.* 908: 99–110.
55. Adelfalk C, Lorenz M, Serra V, von Zglinicki T, Hirsch-Kauffmann M, Schweiger M (2001). Accelerated telomere shortening in Fanconi anemia fibroblasts – a longitudinal study. *FEBS Lett.* 506: 22–6.
56. Serra V, Grune T, Sitte N, Saretzki G, von Zglinicki T (2000). Telomere length as a marker of oxidative stress in primary human fibroblast cultures. *Ann NY Acad Sci.* 908: 327–30.
57. Serra V, von Zglinicki T, Lorenz M, Saretzki G (2003). Extracellular superoxide dismutase is a major antioxidant in human fibroblasts and slows down telomere shortening. *J Biol Chem.* 278: 6824–30.
58. Atamna H, Robinson C, Ingersoll R, Elliott H, Ames BN (2001). N-t-Butyl hydroxylamine is an antioxidant that reverses age-related changes in mitochondria *in vivo* and *in vitro*. *FASEB J.* 15: 2196–204.
59. Chen Q, Fischer A, Reagan JD, Yan LJ, Ames BN (1995). Oxidative DNA damage and senescence of human diploid fibroblast cells. *Proc Natl Acad Sci USA* 92: 4337–41.
60. Furumoto K, Inoue E, Nagao N, Hiyama E, Miwa N (1998). Age-dependent telomere shortening is slowed down by enrichment of intracellular vitamin C via suppression of oxidative stress. *Life Sci.* 63: 935–48.
61. Kashino G, Kodama S, Nakayama Y, *et al.* (2003). Relief of oxidative stress by ascorbic acid delays cellular senescence of normal human and Werner syndrome fibroblast cells. *Free Radic Biol Med.* 35: 438–43.

62. Kelso GF, Porteous CM, Hughes G, *et al.* (2002). Prevention of mitochondrial oxidative damage using targeted antioxidants. *Ann NY Acad Sci.* 959: 263–74.
63. Saretzki G, Murphy MP, von Zglinicki T (2003). MitoQ counteracts telomere shortening and elongates lifespan of fibroblasts under mild oxidative stress. *Ageing Cell* 2: 141–3.
64. von Zglinicki T, Buerkle A, Kirkwood TBL (2001). Stress, DNA damage and ageing – an integrative apporach. *Exp Gerontol.* 36: 1049–62.
65. Kruk PA, Rampino NJ, Bohr VA (1995). DNA damage and repair in telomeres: relation to aging. *Proc Natl Acad Sci USA* 92: 258–62.
66. Campisi J (2003). Cellular senescence and apoptosis: how cellular responses might influence aging phenotypes. *Exp Gerontol.* 38: 5–12.
67. Shelton DN, Chang E, Whittier PS, Choi D, Funk WD (1999). Microarray analysis of replicative senescence. *Curr Biol.* 9: 939–45.
68. Krtolica A, Rarrinello S, Lockett S, Desprez PY, Campisi J (2001). Senescent fibroblasts promote epithelial cell growth and tumorigenesis: a link between cancer and aging. *Proc Natl Acad Sci USA* 98: 12072–7.
69. Brod SA (2000). Unregulated inflammation shortens human functional longevity. *Inflamm Res.* 49: 561–70.
70. Blasco MA (2002). Immunosenescence phenotypes in the telomerase knockout mouse. *Springer Semin Immunopathol.* 24: 75–85.
71. Samani NJ, Boultby R, Butler R, Thompson JR, Goodall AH (2001). Telomere shortening in atherosclerosis. *Lancet* 358: 472–3.
72. Cawthon RM, Smith KR, O'Brien E, Sivatchenko A, Kerber RA (2003). Association between telomere length in blood and mortality in people aged 60 years or older. *Lancet* 361: 393–5.

Probing the *In Vivo* Relevance of Oxidative Stress in Aging Using Knockout and Transgenic Mice

Florian L. Muller[2], James Mele[2], Holly Van Remmen[1,3,4] and Arlan Richardson[1,3,4]

Departments of Physiology[1], and Cellular and Structural Biology[2], and the Barshop Center for Longevity and Aging Studies[3] at the The University of Texas Health Science Center at San Antonio, and the Geriatric Research, Education and Clinical Center[4], South Texas Veterans Health Care System, San Antonio, Texas 78284-7756, USA

Introduction

In 1956, Denham Harman, proposed the free radical theory of aging [1]. The free radical theory has since expanded to include not only strictly free radicals but also other forms of activated oxygen such as peroxy nitrite and peroxides. We shall refer to this slightly modified version of the free radical theory as the oxidative stress theory of aging, and use the term reactive oxygen species (ROS) to denote all forms of activated oxygen, including free radicals. In its current form, the oxidative stress hypothesis of aging can be stated as follows [2, 3]: *A chronic state of oxidative stress exists in cells of aerobic organisms even under normal physiological conditions. This results in a steady-state accumulation of oxidative damage in a variety of macromolecules. Oxidative damage increases during aging, which results in a progressive loss in the functional efficiency of various cellular processes, ultimately leading to age-related disease and death.* The theory has proved remarkably resilient and adaptable: it provides an important mechanistic basis for the rate of living theory, and the two concepts have now essentially merged [2]. Although the oxidative stress theory is currently one of the most popular explanations for how aging occurs at the biochemical level, most of the evidence in its support is correlative.

One consistent line of evidence for the oxidative stress hypothesis of aging has been the large number of studies that have shown an age-related increase in oxidative damage to a variety of molecules (lipid, protein, and DNA) in organisms ranging from invertebrates to humans [2, 3]. Another strong line of evidence in support of this theory is the fact that dietary restriction, which is the most studied experimental

T. von Zglinicki (ed.), Aging at the Molecular Level, 131–144.

manipulation that has been shown to retard aging and extend lifespan in rodents, consistently alters the age-related accumulation of oxidative damage in rodent tissues [3–6]. Dietary restriction has been shown to reduce the level of oxidative damage in liver and other tissues as measured by a decrease in lipofuscin [7, 8], lipid peroxidation [9, 10], protein oxidation [11, 12], and DNA oxidation [13, 14]. Although the effect of dietary restriction on antioxidant enzymes is not clear, the literature is consistent with the observation that dietary restriction decreases the rate of ROS production by mitochondria [12, 15–17].

Currently, the strongest evidence for the oxidative stress theory of aging has come from studies with invertebrates. Because most invertebrates have an open respiratory system, it is possible to manipulate *in situ* oxygen tension (pO_2) by simply changing the ambient oxygen concentration. Because oxygen tension profoundly affects the rate of ROS production [2, 18], changing ambient pO_2 can be seen as a treatment modulating *in vivo* oxidative stress. In *C. elegans* and *Drosophila*, increasing oxygen tension above atmospheric indeed shortens lifespan in a dose-dependent manner [19–23]. In *C. elegans*, decreasing O_2 tension below atmospheric actually increases lifespan [19, 20]; however, results from such studies in *Drosophila* have been less conclusive [21–23]. With regard to the classic "antioxidant" feeding experiments, results generally indicate that supplemental antioxidants effectively extend lifespan in *C. elegans* and to a lesser extent, in *Drosophila* [2]. Most recently, it was shown that feeding small molecule SOD/Catalase mimetics to *C. elegans* increased both mean and maximum lifespan respectively [24]; unfortunately, it does not appear that these compounds are effective in *Drosophila* [25].

In mutant *C. elegans*, Honda and Honda [26, 27] showed that the *daf-2* mutants, which have extended lifespan, are resistant to oxidative stress and have increased levels of Mn superoxide dismutase. Taub *et al.* [28] also showed that mutations in the cytosolic catalase gene (*ctl-1*) eliminated the *daf-c* and *clk-1* mediated extension of adult lifespan in *C. elegans*, suggesting that the extension of lifespan in the *daf-c* and *clk-1* mutants was due to reduced oxidative damage. In addition, Arking's laboratory found that the extended longevity of long-lived *Drosophila* was correlated to an increased expression of superoxide dismutase and catalase [29, 30]. Rose's laboratory found that the longer lifespan in their strains of *Drosophila* co-varies with increased resistance to various kinds of stress and with the increased frequency of expression of a more active form of superoxide dismutase [31].

The most powerful and most persuasive evidence supporting the oxidative stress theory comes from studies in which antioxidant enzymes were overexpressed in *Drosophila*. The observation by Orr and Sohal in 1994 that overexpression of both Cu/Zn-superoxide dismutase (CuZnSOD) and catalase resulted in a significant increase (14 to 34%) in the lifespan of *Drosophila* and resulted in lower levels of protein and DNA oxidation, is the most direct evidence that the age-related accumulation of oxidative damage is responsible for aging [32]. Their previous studies showed that overexpression of CuZnSOD or catalase alone had only a very small positive effect or no effect on lifespan, respectively [33, 34]. However, transgenic studies in *Drosophila* using P-element mediated transformation are problematic due to the fact that the control and experimental lines have different genetic backgrounds, a factor that has been shown to alter lifespan independently from any transgenetic manipulation [35]. Thus, at present it is not clear whether the effect observed was actually caused by the overexpressed transgenes or the genetic background [36]. Parkes *et al.* [37] recently targeted the overexpression of CuZnSOD in *Drosophila* to motorneurons using a yeast UAS element that was regulated by a GAL4 activator and found that overexpression of CuZnSOD in motorneurons resulted in an increase in lifespan (40%) as well as an increase in resistance to paraquat and γ-irradiation.

To overcome these parental genotypic effects, Tower's laboratory used an inducible yeast FLP/FRT recombination system to induce the expression of various antioxidant enzymes [38]. Overexpression of CuZnSOD resulted in an increase in lifespan of up to 48% (albeit the increase in maximum lifespan was not significant), and a more typical increase was around 20%. However, these investigators found no added benefit from overexpressing both catalase and CuZnSOD. The same laboratory recently reported that overexpression of MnSOD also increased average lifespan by up to 35% (in *Drosophila,* [36]). The effect of concomitant overexpression of these genes was additive, but the overexpression of one transgene seemed to downregulate the expression of the other, so that in real numbers, overexpression of both MnSOD and CuZnSOD did not yield a longer lifespan than the overexpression of either transgene alone. On the other hand, a very recent study by Sohal and co-workers showed that only 5% of wild type levels of CuZnSOD activity was necessary to restore normal lifespan to a CuZnSOD mutant [39], casting doubt on a simple dose-response relationship between SOD activity and lifespan.

Very recent transgenic studies provide even more evidence for the oxidative stress theory in *Drosophila.* Overexpression of the protein oxidative damage repair enzyme, peptide-S-methionine sulfoxide reductase (MsrA) was found to increase average lifespan up to 85% [40]. Although this study seems to indicate that maximum lifespan also was increased; however, actual data for maximum lifespan were not provided. This study took great pains to establish that the changes in lifespan were not due to genetic background issues as previously discussed. Hopefully, other investigators will confirm MsrA overexpression and lifespan extension in the near future.

In this chapter, we discuss the application of transgenic/knockout technology to the study of oxidative stress, as well as its relationship to aging, in mice.

Knockouts of antioxidant enzymes in mice

Over the last decade various knockout mouse models have been generated that have alterations in a wide variety of genes involved in antioxidant protection. A discussion of the strengths and weaknesses of using transgenic and knockout mice to study aging are reviewed by Richardson *et al.* [41]. The knockout/transgenic mouse has become the litmus test for any modern biomedical hypothesis. With regard to the pathogenesis and biological toxicity of oxygen free radicals, transgenic/knockout mice have strengthened established views considerably, but they also have provided a string of surprises. In a previous book chapter, we extensively summarized these data [42], and we now provide updated highlights.

The biggest surprise thus far from ablation studies in antioxidant enzymes is that null mutations in the glutathione peroxidase (*Gpx1*) or CuZnSOD (*Sod1*) genes, which code for the major H_2O_2 and $O_2^{\cdot-}$ scavenging activities in the cytoplasm, result in viable mice that live to at least 20 months [43–45]. At present, it is impossible to say exactly why these two knockouts are not lethal. Possibly, this is due to the fact that endogenous levels of $H_2O_2/O_2^{\cdot-}$ are very low in the cytoplasm, or that there are alternative scavenging mechanisms, or that there are simply no viability-critical targets of ROS in the cytoplasm. Regardless, this experimental result further solidifies Harman's postulate that the mitochondrion is the main source, as well as the main target of ROS [46].

Although null mutations in either *Sod1* or *Gpx1* are not lethal, these mice are more sensitive to exogenous oxidative stress and have some defects, indicating that cytoplasmic $O_2^{\cdot-}$ and H_2O_2 do have some pathogenic potential. For example, *Sod1* knockout females are almost completely infertile [47, 48] and *Gpx1* knockout mice develop cataracts at a much earlier age than do wild type animals [45]. While the exact lifespan of these animals has not been determined, the literature indicates that *Gpx1* knockout mice are viable to at least 22 months [45], and *Sod1* knockout animals are viable to at least 19 months [43].

While knocking out the most widely expressed antioxidant enzymes (*Sod1* and *Gpx1*) fails to yield a dramatic phenotype, it is surprising that null mutations in an overlooked member of the glutathione peroxidase family, phospholipid hydroperoxidase (PHGPx, the product of the *Gpx4* gene), results in early embryonic lethality [49]. Interestingly, this corresponds to about the time when embryos from vitamin E deficient animals are re-absorbed (see below) [50, 51]. PHGPx is a unique glutathione peroxidase in that it reduces lipid hydroperoxides to their respective alcohols, which in the absence of this catalytic activity could decompose to $OH^{\cdot}$ and propagate lipid peroxidation or damage nearby macromolecules in cellular membranes [49, 52]. *Gpx4* is widely distributed in tissues at low levels, except in testis where it is found in relatively high levels. PHGPx is found in various cellular compartments, including the mitochondria [52]. It is impossible to say as yet whether the lethality of *Gpx4*$^{-/-}$ is due to lipid peroxidation or some unrelated function of *Gpx4*. Oxidative damage was not measured in *Gpx4* null embryos [49], and it is known that *Gpx4* has a non-antioxidant structural role in the capsid of sperms [53]. On the other hand, cells from the *Gpx4*$^{+/-}$ mice are sensitive to a variety of oxidative stressors [49], emphasizing the importance of the antioxidant function of this enzyme and the physiological importance of oxidative damage to membranes. Regardless, it must first be established that it is the lack of antioxidant function that is truly responsible for the death of the *Gpx4*$^{-/-}$ mice; failing that, it is regretful that the strongest possible conclusion, namely that lipid hydroperoxides are incompatible with life, can not be drawn from the present work.

In a related matter, the *Gpx4* knockout mice may help elucidate the complex nutritional relationship that exists between selenium and vitamin E. It is well known that these two nutrients "spare" one another, a phenomenon previously interpreted to be due to the synergistic action of glutathione peroxidase 1 and vitamin E [51, 54]. It is now known, thanks to the *Gpx1* knockout mice, that this is incorrect [55]. This view was questionable because H_2O_2, the main substrate of *Gpx1*, is not particularly potent in the induction of lipid peroxidation [51, 56]. However, it is much more probable that the selenium/vitamin E nutritional interaction is due to PHGPx. The pathway can be described as follows: vitamin E first donates a hydrogen atom to a lipid hydroperoxide radical (LOO$^{\cdot}$), resulting in a lipid hydroperoxide (LOOH) that can then be reduced by PHGPx to LOH and H_2O [51, 52]. The generation of the *Gpx4* knockout mouse should allow the testing of this interaction *in vivo*, i.e., *Gpx4*$^{+/-}$ mice should be more susceptible to vitamin E deficiency than wild type.

A large body of evidence in model organisms and in purified organelles suggests that the mitochondrial electron transport chain is one of the main sources of superoxide within the cell [reviewed in ref. 57]. That this is true in mammalian cells as well, was corroborated by the finding that the null mutation in the mitochondrial matrix MnSOD (*Sod2*) is neonatally lethal [58, 59]. The cause of death is strain specific, cardiomyopathy, metabolic acidosis and neurodegeneration in C57B6, DBA and C57DBA F1 respectively [60].

Knockout mice heterozygous for the *Sod2* gene (*Sod2*$^{+/-}$) show reduced (30–80%) MnSOD activity in all tissues studied without any compensatory up-expression in other antioxidant enzymes [61]. Thus, these mice have a compromised antioxidant defense system. Because MnSOD is located in the mitochondrial matrix, it plays a critical role in protecting the mitochondria from oxidant stress by enzymatically scavenging superoxide anions that are produced as a by-product of the respiratory chain. Studies from our laboratory show that the *Sod2*$^{+/-}$ mice have altered mitochondrial function (e.g., reduced respiration and an increased sensitivity of the mitochondrial permeability transition to oxidative stress) and increased oxidative damage [62, 63]. Studies by our laboratory also indicate that *Sod2*$^{+/-}$ mice are sensitive to oxidative stress, e.g., paraquat, and have increased oxidative damage to DNA over the lifespan of the mice. However, our preliminary studies indicate that these mice have no change in lifespan and no evidence of accelerated aging.

Knockouts of oxidative damage repair enzymes

While the biochemistry and biological significance of antioxidant enzymes has received considerable experimental attention, far fewer studies exist as to the biochemistry and physiological significance of oxidative damage repair pathways. Two such systems, namely the repair of oxidized methionine (methionine sulfoxide) and the repair of oxidized guanine (8-oxo-dG), have been probed using gene knockout techniques.

In the case of the methionine sulfoxide repair pathway, the results have proved quite exciting. The null mutation of methionine sulfoxide reductase A (methionine-S-sulfoxide reductase, from the *MsrA* gene) in mice resulted in a considerably shortened lifespan as compared to heterozygote and wild type [64]. The knockout mice were also more sensitive to hyperoxia. Although H_2O_2, $O_2^{\cdot -}$, OH$^{\cdot}$ and even molecular O_2 each can oxidize methionine to its sulfoxide *in vitro*, it is not clear which species predominantly oxidizes methionine *in vivo*. It is nevertheless remarkable that enough methionine sulfoxide can form, in the presence of wild type levels of antioxidant enzymes and low *in situ* concentrations of O_2, for this oxidative modification to become lifespan-limiting. This finding is obviously in agreement with the oxidative stress theory of aging, but the question still remains as to whether this phenomenon represents accelerated aging or simply premature death unrelated to aging. Analyzing the pathology of *MsrA* knockout mice and comparing it to age-related pathology of wild type mice should offer some clarification of this issue. An additional problem with these studies is that the lifespan of the wild type mice was considerably shorter (mean of 22–23 months) than the mean lifespan of 27 to 29 months normally seen for most mouse strains. This could be indicative of less than optimal husbandry conditions, and it is not yet clear whether in a pure genetic background and with better husbandry conditions, the short lifespan of the $MsrA^{-/-}$ mice would still be observed.

In stark contrast to the knockout of the *MsrA* gene, genetic ablation of *Ogg1*, the major glycosylase responsible for the removal of the mutagenic base 8-oxo-dG, results in mice with no overt pathologies [65, 66], which are viable to at least 27 months [67]. Steady-state levels of 8-oxo-dG are increased 2- to 3-fold in these mice and the rate of accumulation of this lesion with age is also accelerated [67, 68]. The spontaneous mutagenesis is increased in these mice, albeit only in the liver (approximately 2-fold). Quite surprisingly, spontaneous carcinogenesis was not increased in the $Ogg1^{-/-}$ mice [69]. On the other hand, the null mutation in the *Mth1* gene, which codes for a protein that hydrolyzes 8-oxo-dGTP and prevents 8-oxo-dG incorporation into DNA by rapidly dividing cells, results in an increased tumor burden by 18 months of age [70]. Whether this actually leads to a decrease in lifespan has not been determined.

Effect of antioxidant enzyme overexpression on lifespan in mice

The most obvious way of validating the oxidative stress theory is through the overexpression of antioxidant enzymes. Indeed, since the publication of positive results in *Drosophila*, many investigators have overexpressed antioxidant enzymes (mainly *Sod1*, *Sod2*, *Gpx1*) and tested the potential benefits of these modifications on a variety of pathophysiological conditions. The results have been less than equivocal, sometimes a small benefit is seen, but most often no real effect is observed [reviewed in ref. 42]. In a few instances, the effect of antioxidant enzyme overexpression is outright pathological [71, 72].

With regards to aging, there is only one published study investigating the effect of antioxidant enzyme overexpression on longevity. Transgenic mice overexpressing CuZnSOD, which were generated by Epstein's and Groner's laboratories using the 15-kb genomic fragment containing the human SOD1 gene [73], have been extensively studied in respect to resistance to oxidative stress and a variety of other biological parameters [see Table 1 in ref. 42]. In general, these mice, show increased resistance to oxidative stress, e.g., focal cerebral ischemia, pulmonary oxygen toxicity, and reactive oxygen species such as peroxynitrite. However, several negative effects have also been reported in these transgenic mice, e.g., reduced macrophage function and thymus envolution [72]. Recently, Huang *et al.* [73] reported the results of a detailed study of the lifespan of CuZnSOD hemizygous and homozygous transgenic mice, which showed 1.5- to 3-fold and 2- to 5-fold, respectively, higher activities of CuZnSOD in various tissues compared to nontransgenic, control mice. In a pilot study, in which a total of 45 mice (15 nontransgenic, 13 hemizygous, and 17 homozygous) were studied in conventional animal housing, the lifespan (mean survival) of the homozygous transgenic mice was observed to be significantly shorter (24%) compared to the control, nontransgenic mice. In a larger lifespan study, in which a total of 417 mice were studied (198 nontransgenic mice, 200 hemizygous mice, and 98 homozygous mice) under barrier conditions, no significant difference in lifespan was observed between the control, nontransgenic mice and the hemizygous and homozygous transgenic mice. The mean ($\pm$SEM) and the maximum (when 95% had died) survival for the nontransgenic, hemizygous, and homozygous transgenic mice were: 20.1 ± 0.6 and 31 months; 19.8 ± 0.5 and 31 months; 18.5 ± 0.7 and 30 months, respectively. It should be noted that the lifespan of these mice is shorter than that normally observed for that strain. Thus, in contrast to what has been reported in *Drosophila*, overexpression of CuZnSOD does not appear to enhance the lifespan of mice. In fact, under conventional housing conditions, when the mice are exposed to more pathogenic insults, the lifespan of mice overexpressing CuZnSOD is significantly reduced. Although these studies are not supportive of the oxidative stress theory, one should note that no parameter of oxidative damage was actually measured in these studies; therefore, it is not clear if the increased expression of CuZnSOD had any effect on the age-related accumulation of oxidative damage in tissues of the transgenic mice.

Knockout mice with altered oxidative stress response signaling

Migliaccio *et al.* [74] reported that knockout mice null for the *p66^{shc}* gene showed increased resistance to oxidative stress and an increase in lifespan. The *Shc* locus encodes three proteins distinguished by differential splicing: the 46, 52, and 66kD forms. The 46kD and 52kD forms are involved in signal transduction through the Ras and MAP kinase pathways, while the 66kD form, p66^{shc}, appears to play a role in response of cells to oxidative stress, e.g., the protein becomes phosphorylated when cells are exposed to UV light or hydrogen peroxide. Cells from the mice homozygous for the null mutation in *p66^{shc}* were observed to be more resistant to apoptosis induced by either UV irradiation or hydrogen peroxide. In addition, mice with the disrupted *p66^{shc}* gene were found to be resistant to paraquat toxicity. The mice lacking *p66^{shc}* lived approximately 30% longer than wild type littermates i.e., the survival increased from approximately 850 days in the wild type mice to almost 1100 days in the homozygous knockout mice for p66^{shc}. Although this exciting observation is consistent with the oxidative stress theory of aging, it should be noted that the survival data were generated with only a very limited number of mice (14 wild type and 15 homozygous *p66^{shc}* knockout mice) and that the mean survival of the mice was relatively short, i.e., the quality of the animal husbandry conditions might be poor.

Synopsis, conclusions, and perspectives

Before discussing the contrasting effects of antioxidant enzymes on lifespan in mice and invertebrates, it is worth emphasizing that there are fundamental differences that make *Drosophila* intrinsically more sensitive or prone to ROS damage than mice (and mammals in general). Insects have a so-called "open" respiratory system such that gas exchange occurs through a tracheal system that puts cells in direct (diffusion) contact with atmospheric (21%) oxygen. By contrast, mammals use a heme oxygen carrying system, which tightly regulates *in situ* pO_2, to less than 2% O_2 [75]. Also worth remembering is that fly mitochondria produce ROS at a roughly 6-fold faster rate than do mice mitochondria [76]. Taking these facts together, it becomes obvious that *Drosophila* are exposed to a considerably higher steady-state level of oxidative stress than mice. With this in mind, we will consider the contrasting results of overexpressing antioxidant enzymes in *Drosophila* and mice.

Contrary to what has been observed in *Drosophila* by two laboratories, overexpression of CuZnSOD does not increase lifespan in mice [73]. However, it should be noted that the initial results in *Drosophila* of overexpressing CuZn SOD were also negative, i.e., the first studies to overexpress CuZnSOD found no change in lifespan [77]. Indeed, tissue-specific and conditional promoters were required to obtain lifespan extension via CuZnSOD overexpression [37, 38]. However, comparing knockout phenotypes of CuZnSOD between mice and *Drosophila* suggests that positive data obtained with the follow-up *Drosophila* experiments will not hold in mammals. For example, the homozygous knockout of CuZnSOD in mice is relatively harmless: the mice are viable to at least 20 months and display a few minor pathologies, among those, sub-fertility in females [47, 48]. This is in stark contrast to the CuZnSOD null mutant in *Drosophila*, which shows an 80% decrease in lifespan, male and female infertility, and several other acute pathologies [78, 79]. Thus, CuZnSOD activity does not appear to be as important for viability (and lifespan) in mice as it does in *Drosophila*. In other words, if a full knockout of CuZnSOD barely decreases lifespan (in mice), why would increasing CuZnSOD activity increase lifespan? The same argument could be made with respect to overexpressing glutathione peroxidase 1. Because mice null for *Gpx1* are viable to at least 22 months [45], overexpression of *Gpx1* is unlikely to have a major effect on lifespan.

In contrast to CuZnSOD, the null mutant of MnSOD is lethal in both *Drosophila* and mice [58, 59, 80], strongly emphasizing the importance of the mitochondrion as a source of ROS. Cells from mice deficient (50% of wild type) in MnSOD show increased sensitivity to a variety of oxidative stresses, and the mice are more sensitive to paraquat toxicity. With this in mind, our laboratory asked the question of whether 50% down-regulation of MnSOD would increase oxidative damage and decrease longevity, i.e., we have investigated whether there was a dose-response relationship between MnSOD activity and lifespan. Although oxidative DNA damage and pathology (tumor burden) are increased, there was no significant decrease in lifespan. These data give support to the oxidative stress theory in its weak form (which states that oxidative stress is associated with aging and age-related disorders), but not in its strongest form (which states that oxidative stress determines lifespan). It is too early to say whether these data simply represent an exception, or are the beginning of trend away from oxidative stress as a lifespan-determining factor.

From the above data, it is also obvious that a simple dose-response relationship between antioxidant enzymes (at least the "classical" antioxidant enzymes) and lifespan is very unlikely. However, this does not mean that the oxidative stress theory of aging has been proven wrong. It should be pointed out that phylogenic comparisons of antioxidant enzymes have, if anything, shown a negative correlation between lifespan and concentration of antioxidant enzymes [reviewed in refs. 15, 81]. However, the steady-state level of oxidative damage, as measured by a diverse series of techniques, negatively correlates with lifespan, just as the oxidative stress theory would predict [2, 15, 81, 82]. By the same token, the rates of ROS generation also negatively correlate with lifespan, again consistent with the oxidative stress theory [15, 76, 83]. While tentative explanations have been advanced to explain this contradictory observation [57, 81], the simple truth is that the complexity of the *in vivo* redox situation is incompletely understood and that much more work is needed. Incidentally, preliminary experiments in our laboratory show that H_2O_2 production rates are not changed in *Sod2*$^{+/-}$ mice (unpublished).

On the other hand, "non-traditional" genetic manipulations have provided support, albeit not proof, for the oxidative stress theory of aging. For example, a null mutation in methionine-S-sulfoxide [64] results in shortened lifespan, just as the "strong" oxidative stress theory would predict (not withstanding the potential problems we have raised above with this study). The next obvious experiment is to overexpress *MsrA* as well as to knockout *MsrB*, which encodes the complementary methionine-R-sulfoxide reductase enzyme.

The extended lifespan seen in the *p66*$^{shc-/-}$ mice would also support the strong oxidative stress theory of aging [74]. However, as noted, it is critical that this study be repeated and that pathological data and data on biomarkers be obtained on these mice to demonstrate that the extended longevity is due to retarded aging. Nevertheless, this study does illustrate a new strategy for testing the role of oxidative stress in aging and disease: instead of manipulating the levels of antioxidant enzymes that mitigate free radicals, this new approach involves manipulating "master" genes that, via signal transduction control a battery of antioxidant genes to maintain the cell in the proper oxidative state. This strategy is in its infancy. In *Drosophila*, genes such as *Methusela* [84] may fall in this category. Perhaps in our ignorance of the true *in vivo* complexity of free radicals and free radical scavenging enzymes, we simply have not yet posed the questions correctly.

Acknowledgments

This work was supported by grants (Merit Review AR and HVR) and an Environmental Hazards Center (AR, HVR) from the Department of Veteran Affairs, NIH grants R01-AG015908 (AR), P01-AG19316 (AR) and the San Antonio Nathan Shock Aging Center (1P30-AG13319), and a grant from the American Cancer Society (HVR).

References

1. Harman D (1956). Aging: a theory based on free radical and radiation chemistry. *J Gerontol.* 11: 298–300.
2. Beckman KB, Ames BN (1998). The free radical theory of aging matures. *Physiol Rev.* 78: 547–81.
3. Sohal RS, Weindruch R (1996). Oxidative stress, caloric restriction, and aging. *Science* 273: 59–63.
4. Masoro EJ (2000). Caloric restriction and aging: an update. *Exp Gerontol.* 35: 299–305.
5. Yu BP, Lim BO, Sugano M (2002). Dietary restriction downregulates free radical and lipid peroxide production: plausible mechanism for elongation of life span. *J Nutr Sci Vitaminol. (Tokyo)* 48: 257–64.
6. Yu BP, Yang R (1996). Critical evaluation of the free radical theory of aging. A proposal for the oxidative stress hypothesis. *Ann NY Acad Sci.* 786: 1–11.
7. Brunk UT, Terman A (2002). Lipofuscin: mechanisms of age-related accumulation and influence on cell function. *Free Radic Biol Med.* 33: 611–19.
8. Brunk UT, Terman A (2002). The mitochondrial-lysosomal axis theory of aging: accumulation of damaged mitochondria as a result of imperfect autophagocytosis. *Eur J Biochem.* 269: 1996–2002.
9. Yu BP, Laganiere S, Kim JW (1988). Influence of life-prolonging food restriction on membrane lipoperoxidation and antioxidant status. *Basic Life Sci.* 49: 1067–73.
10. Yu BP (1996). Aging and oxidative stress: modulation by dietary restriction. *Free Radic Biol Med.* 21: 651–68.
11. Youngman LD, Park JY, Ames BN (1992). Protein oxidation associated with aging is reduced by dietary restriction of protein or calories. *Proc Natl Acad Sci USA* 89: 9112–16.
12. Sohal RS, Ku HH, Agarwal S, Forster MJ, Lal H (1994). Oxidative damage, mitochondrial oxidant generation and antioxidant defenses during aging and in response to food restriction in the mouse. *Mech Ageing Dev.* 74: 121–33.
13. Shigenaga MK, Hagen TM, Ames BN (1994). Oxidative damage and mitochondrial decay in aging. *Proc Natl Acad Sci USA* 91: 10771–8.
14. Beckman KB, Ames BN (1999). Endogenous oxidative damage of mtDNA. *Mutat Res.* 424: 51–8.
15. Barja G (2002). Rate of generation of oxidative stress-related damage and animal longevity. *Free Radic Biol Med.* 33: 1167–72.
16. Merry BJ (2000). Calorie restriction and age-related oxidative stress. *Ann NY Acad Sci.* 908: 180–98.
17. Gredilla R, Lopez-Torres M, Barja G (2002). Effect of time of restriction on the decrease in mitochondrial H_2O_2 production and oxidative DNA damage in the heart of food-restricted rats. *Microsc Res Tech.* 59: 273–7.
18. Chance B, Sies H, Boveris A (1979). Hydroperoxide metabolism in mammalian organs. *Physiol Rev.* 59: 527–605.
19. Honda S, Ishii N, Suzuki K, Matsuo M (1993). Oxygen-dependent perturbation of life span and aging rate in the nematode. *J Gerontol.* 48: B57–61.
20. Honda S, Matsuo M (1992). Lifespan shortening of the nematode *Caenorhabditis elegans* under higher concentrations of oxygen. *Mech Ageing Dev.* 63: 235–46.
21. Strehler BL (1977). *Time, Cells, and Aging*. New York: Academic Press.
22. Baret P, Fouarge A, Bullens P, Lints FA (1994). Life-span of *Drosophila melanogaster* in highly oxygenated atmospheres. *Mech Ageing Dev.* 76: 25–31.

23. Miquel J, Lundgren PR, Bensch KG (1975). Effects of oxygen-nitrogen (1: 1) at 760 Torr on the life span and fine structure of *Drosophila melanogaster. Mech Ageing Dev.* 4: 41–57.
24. Melov S, Coskun P, Patel M, *et al.* (1999). Mitochondrial disease in superoxide dismutase 2 mutant mice. *Proc Natl Acad Sci USA* 96: 846–51.
25. Bayne AC, Sohal RS (2002). Effects of superoxide dismutase/catalase mimetics on life span and oxidative stress resistance in the housefly, *Musca domestica. Free Radic Biol Med.* 32: 1229–34.
26. Honda Y, Honda S (1999). The daf-2 gene network for longevity regulates oxidative stress resistance and Mn-superoxide dismutase gene expression in *Caenorhabditis elegans. FASEB J.* 13: 1385–93.
27. Honda Y, Honda S (2002). Oxidative stress and life span determination in the nematode *Caenorhabditis elegans.* Ann NY Acad Sci. 959: 466–74.
28. Taub J, Lau JF, Ma C, *et al.* (1999). A cytosolic catalase is needed to extend adult lifespan in *C. elegans* daf-C and clk-1 mutants. *Nature* 399: 162–6.
29. Arking R, Burde V, Graves K, *et al.* (2000). Forward and reverse selection for longevity in *Drosophila* is characterized by alteration of antioxidant gene expression and oxidative damage patterns. *Exp Gerontol.* 35: 167–85.
30. Hari R, Burde V, Arking R (1998). Immunological confirmation of elevated levels of CuZn superoxide dismutase protein in an artificially selected long-lived strain of *Drosophila melanogaster. Exp Gerontol.* 33: 227–37.
31. Tyler RH, Brar H, Singh M, *et al.* (1993). The effect of superoxide dismutase alleles on aging in *Drosophila. Genetica* 91: 143–9.
32. Sohal RS, Agarwal A, Agarwal S, Orr WC (1995). Simultaneous overexpression of copper- and zinc-containing superoxide dismutase and catalase retards age-related oxidative damage and increases metabolic potential in *Drosophila melanogaster. J Biol Chem.* 270: 15671–4.
33. Orr WC, Sohal RS (1993). Effects of Cu-Zn superoxide dismutase overexpression of life span and resistance to oxidative stress in transgenic *Drosophila melanogaster. Arch Biochem Biophys.* 301: 34–40.
34. Orr WC, Sohal RS (1992). The effects of catalase gene overexpression on life span and resistance to oxidative stress in transgenic *Drosophila melanogaster. Arch Biochem Biophys.* 297: 35–41.
35. Kaiser M, Gasser M, Ackermann R, Stearns SC (1997). P-element inserts in transgenic flies: a cautionary tale. *Heredity* 78(Pt 1): 1–11.
36. Sun J, Folk D, Bradley TJ, Tower J (2002). Induced overexpression of mitochondrial Mn-superoxide dismutase extends the life span of adult *Drosophila melanogaster. Genetics* 161: 661–72.
37. Parkes TL, Elia AJ, Dickinson D, Hilliker AJ, Phillips JP, Boulianne GL (1998). Extension of Drosophila lifespan by overexpression of human SOD1 in motorneurons. *Nat Genet.* 19: 171–4.
38. Sun J, Tower J (1999). FLP recombinase-mediated induction of Cu/Zn-superoxide dismutase transgene expression can extend the life span of adult *Drosophila melanogaster* flies. *Mol Cell Biol.* 19: 216–28.
39. Mockett RJ, Radyuk SN, Benes JJ, Orr WC, Sohal RS (2003). Phenotypic effects of familial amyotrophic lateral sclerosis mutant Sod alleles in transgenic *Drosophila. Proc Natl Acad Sci USA* 100: 301–6.
40. Ruan H, Tang XD, Chen ML, *et al.* (2002). High-quality life extension by the enzyme peptide methionine sulfoxide reductase. *Proc Natl Acad Sci USA* 99: 2748–53.

41. Richardson A, Heydari AR, Morgan WW, Nelson JF, Sharp ZD, Walter CA (1997). Use of transgenic mice in aging research. *Inst Lab Animal Res J.* 38: 125–36.
42. Van Remmen H, Chen X, Mele J, Richardson A (2002). Aging and oxidative stress in transgenic mice. In: Cutler RG, Rodriguez H, eds. *Oxidative Stress and Aging: Advances in Basic Science, Diagnostics, and Interventions.* Singapore: World Scientific Publishing Co.Pte. Ltd, pp. 1029–69.
43. McFadden SL, Ding D, Reaume AG, Flood DG, Salvi RJ (1999). Age-related cochlear hair cell loss is enhanced in mice lacking copper/zinc superoxide dismutase. *Neurobiol Aging* 20: 1–8.
44. Reaume AG, Elliott JL, Hoffman EK, *et al.* WD (1996). Motor neurons in Cu/Zn superoxide dismutase-deficient mice develop normally but exhibit enhanced cell death after axonal injury. *Nat Genet.* 13: 43–7.
45. Reddy VN, Giblin FJ, Lin LR, *et al.* (2001). Glutathione peroxidase-1 deficiency leads to increased nuclear light scattering, membrane damage, and cataract formation in gene-knockout mice. *Invest Ophthalmol Vis Sci.* 42: 3247–55.
46. Harman D (1972). The biologic clock: the mitochondria? *J Am Geriatr Soc.* 20: 145–7.
47. Ho YS, Gargano M, Cao J, Bronson RT, Heimler I, Hutz RJ (1998). Reduced fertility in female mice lacking copper-zinc superoxide dismutase. *J Biol Chem.* 273: 7765–9.
48. Matzuk MM, Dionne L, Guo Q, Kumar TR, Lebovitz RM (1998). Ovarian function in superoxide dismutase 1 and 2 knockout mice. *Endocrinology* 139: 4008–11.
49. Yant LJ, Ran Q, Rao L, *et al.* (2003). The selenoprotein GPX4 is essential for mouse development and protects from radiation and oxidative damage insults. *Free Radic Biol Med.* 34: 496–502.
50. Evans HM, Bishop KS (1922). On the existence of a hithero unrecognized dietary factor essential for reproduction. *Science* 56: 650–1.
51. Halliwell B, Gutteridge J (1999). *Free Radicals in Biology and Medicine.* New York: Oxford University Press.
52. Imai H, Nakagawa Y (2003). Biological significance of phospholipid hydroperoxide glutathione peroxidase (PHGPx, GPx4) in mammalian cells. *Free Radic Biol Med.* 34: 145–69.
53. Ursini F, Heim S, Kiess M, *et al.* (1999). Dual function of the selenoprotein PHGPx during sperm maturation. *Science* 285: 1393–6.
54. Cheng WH, Valentine BA, Lei XG (1999). High levels of dietary vitamin E do not replace cellular glutathione peroxidase in protecting mice from acute oxidative stress. *J Nutr.* 129: 1951–7.
55. Cheng WH, Ho YS, Ross DA, Valentine BA, Combs GF, Lei XG (1997). Cellular glutathione peroxidase knockout mice express normal levels of selenium-dependent plasma and phospholipid hydroperoxide glutathione peroxidases in various tissues. *J Nutr.* 127: 1445–50.
56. Antunes F, Salvador A, Marinho HS, Alves R, Pinto RE (1996). Lipid peroxidation in mitochondrial inner membranes. I. An integrative kinetic model. *Free Radic Biol Med.* 21: 917–43.
57. Muller F (2000). The nature and mechanism of superoxide production by the electron transport chain: its relevance to aging. *J Am Aging Assoc.* 23: 227–53.
58. Li Y, Huang TT, Carlson EJ, *et al.* (1995). Dilated cardiomyopathy and neonatal lethality in mutant mice lacking manganese superoxide dismutase. *Nat Genet.* 11: 376–81.
59. Lebovitz RM, Zhang H, Vogel H, *et al.* (1996). Neurodegeneration, myocardial injury, and perinatal death in mitochondrial superoxide dismutase-deficient mice. *Proc Natl Acad Sci USA* 93: 9782–7.

60. Huang TT, Carlson EJ, Kozy HM, *et al.* (2001). Genetic modification of prenatal lethality and dilated cardiomyopathy in Mn superoxide dismutase mutant mice. *Free Radic Biol Med.* 31: 1101–10.
61. Van Remmen H, Salvador C, Yang H, Huang TT, Epstein CJ, Richardson A (1999). Characterization of the antioxidant status of the heterozygous manganese superoxide dismutase knockout mouse. *Arch Biochem Biophys.* 363: 91–7.
62. Van Remmen H, Williams MD, Guo Z, *et al.* (2001). Knockout mice heterozygous for Sod2 show alterations in cardiac mitochondrial function and apoptosis. *Am J Physiol Heart Circ Physiol.* 281: H1422–32.
63. Williams MD, Van Remmen H, Conrad CC, Huang TT, Epstein CJ, Richardson A (1998). Increased oxidative damage is correlated to altered mitochondrial function in heterozygous manganese superoxide dismutase knockout mice. *J Biol Chem.* 273: 28510–15.
64. Moskovitz J, Bar-Noy S, Williams WM, Requena J, Berlett BS, Stadtman ER (2001). Methionine sulfoxide reductase (MsrA) is a regulator of antioxidant defense and lifespan in mammals. *Proc Natl Acad Sci USA* 98: 12920–5.
65. Klungland A, Rosewell I, Hollenbach S, *et al.* (1999).. Accumulation of premutagenic DNA lesions in mice defective in removal of oxidative base damage. *Proc Natl Acad Sci USA* 96: 13300–5.
66. Minowa O, Arai T, Hirano M, *et al.* (2000). Mmh/Oggl gene inactivation results in accumulation of 8-hydroxyguanine in mice. *Proc Natl Acad Sci USA* 97: 4156–61.
67. Osterod M, Hollenbach S, Hengstler JG, Barnes DE, Lindahl T, Epe B (2001). Age-related and tissue-specific accumulation of oxidative DNA base damage in 7,8-dihydro-8-oxoguanine-DNA glycosylase (Oggl) deficient mice. *Carcinogenesis* 22: 1459–63.
68. Nishimura S (2002). Involvement of mammalian OGG1(MMH) in excision of the 8-hydroxyguanine residue in DNA. *Free Radic Biol Med.* 32: 813–21.
69. Epe B (2002). Role of endogenous oxidative DNA damage in carcinogenesis: what can we learn from repair-deficient mice? *Biol Chem.* 383: 467–75.
70. Tsuzuki T, Egashira A, Igarashi H, *et al.* (2001). Spontaneous tumorigenesis in mice defective in the MTH1 gene encoding 8-oxo-dGTPase. *Proc Natl Acad Sci USA* 98: 11456–61.
71. Raineri I, Carlson EJ, Gacayan R, *et al.* (2001). Strain-dependent high-level expression of a transgene for manganese superoxide dismutase is associated with growth retardation and decreased fertility. *Free Radic Biol Med.* 31: 1018–30.
72. Rando TA, Epstein CJ (1999). Copper/zinc superoxide dismutase: more is not necessarily better! *Ann Neurol.* 46: 135–6.
73. Huang TT, Carlson EJ, Gillespie AM, Shi Y, Epstein CJ (2000). Ubiquitous overexpression of Cu,Zn superoxide dismutase does not extend life span in mice. *J Gerontol A Biol Sci Med Sci.* 55: B5–9.
74. Migliaccio E, Giorgio M, Mele S, *et al.* (1999). The p66shc adaptor protein controls oxidative stress response and life span in mammals. *Nature* 402: 309–13.
75. Jones DP, Mason HS (1978). Gradients of O_2 concentration in hepatocytes. *J Biol Chem.* 253: 4874–80.
76. Sohal RS, Svensson I, Sohal BH, Brunk UT (1989). Superoxide anion radical production in different animal species. *Mech Ageing Dev.* 49: 129–35.
77. Seto NO, Hayashi S, Tener GM (1990). Overexpression of Cu-Zn superoxide dismutase in *Drosophila* does not affect life-span. *Proc Natl Acad Sci USA* 87: 4270–4.
78. Phillips JP, Campbell SD, Michaud D, Charbonneau M, Hilliker AJ (1989). Null mutation of copper/zinc superoxide dismutase in *Drosophila* confers hypersensitivity to paraquat and reduced longevity. *Proc Natl Acad Sci USA* 86: 2761–5.

79. Parkes TL, Kirby K, Phillips JP, Hilliker AJ (1998). Transgenic analysis of the cSOD-null phenotypic syndrome in *Drosophila. Genome* 41: 642–51.
80. Kirby K, Hu J, Hilliker AJ, Phillips JP (2002). RNA interference-mediated silencing of Sod2 in *Drosophila* leads to early adult-onset mortality and elevated endogenous oxidative stress. *Proc Natl Acad Sci USA* 99: 16162–7.
81. de Grey ADNJ (2000). Non-correlation between maximum life span and antioxidant enzyme levels among homotherms: implications for retarding human aging. *J Antiaging Med.* 3: 25–36.
82. Sohal RS, Ku HH, Agarwal S (1993). Biochemical correlates of longevity in two closely related rodent species. *Biochem Biophys Res Commun.* 196: 7–11.
83. Sohal RS, Svensson I, Brunk UT (1990). Hydrogen peroxide production by liver mitochondria in different species. *Mech Ageing Dev.* 53: 209–15.
84. Lin YJ, Seroude L, Benzer S (1998). Extended life-span and stress resistance in the Drosophila mutant methuselah. *Science* 282: 943–6.

Non-oxidative Modification of DNA and Proteins

Alan R. Hipkiss
Department of Experimental Therapeutics, William Harvey Research Institute, Charterhouse Square, London EC1M 6BQ, UK

Introduction

Aging is associated with progressive cellular dysfunction which in all probability derives from altered forms of both protein and DNA. There are many potential sources of macromolecular instability both oxidative and non-oxidative. In recent years much attention has been given to the role of reactive oxygen species (ROS) as potential sources macromolecular alteration reflecting the popularity of the free-radical theory of aging. In an attempt to restore a more balanced view and in keeping with the obvious multifactorial nature of aging, this chapter will outline those non-oxidative phenomena that can and do influence the structure and function of DNA and proteins, some of which are listed in Table 1. Oxidative events that affect DNA and proteins will be dealt with elsewhere in this volume.

DNA

Spontaneous changes

DNA is not stable. It is subject to a surprising amount of change to the bases, as well as cleavage of phosphodiester bonds, apparently in complete absence of any exogenous agents, including reactive oxygen species. These spontaneous events are sometimes ignored when possible sources of DNA alteration and age-related somatic mutation are considered. The review by Thomas Lindahl [1], although a decade old, summarizes well the possible contribution of slow spontaneous DNA damage to the phenomenon of aging. In his article [1] Lindahl argues that in developed societies with their recent much extended average lifespan, slow, low-frequency changes are likely to be important (for human aging), because their effects have not been suffiently deleterious to provoke evolution of compensating correction/repair processes. Should this be proven to be the case, it is possible that many of the changes

T. von Zglinicki (ed.), Aging at the Molecular Level, 145–177.

Table 1. *Spontaneous non-oxidative, non-enzymic changes to DNA and proteins that may impact macromolecular aging*

Source of damage	Change to DNA	Change to protein
Spontaneous changes to primary structure	Depurination (depyrimidation) Strand breakage	Amino acid residue Isomerization Racemization Dephosphorylation
	Deamination 5-methylcytosine to thymine Cytosine to uracil Guanine to hypoxanthine Adenine to xanthine	Deamidation Asparagine to aspartic acid (L, D or iso-forms) Glutamine to glutamic acid (L, D or iso-forms)
Non-oxidative, non-enzymic modifications		
Glycation by reactive metabolites (aldehydes and ketones)	Adducts to guanine Cross-linked structures	Adducts to lysine, arginine, cysteine, histidine Cross- linked structures
Methylation by SAM	O^6-methylguanine	

under current and highly active investigation may be irrelevant as major causative determinants of human aging, because processes that correct those macromolecular alterations have evolved. Conversely, perhaps it is the infrequent but uncorrected imperfections in DNA and perhaps its maintenance system, which permitted evolution, that cause the ultimate demise of the individual, via the "process" that we call aging.

Depurination and depyrimidination

Loss of purine bases (depurination) from DNA can occur spontaneously at a frequency of approximately 580 per hour per (human) cell and is thought [2] to be the result of merely living at 37°C. In absence of their repair, such losses produce abasic sites that can cause replication faults or transcription errors. Depyrimidation is much less frequent, occuring at about one tenth of rate of depurination. Additional base losses are enzymically-mediated by various DNA-glycosylases which specifically eliminate modified bases after oxidative attack or alkylation (see Chapter 2, this volume). An increase in abasic sites has been found to accompany senescence of cultured human fibroblasts as well as in leukocytes from old human donors. It is thought, however, that the elevated apurinic sites levels are due to a lowered rate of repair rather than any increase in formation [3].

Deamination

Deamination of the pyrimidine cytosine is thought [2] to occur at a rate of 8 events per hour per (human) cell at 37°C. Cytosine's deamination results in uracil, an illegitimate DNA base which is normally removed by DNA-uracil glycosylase (see Chapter 2, this volume). Adenine and guanine's deamination to xanthine and hypoxanthine respectively are both highly mutagenic, but occur predominantly as a consequence of attack by reactive oxygen and nitrogen species [4] (again see Chapter 2, this volume).

Providing repair is 100% efficient (but see comment above and at the conclusion of this chapter) and occurs before gene replication or expression, there should be no effects of cytosine deamination on gene compositon. Effects on gene regulation are possible however. This is because some cytosines are methylated at the 5-position, especially in CpG islands which play a role in gene silencing (see Chapters 8 and 13, this volume). Spontaneous deamination of 5-methylcytosine creates thymine, a perfectly legitimate component of DNA which is consequently not eliminated by any DNA glycosylase. The consequent mismatch, however, should be corrected by a special T-G mismatch repair enzyme which replaces the offending thymine with a cytosine.

Single-strand breaks

According to Setlow [2] the most common form of spontaneous damage to DNA is the single-strand break. He reports an estimated rate of 2300 single-strand breaks per hour in human cells. The origin of the the breaks is not clear, but they are thought to derive from depurination and depyrimidination events as well as enzyme-mediated base excision, all of which render the phosphodiester bond between the deoxyribose moieties highly labile. They are readily cleaved in the presence of basic agents such as lysine residues on histones, should the lysine's ε-amino group react with the free aldehyde function of the abasic deoxyribose (see sections on non-enzymic glycosylation of proteins and DNA).

Spontaneous methylation
The reactive metabolite S-adenosylmethionine (SAM) can spontaneously methylate guanine in DNA to produce O^6-methylguanine which affects the bases's hydrogen bonding ability. This modification is repaired by a specific protein, a suicidal O^6-methylguanine-DNA methyl transferase (MGMT), in which a specific cysteine residue receives the offending methyl group [5]. This completely inactivates the protein such that it has to be ubiquitinated and then degraded by the proteasome system [6]. This may partly explain why maintenance of cellular protein degradative function is important for DNA integrity [see ref. 7 and refs. therein]. Methylation of the specific cysteine residue on the MGMT protein, however, also turns it into a transcriptional activator thereby promoting a compensating increase in repair protein expression [8].

Glycation
There have been some reports showing that DNA is subject to non-enzymic glycosylation where an aldehyde or ketone group of the glycating agent reacts with an amino group on DNA, although the purine and pyrimidine rings make DNA amino groups less reactive than those on proteins [9, 10]. The occurrence of glycated DNA lesions *in vivo* have been estimated to be very low (less than 1% of oxygen-mediated damage) [11] and therefore are almost indetectable [12]. Because of the longevity of DNA in long-lived species however, they should not be disregarded [13].

In order to create detectable DNA glycation in the laboratory, aldehydes more active than glucose are required, such as the unphosphorylated forms of the glycolytic intermediates dihydroxyacetone phosphate and glyceraldehyde-3-phosphate, and methylglyoxal (pyruvaldehyde) which is a by-product of normal metabolic intermediates [11]. The major product of DNA glycation is N^2-(1-carboxyethyl)-deoxyguanine which has also been isolated following treatment of plasmid DNA with dihydroxyacetone [14]. The effect of glycation of plasmid DNA with this agent is an increase in mutagenicity, most likely due to increased spontaneous depurination of the glycated DNA [14].

A major source of reactive aldehydes that can react with DNA is lipid peroxidation [15]. Although the origin of the reative agents such as 4-hydroxynonenal and malondialdehyde is oxygen-dependent, their reaction with DNA does not require oxygen. The major point of attack on DNA again appears to be the deoxyguanine amino group followed by cyclization of the adduct [15]. Such adducts have also been detected *in vivo*, which if not repaired could induce mutations etc. [15]. This may partly explain the genotoxicity of 4-hydroxynonenal and related aldehydes [16]. Ethanol treatment of rats induced a 2-fold increase in generation of DNA adducts (ethenodeoxyadenosine and ethenodeoxycytidine) in rat liver [17] possible via increased lipid peroxidation. Glyoxal treatment of human fibroblasts and keratinocytes induced DNA strand breakage but mediated indirectly via the carbonyl agent's initial attack of nuclear protein [18].

Cross-linking
DNA-protein cross-links have been found to increase with age in mouse brain and heart and may be considered as another possible aging marker, especially in non-mitotic tissue [19]. Most DNA-protein cross-links are secondary consequences of protein oxidation especially if polypeptide alkoxyl-, peroxyl- or carbon-centred-radical species are generated. Reactive metabolites such as aldehydes may, however, also induce DNA cross-linking to protein via a non-oxidative mechanism. For example acetaldehyde can induce DNA-protein cross-linking most likely reacting first with protein amino groups (e.g., on lysine side chains) forming polypeptide bound carbonyl groups which then cross-link to guanine amino groups [20, 21]. This is likely to occur in chromatin where lysine-rich histones are available targets for reactive aldehydes before reacting with DNA.

DNA itself is another potential source of reactive aldehydes. Deoxyribose is a reactive aldehyde and therefore potent glycating agent in its non-cyclic (straight chain) form. Whilst incorporated in DNA the sugar is locked into its cyclic conformation by attachment of the purine or pyrimidine base to the 1′-carbon atom. Should, however, the base be lost by spontaneous depurination or the action of any base-excision (repair) glycosylase, then the deoxyribose is a glycating agent with the potential for reacting with any available amino group, on a protein for example. During DNA repair this may occur during elimination of the sugar residue to promote cleavage of the adjacent phosphodiester bond inducing strand breakage. Once eliminated from the polynucleotide backbone the released sugar residue may possess one or even two reactive carbonyls, but its fate is uncertain. It has been shown, however, that the reaction of oxygen free-radical species with DNA can result in as many as 12 different deoxyribose products each bearing at least one reactive aldehyde group [22], all of which could subsequently cause cross-linking phenomena. Interestingly it has been shown that aggregated DNA-histone crosslinks, formed following hydrogen peroxide treatment, can inhibit proteasome function [23]. It is therefore at least conceivable that DNA-histone cross-links, formed via any mechanism, could compromise the proteasome-mediated removal of the O^6-methylguanine repair protein (MGMT) after its methyation as described above, as well as other ubiquitin/proteasome-dependent processes involved in DNA repair [7].

Proteins

One of the most common biochemical signals of aging is the accumulation of aberrant protein forms within cells and tissues [24, 25]. The aberrant proteins are often aggregated, sometimes in association with chaperone or heat shock (stress) proteins and may be conjugated with ubiquitin especially when they accompany age-related pathologies. The source of aberrant polypeptide that accumulate and their possible involvement as either causes or effects of aging and related conditions have been debated for many decades [ref. 26 and refs therein]. This review will cover the various non-oxidative sources of aberrant polypeptide forms and their relationship to the processes of aging. The possible impact of these modifications on other areas of protein biochemistry, which may be more than a little speculative or controversial, will nevertheless be explored.

Biosynthetic errors

One of the earliest molecular theories of aging was the "error catastophe theory" proposed by Leslie Orgel [27,.28]. This theory suggests that the aberrant proteins known to accumulate in aging cells and tissues arise as a result of an age-related increase in error frequency during gene expression; the proposed increase in erroneous polypeptides further compromising the fidelity of gene expression to induce, eventually, an error catastrophe, senescence and death (see also Chapter 12, this volume). One major advantage of the error catastrophe theory was that it was testable, and in the 1970s and 1980s many attempts were made to test it [see ref. 26 for detailed discussion]. It was established at that time that mRNA translation was the most error-prone step in gene expression. Consequently most experimentalists sought to find out whether the translational error rate increased during aging. As documented below, little or no unequivocable evidence was obtained for any age-related increase in the production of translation-induced error proteins and the error catastrophe theory was abandonded by most biogerontologists. More recent studies however show that the decision to search for changes in translational error rates, whilst neglecting any search for alterations in transcriptional errors, the more accurate process, may have been erroneous; an age-related increase in transcriptional errors has been recently obtained [29, 30]. These findings suggest that the demise of the error catastrophe theory of aging may have been a little premature.

Transcriptional errors

As stated above, transcriptional errors are comparatively rare occurring at an estimated rate of one error in every 10^4–10^5 bases transcribed [26]. Evidence suggesting an alteration in transcriptional accuracy with age was first found [31] in the Brattleboro rat which possesses a single base (G) deletion in the vasopressin gene. The young rats produced, as expected, no functional vasopressin, whereas in older animals functional vasopressin was detected. Examination of the older animals indicated no change in the vasopressin gene. When, however, the old animals' vasopressin mRNA was examined a 2-base deletion was detected, the effect of which was to put the reading frame back into its correct alignment, thereby generating a quasi vasopressin polypeptide with just one amino acid residue missing roughly in the middle. It has subsequently been shown that the RNA polymerase in old animals makes errors when transcribing CT repeat sequences in DNA, producing GAGAG motifs of mRNA; it can omit a GA in the transcript producing a frameshift error of +1 base, generating a so-called +1 protein [32, 33]. The reason for this is uncertain, either the intrinsic error rate of the RNA polymerase producing the frameshift increases with age, or the detection and elimination of the aberrant transcripts is compromised in old animals. CT sequences are not uncommon in mammalian genes; they have been detected not only in the vasopressin gene but are also found in the genes coding for amyloid precursor protein (APP) and ubiquitin [32]. Interestingly, the corresponding +1 versions of both APP and ubiquitin have been detected in the aberrant proteins that are associated with Alzheimer's disease, a major human age-related pathology [32]. Moreover the aberrant (+1) form of ubiquitin has been shown to be toxic to cells and compromises the elimination of altered protein forms by the proteasome [34, 35]. The latter observation is consistent with the general predictions of the error catastrophe theory i.e., that erroneous gene products could further compromise protein quality control, thereby inducing an increased accumulation of aberrant proteins characteristic of the senescent state.

Translational errors

It has long been shown that mRNA translation at the ribosome is the most error-prone step in gene expression. The actual mistranslation rate varies between codons but the average is around 3 errors in every 10 000 codons [36]. Many attempts were made 20–30 years ago to show whether translational error rates increase with age by assessing the incidence of protein "stuttering" following amino acid misincorporation during polypeptide synthesis [26]. Although some evidence for stuttering was obtained, there was, however, no clear evidence for any increase in translational errors, suggesting the invalidity of the error catastrophe theory of aging in general [37] but see Holliday [26].

Nevertheless, attempts were made to find out if raising the general translational error rate could lead to cell senescence using an immortal organism *Escherichia coli* [38–40]. These experiments used very low concentrations (0.5–2.5 μg/ml) of the miscoding antibiotics streptomycin or dihydrostreptomycin to raise translational error rates. The experiments showed that bacterial growth in very low concentrations of miscoding antibiotics continued exponentially for up to ten or more cell doublings; growth did, however, eventually slow and the cells finally died. Evidence for increased error rates was indicated by increased phenotypic correction of a nonsense mutation in β-galactosidase [39] and transiently raised and then suppressed proteolyis of pulse-labeled protein [40].

Although there may be a variety of interpretations of these observations, they do superficially suggest that increasing the intrinsic error rate can eventually induce the cessation of growth and finally death (senescence ?) in a supposedly immortal organism, accompanied by the failure to eliminate aberrant protein forms, an error catastrophe perhaps. That both erroneously synthesized proteins and oxidized proteins appear to be degraded by the same proteolytic system in bacteria suggests that overproduction of either of these aberrant protein forms could compromise destruction of both of them. Furthermore, a compromised ability to eliminate altered proteins is now being recognized as possibly central to cellular senescence (see Chapters 14 and 15, this volume), although such proposals have been extant for decades in biogerontology. The recent observation by Holbrook and Menninger [41] is very pertinent here. These authors showed that treatment of yeast with erythromycin, an antibiotic that increases the accuracy of protein synthesis in bacteria, increases cellular lifespan by up to 27%. That this effect was not detected in petite yeast, which lack detectable mitochondrial DNA, suggests that increased accuracy of mitochondrial protein synthesis could be responsible for the effects of cell longevity, which is consistent with an error theory of aging. Consequently, could the relative inaccuracy of mitochondrial protein synthesis (the error rate of bacterial [70S] ribosomes being about 5 times that of cytoplasmic [80S] ribosomes) eventually provoke an intra-mitochondrial error catastrophe, contribute to age-related increases in mitochondrial dysfunction and secondary degenerative events such as apoptosis? That erroneously-synthesized proteins are more susceptible to oxidation than normal gene products [42], and within mitochochondria both forms of aberrant polypeptide are degraded by protease Lon, which also shows an age-related decline [43], provides obvious links between error proteins, oxidative damage, proteolytic decline and age-related mitochondrial dysfunction. That ubiquitin is required for the proteolytic elimination of mitochondria [44–47] suggests links to extra-mitochondrial degradative failure, e.g., proteasomes and lysosomes (see Chapters 14 and 15, this volume), the age-related accumulation of aberrant mitochondria [48], the release of cytochrome c from mitochondria and induction of apoptosis.

In this context one has to at least question the use of streptomycin in cell culture media to help maintain sterility, especially in long-term experiments studying fibroblast senescence. Although the antibiotic should not affect cytoplasmic ribosome function, streptomycin could theoretically affect the error rate of mitochondrial ribosomes in the long-term, similar to that shown for *E. coli* [38–40]. Could streptomycin contribute to mitochondrial instability, or even, at the extreme case, provoke an error catastrophe in mitochondria of cultured fibroblasts grown in the presence of the antibiotic? Additionally could the use of therapeutic levels of streptomycin and similar miscoding antibiotics contribute to mitochondrial instability in human and animal populations in general?

Post-synthetic protein modifications and aging

Spontaneous changes

Racemization and isomerization

All amino acids (except glycine) incorporated into proteins are in the L-configuration. There is, however, a finite possibility for spontaneous racemization of most amino acids into the alternative D-configuration [49]. The process is very slow however (the racemization rate of the fastest, aspartic acid residues, is 0.1% per annum) but D-amino acids can be detected in long-lived proteins such as the crystallins found in the lens core, the enamel and dentine of teeth, and long-lived proteins found in cartilage, bone, skin, arterial walls and red cell membranes [49, 50]. D-aspartic acid and D-serine residues are the most common, whilst D-tyrosine has also been detected in lens tissue [51]. D-amino acids are enriched in lenticular cataracts which are aggregates of improperly folded protein; either their presence promotes the initial aggregation or the long half-life of the aggregated polypeptide permits racemization. It is uncertain what effects the presence of D-amino acid residues induce biologically, but it is likely that the half-life of the protein could be increased, as D-amino acid residues are known to increase the half-lives of synthetic peptides [52].

Aspartic acid residues are the most commonly studied example of amino acid racemization. The transformation between the L- and D-forms is thought to occur via a succinimide intermediate [53]. Whilst aspartic acid residues can and do undergo racemization, the vast majority of D-aspartic acid residues found in proteins arise following the non-enzymic deamidation of asparagine residues (see below).

Whereas amino acid racemization in proteins usually occurs non-enzymically, there are examples of specific enzymically-mediated transformations [52] as shown by the funnel-web spider which can convert L-serine at position 46 of the agatoxin to the D-form; opioid peptides containing D-alanine have been found in the skin of a South American tree frog, whilst peptides containing D-methionine, D-leucine and D-phenylalanine have been obtained from invertebrates, other amphibia and molluscs [52]. Nothing is known about any relationship (if any) between these observations and the slow accumulation of D-amino acid residues in proteins during aging. Could a chronic infection by an organism possessing the ability to convert L-amino acid residues into the D-form accelerate aging in some tissues? Perhaps this is a speculation too far.

Aspartic acid residues can also undergo spontaneous isomerization. Studies show a half-life of 53 days for L-aspartyl isomerization using a model hexapeptide [53], but about 38 times slower than the corresponding asparaginyl peptide. At Asp 151 of human α-A-crystallin obtained from elderly subjects (mean age 80 years) about 50% was present as the iso-peptide, whereas Asp at the same position in the protein obtained from an 11 month old child showed about 5% of the molecules possessed this modification, indicating that position 151 is a site that is highly susceptible to isomerization [54].

Deamidation of asparagine and glutamine residues

Asparagine and glutamine residues are also prone to spontaneous deamidation resulting in the generation of aspartic acid and glutamic acid residues respectively and thereby creating altered protein forms bearing negative charges in place of the amide functions [55]. Asparagine residues appear to deamidate about 10–100-fold faster than glutamine residues, consequently the former have been far more extensively studied [56]. Though generally regarded as slow, studies with model hexapeptides show half-lives as short as two days or less are possible [56] and in most proteins primary structure and conformation affect deamidation rates. Spontaneous deamidation in a recombinant human protein has been shown to produce a 50-fold decrease in biological activity [57].

Deamidation of asparagines is known to procede via a cyclic succinimide intermediate (see Figure 1) which can spontaneously cleave to form one of four possible products, L-aspartic acid, D-aspartic acid, L-iso-aspartic acid and D-iso-aspartic acid [53]. The latter two forms are β-amino acids residues with an extra methylene group present in the polypeptide backbone, whilst the original α-carbonyl group now forms the carboxyl side chain. It appears that the β-peptide bond is not cleaved by the regular proteolytic apparatus (proteasomes and lysosomes) such that β-peptides are found in urine [58, 59]. Whether β-peptides affect the cell's ability to cleave other aberrant protein forms is unknown, but many protease inhibitors appear to contain iso-peptide bonds.

Asparagine residue deamidation is very dependent on its context as adjacent residues can profoundly influence deamidation rates, those on the carboxyl side having the greater influence [53, 56]. It appears that deamidation is fastest when glycine is present in the carboxyl position; histidine, serine and alanine also permit

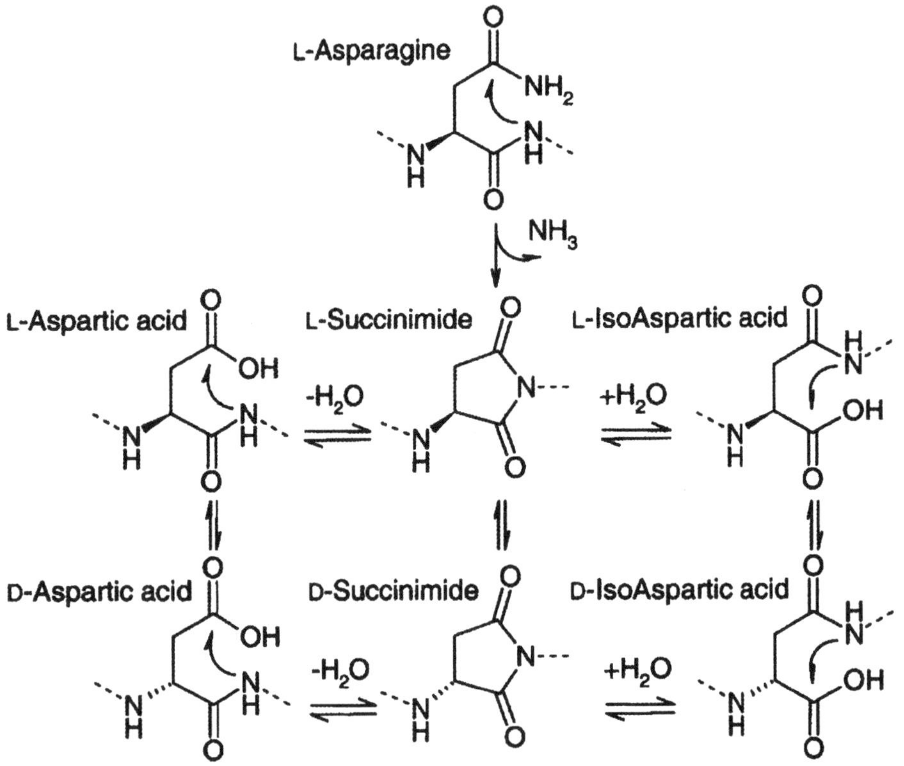

Figure 1. Instability of asparagine and aspartic acid residues: succinimide intermediates and possible products. Dotted lines indicate the polypeptide backbone. (Figure from Cloos and Jensen (2000) Biogerontology 1: 341–56).

relatively rapid deamidation of adjacent asparagine residues. Protein secondary and tertiary structures and concentration also affect deamidation rates [60–62]; within the same protein some asparagine residues appear to be far more resistant to deamidation than others. There is one example (human hypoxanthine guanine phosphoribosyltransferase) of where a protein's 3-dimensional structure accelerates deamidation [63]. Robinson and Robinson [56] list 126 proteins and the calculated rate constants for 1371 different asparagine residues, their half-lives ranging from as low as one day (uracil-DNA-glycosylase) to more than 6000 days; experimental studies also showed half-lives as short as 1.4 days at 37°C and pH 7.4 when glycine followed asparagine in a model peptide [53].

Glutamine residues also deamidate but at much slower rates than asparagine. Even when the process is most favored, i.e., when glycine is on the carboxyl side of glutamine, the rate is at least 400-fold slower than asparagine deamidation [56]. The mechanism is assumed to be analogous to asparagine deamidation but proceeding via a glutaride intermediate [53]. Differential deamidation rates for glutamine residues within the same polypeptide species have been reported, for example in α-A-crystallin glutamine at position 50 undergoes 30% deamidation within 64 years, whilst glutamines at positions 6 and 147 showed no detectable deamidation during the same time period [64].

The specific biological effects of deamidation have not been extensively studied, except in long-lived proteins such as in the lens where it has been clearly shown to be associated with aging and cataractogenesis [55, 64, 65]. In general terms the presence of iso-aspartic acid residues in proteins can alter proteolytic susceptibility and elicit autoimmunity [66], both features of the aged phenotype. Age-related deamidation has been observed in histones in chromatin in rodent liver [67] and brain [68], which could affect gene activity. D-iso-aspartic has been found in elastin fibres of sun-damaged skin [69] suggesting that ultra-violet irradiation may influence asparagine/aspartic acid deamidation and isomerization during sun-induced aging of skin.

Asparagine instability may play a role in age-related neurodegenerative diseases. For example, iso-aspartic acid residues have been detected in β-amyloid plaque [70]; however whether the presence of the altered amino acid form is a cause or effect of amyloid stability in Alzheimer's disease is uncertain. Iso-aspartic acid residues have been detected in aggregated tau proteins [71], which are also associated with Alzheimer's disease. Spontaneous deamidation and isomerization have, furthermore, been observed in a murine prion protein and in a fragment corresponding to the human prion protein [72]. Sandmeier *et al.* [72] suggest that isomerization of asparagine 108 might be an integral feature of the conversion of the non-toxic prion protein PrP^{C} into the toxic form PrP^{Sc} . Small amounts of asparagine deamidation (less than 5%) in an amylin fragment can promote formation of aggregates that have the hallmarks of amyloid [73]. It is, therefore, not inconceivable that asparagine deamidation could play an initiating role in many of the aggregation phenomena associated with age-related disease, especially should the activity of the enzyme, PIMT (see below) that partially repairs the protein, and proteolytic activities in general, decline with age.

It is interesting to note that uracil DNA-glycosylase has the shortest deamidation half-life (one day) according to Robinson and Robinson [56]. The enzyme, which has three consecutive asparagine residues at its active site, is important for DNA repair, another activity that apparently declines during aging [3]. Could asparagine instability contribute to the age-related decline in DNA repair when protein synthesis rates decline? A further indication of the interrelationship between various possible causal events in aging is the observation that the deamidation potential is increased in oxidized proteins [74]; that histidine residues are converted to aspartic acid and aspargine residues upon oxidation [75], which have the potential for isomerization and racemization, could partly explain this observation.

Aging is also associated with apoptotic dysfunction; the increased incidence of cancer may be associated with a failure of apoptosis, whereas degenerative disorders may result in part from excessive apoptosis. It may be relevant that aspartic acid residues are central to the action of caspases involved in apoptosis: aspartic acid residues are present at the active sites of these enzymes, and caspases cleave their substrates at aspartic acid residues [76]. Consequently could aspartic acid residue racemization and isomerization in the enzyme or a substrate affect caspase activity, and could deamidation of asparagine residues or oxidation of histidine residues to aspartic acid residues create illegitimate caspase binding sites? These are theoretical possibilities that should be investigated, particularly as aging is associated with increased protein half-lives which increases the likelyhood for these changes to occur.

Repair of iso-aspartic acid residues in proteins

Asparagine deamidation must be biologically important because an enzyme exists that partially repairs proteins that contain iso-aspartic acid residues. This enzyme is called protein iso-aspartate methyltransferase (PIMT) and it initiates repair by specifially methylating the free carboxyl group of iso-aspartic acid residues [77]. The resultant methyl ester can then regenerate the succinimide intermediate, the spontaneous cleavage of which has a 50% possibility of forming the normo-aspartic acid derivative, thereby eliminating the extra methylene group from the polypeptide backbone; that which reforms the β-peptide bond is remethylated by PIMT and the process repeats. Knock-out mice that lack this enzyme suffer from epileptic seisures, have shortened life-spans and accumulate much aberrant protein in their tissues especially their brains [78–80]. PIMT-deficient mice showed an 80-fold accumulation of iso-aspartic acid residues in histone H2B from brain, and inhibition of PIMT by adenosine dialdehyde promoted an 18-fold increase in iso-aspartic acid residue content in histone H2B of cultured cells [81]. Down regulation of PIMT in humans suffering from mesial temporal lobe epilepsy, appears to provoke elevated levels of iso-aspartic acid residues in β-tubulin and several other proteins [82]. Folate deficiency has been reported to suppress PIMT activity and thereby stimulate iso-aspartic acid residue accumulation in young rat liver [67].

The PIMT protein may undergo automethylation resulting in a partial self repair and thereby help maintain PIMT activity during aging [83]. It is uncertain whether PIMT activity declines with age, whilst PIMT overexpression can extend lifespan in *Drosophila* under conditions of heat stress (29°C) but not in unstressed (25°C) flies however [84]. This observation may at least question the supremacy of oxidation processes as the rate-limiting steps in aging in all cases.

Whilst the repair of iso-aspartyl residues by PIMT appears to be a well distributed phenomenon, being present in bacteria and plants as well as mammals, no comparable enzyme has been detected that is capable of repairing the likely products of glutamine deamidation such as γ-glutamic acid residues, where two extra methylene groups would be inserted into the polypeptide backbone. This could mean that cleavage of the cyclic intermediate is invariably "correct" yielding a free γ-carboxyl group and no "incorrect" form is generated with a free α-carboxyl group and the β- and γ-carbon atoms inserted into the polypeptide backbone. Alternatively such incorrect rearrangements could occur so slowly that they have not been biologically significant so evolution of a repair function was unnecessary. However, perhaps humankind's recently acquired "unnatural" longevity has exposed glutamine deamidation, a low frequency time-dependent change in protein chemistry, as another factor compromising biochemical function in elderly individuals but against which no defense/repair system has evolved.

Polyglutamine diseases

There are number of conditions such as Huntington's disease, dentatorubral pallidoluysian atrophy, spinocerebellar ataxia, spinobulbar muscular atrophy and Machado-Joseph disease where polypeptides accumulate with long polyglutamine tails and promote gain of function effects. The onset of some of the associated pathologies in, for example Huntington's disease, is invariably age-related where proteolytically-resistant aggregates of the polyglutamine-containing peptides appear to be causal to the associated pathologies [85]. In humans [86], in an animal model [87] and in a bacterium [88], the onset of pathology or toxicity is dependent upon the length of the polyglutamine tail and age (time), both of which could theoretically increase the probability of deamidation. It is curious that there are no reported studies on whether glutamine deamidation could influence the aggregation of the polyglutamine peptides. The recent observation that low levels of asparagine deamidation can dramatically affect aggregation of amyloidogenic peptides [73] as well as their proteolytic susceptibility (see above), suggest that glutamine deamidation should be investigated as a possible initiating factor in the aggregation of polyglutamine proteins.

Deamidation and carnosine

Deamidation of asparagine and glutamine may be related to carnosine (β-alanyl-L-histidine), a dipeptide that appears to suppress senescence in cultured human fibroblasts and even rejuvenate senescent cells [89, 90]. Studies in food chemistry have shown that γ-glutamyl-carnosine is present in the high molecular weight fraction (protein) of beef preparations [91, 92]. It was shown that such an adduct could be formed upon the heat-induced deamidation of glutamine residues in beef proteins [93]. Furthermore the adduct was detected when free glutamine deamidated in the presence of carnosine, whilst β-aspartyl-carnosine was formed when free asparagine was deamidated in the presence of the dipeptide [93]. It is curious that the latter adduct was not detected in any of the beef preparations (heated or otherwise) despite the fact that when free asparagine deamidated in the presence of carnosine, formation of the adduct proceded at 5-times that rate at which the analogous glutamine-derived adduct was formed. One possible explanation is that reaction between the asparagine-derived moiety does occur but the products are unstable resulting in cleavage of the β-peptide bond between carnosine and the aspartic acid residue. This proposal is supported by the detection of β-aspartyl-β-alanine adducts in the beef preparations, which appear to be increased in amount compared to γ-glutamyl-β-alanine [93]. It is uncertain, however, whether carnosine has any effects on deamidation of asparagine and glutamine residues *in vivo*, but the dipeptide is present in some tissues at concentrations as high as 20 mM. Perhaps carnosine is a protective agent against glutamine residue deamidation, whilst PIMT looks after deamidated asparagine residues.

Generation of reactive carbonyl groups on proteins under non-oxidative conditions: glycation

There is much evidence that shows that proteins are subject to oxidation and that aging is associated with an accumulation of oxidized polypeptide forms (see Chapter 3, this volume). Many of the products of oxidative attack on proteins appear to have one feature in common, the presence of an aldehyde or carbonyl group on the polypeptide [75]. Indeed the presence of increased amounts of protein-associated carbonyls is a frequent indicator of oxidative damage, the amount of which has been shown to increase with age. As will be discussed elsewhere in this volume the presence of carbonyl groups on a polypeptide may be signal for its destruction by the proteasome and/or the lysosome. However the presence of carbonyl groups on proteins can facilitate protein-protein cross-linking [94] or protein-DNA cross-linking [18]. Cells appear to have limited defense against cross-linked structures which seem to compromise DNA repair and proteasome function. Suppressed proteasome function may be partly responsible for the well known accumulation of aberrant protein forms that characterize many aged cells and tissues. Chapters 14 and 15 (this volume) will deal with the inhibitory role of cross-linked proteins on proteasome and lysosomal function respectively.

Non-enzymic glycosylation, now termed glycation, is another source of protein carbonyls [95]. This process was first discovered in food chemistry (called the Maillard or browning reaction) where glucose reacts with proteins eventually producing yellow/brown products that are frequently autofluorescent and contain cross-links [12, 13 and refs. therein]. At its simplest (see Figure 2) glycation is thought to proceed first by formation of a Schiff's base between the target amino group (usually on a protein) and the reactive aldehyde or ketonic group of the glycating agent (e.g., glucose). The Schiff's base then spontaneously rearranges to form an Amadori product which, importantly, has regained a carbonyl group. Up to this stage the reactions are thought to be reversible. The Amadori product can, however, undergo a variety of reactions. First it can simply reverse to reform the Schiff's base, this is the least dangerous route. Secondly, the carbonyl group on the Amadori product may react with another amino group to form an intra- or intermolecular cross-link. Thirdly, the amino group can be regenerated directly with the release of a dicarbonyl deoxyglucosone, which, now bearing a second carbonyl group adjacent to the first, is a far more reactive glycating agent (called a propagator) than the initial sugar. Fourthly, the sugar may fragment leaving, for example, a carboxylmethyllysine and an unreactive four-carbon fragmentation product. The chemistry of these changes is still ill-defined, although many structures have been characterized. These include polycyclic fluorescent forms derived from cross-linking reactions. Collectively the structures that are finally formed are termed advanced glycation end-products (AGEs) [96]. AGEs have been detected in many aged tissues [12, 13, 97] and during senescence of cultured cells [98]. They are often associated with age-related pathologies [99] such as cataracts, nephropathies [100], neuropathies including Alzheimer's disease [101–103] and atherosclerosis [95]. The incidence of protein AGEs is also raised in uncontrolled diabetes [104, 105], a condition which, to some extent, mimics aspects of early senescence.

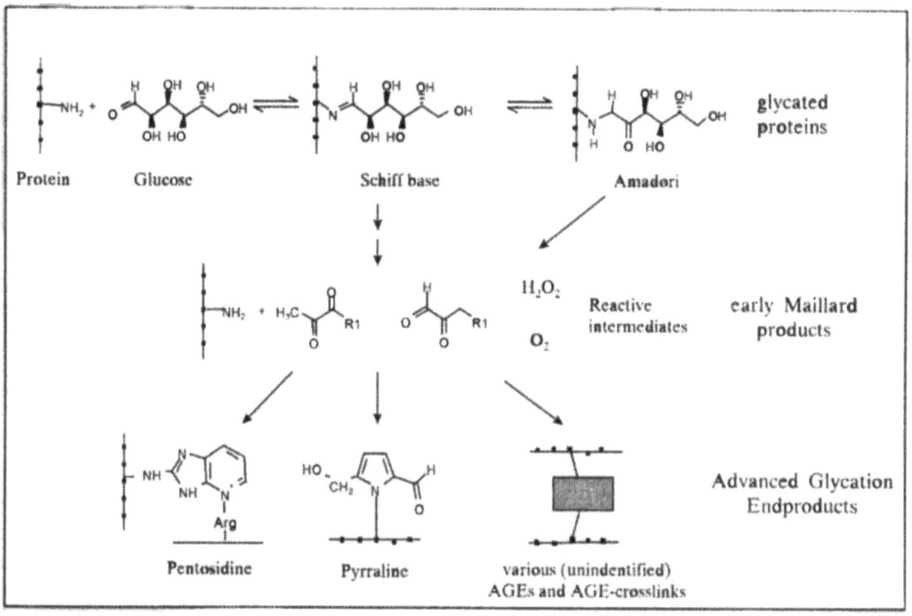

***Figure** 2. Non-enzymic glycosylation of a protein amino group with glucose; intermediates and formation of advanced glycosylation end products (AGEs). (Figure from Dukic-Stefanovic et al. (2001) Biogerontology 2: 19–34).*

Because of its obvious implications for diabetes where circulating glucose levels are markedly raised, much work initially considered glucose to be the primary glycating agent responsible for protein modification. Glucose is the least reactive of the common sugars because it exists predominantly in the unreactive cyclic form; other sugars such as fructose, galactose, ribose and deoxyribose, free or phosphorylated, are far more reactive. Not only is extracellular sugar a source of extracellular glycation especially in diabetics, but excessive glycolysis is potentially deleterious because many glycolytic intermediates such as glyceraldehyde-3-phosphate and dihydroxyacetone-phosphate are highly reactive towards proteins and can glycate them much more rapidly than glucose. Furthermore a spontaneous by-product of glyceraldehyde-3-phosphate and dihydroxyacetone-phosphate is methylglyoxal (pyruvaldehyde) which, being a bicarbonyl, is even more reactive towards proteins, and again generates proteins carbonyl groups and cross-links. Methylglyoxal is also a product of threonine and glycine metabolism. Catabolism of these amino acids generates aminoacetone, which monoamino oxidase converts into methylglyoxal. Fatty acid metabolism is another source of methylglyoxal [106]. Methylglyoxal can react with lysine, arginine and cysteine residues in proteins and possibly guanidine in DNA [106]. Metabolic elimination of methylglyoxal normally occurs by aldose reductase activity which produces 1,2-propanediol and requires NADPH; alternatively glyoxalases I and II employ glutathione catalytically to convert methylglyoxal to lactate [106]. Consequently any overproduction of methylglyoxal could directly damage proteins and possibly DNA especially if glutathione and NADPH levels are low. Additionally it has recently been shown that exogenous methylglyoxal promotes mitochondrial damage in toad embryos [107], and methylglyoxal's reaction with arginine and other guanidino compounds can induce production of superoxide anions [108]. The arbitrary separation between of oxidative and non-oxidative protein modification is therefore difficult to justify. Mitochondrial dysfunction and apoptosis have also been shown to be induced when glycated-oxidized-HDL was added to cultured endothelial cells, although glycated-but-un-oxidized HDL showed little toxicity [109].

As outlined in Chaplen [106] excess glycolysis can increase intracellular levels of glycolytic intermediates and thereby methylglyoxal, which can not only directly damage protein but also indirectly damage them by inducing ROS production. Given the likelyhood that increased glycolytic flux increases intracellular methylglyoxal production [106] one can immediately begin to understanding the beneficial effects of caloric restriction especially where the majority of the calories are supplied as carbohydrate as in most diets given to experimental animals. Suppression of glycolysis rather than any decrease of mitochondrial function may explain the anti-aging effects of caloric restriction. Futhermore, glucose itself is subject to metal catalyzed auto-oxidation with concomittant generation of ROS and reactive bicarbonyls [110].

Oxidative events are also an indirect source of protein carbonyls. Lipid peroxidation can generate reactive products such as malondialdehyde, hydroxynonenal and acrolein [111–114] all of which damage proteins by reacting non-oxidatively with lysine, histidine or arginine side chains to produce polypeptide carbonyl groups with the potential for deleterious cross-linking as described above. Such products have been called Advanced Lipid peroxidation Endproducts or ALEs [115]. Furthermore it has recently been shown that free-radical mediated arachidonic acid oxidation can produce a mixture of γ-isoketals capable of reacting with proteins to produce structures that are not only resistant to proteolysis but inhibit proteasome function [116].

As a result of the reaction of various amino acid side chains with a variety of glycating agents (from glucose to acrolein) a large number of structures have been described [96], some of which are listed in Table 2. Non-oxidative glycation induces a number of effects on proteins such as decreased proteolytic susceptibility, induction of autoantibodies, decreased protein solubility, cross-linking to other proteins and increased aggregation. Protein glycation may also create catalytic sites for free-radical generation [117] thereby again linking oxidative and non-oxidative aging theories. The extent to which any of these modifications might be causal to aging rather than its concommitants remains to be determined. Nevertheless it is clear that much protein modification that characterizes the aged phenotype can occur in absence of overt oxidative damage.

***Table 2**. Possible sources of some AGEs and ALEs*

AGEs	Origin
Pentosidine	Cross-linked arginine and lysine
Carboxymethyl-lysine	Lysine and glyoxal
Carboxyethyl-lysine	Lysine and methylglyoxal
Pyrraline	Lysine and 3-deoxyglucosone
GOLD	Lysine-lysine cross-linked with glyoxal
MOLD	Lysine-lysine cross-linked with methylglyoxal
DOLD	Lysine-lysine cross-linked with 3-deoxyglucosone
Hydroimidazolones	Arginine and glyoxal, methylglyoxal and 3-deoxyglucosone
Argpyrimidine	Arginine and 3-deoxyglucosone
ALEs	Lysine and malondialdehyde
	Lysine-lysine cross-linked by malondialdehyde
	Lysine and 4-hydroxynonenal
	Histidine and 4-hydroxynonenal
	Cysteine and 4-hydroxynonenal

GOLD = glyoxal derived lysine dimer.
MOLD = methylglyoxallysine dimer.
DOLD = deoxyglucosone lysine dimer.

ADP-ribose and glycation

A major response to DNA damage is the poly(ADP-ribosyl)ation of nuclear proteins catalyzed by poly(ADP-ribose) polymerase-1 (PARP-1) [118, 119]. There is still much debate about the role that this modification plays. For example ADP-ribose which is produced in the nucleus following DNA damage is a highly reactive glycating agent [120] capable of reacting with histone H1, most likely at lysine residues [121], producing a ketoamine. Synthesis of the ADP-ribose is carried out by the enzymes polyADP-ribose synthetase and polyADP-ribose glycohydrolase whose co-ordinate activies generate large amounts of ADP-ribose following DNA damage. One possible explanation is that ADP-ribose facilitates histone lysine residue deacetylation forming 1-O-acetyl-ADP-ribose as product [122] and hence possibly playing a role in gene regulation. Another explanation is that the release of free ADP-ribose within the vicinity of basic histones prevents lysine residue amino groups reacting with the free ribose moieties that would be generated on DNA following depurination and depyrimidation events, occurring either spontaneously or following enzyme-mediated elimination of modified bases during base-excision repair.This could thereby inhibit formation of DNA-protein cross-links which have not only decreased susceptibility to proteasome activity [123] but may even be inhibitory. The ability of polyADP-ribose polymerase to activate nuclear proteasomes to degrade oxidized proteins [23, 124] is consistent with this proposal. It would be interesting to determine directly if the ketoamine formed as a result of histone glycation with ADP-ribose is a substrate for the proteasome system.

Serine dephosphorylation

Post-synthetic phosphorylation of serine residues is a form of protein modification frequently found in phosphoproteins with high affinity for divalent cations, mostly calcium. Many phosphoproteins are found in long-lived tissues such as bone (osteonectin, osteopontin and bone sialoprotein) and teeth (dentin sialoprotein and dentin phosphophoryn [125]. Human dentin shows a large age-related decline in phosphoserine content, falling at 75 years to about 5% of the level found in dentin phosphoprotein of an 11 year old. The authors [125] propose (see Figure 3) that a phosphoserine residue can undergo loss of its phosphate group, at a rate around 4% per year, which proceeds via two possible mechanisms yielding either, via hydrolysis, serine and eventually glycine, or via β-elimination, a highly reactive dehydroalanine which can react with a histidine residue to form a histidinoalanine cross-link. Histidine and lysine residues show age-related declines in human dentin phosphoprotein at around 0.9% and 0.3% per year respectively [125]. It is further suggested that serine racemization may accompany dephosphorylation and that phosphothreonine residues could also undergo dephosphorylation via a similar series of reactions [125].

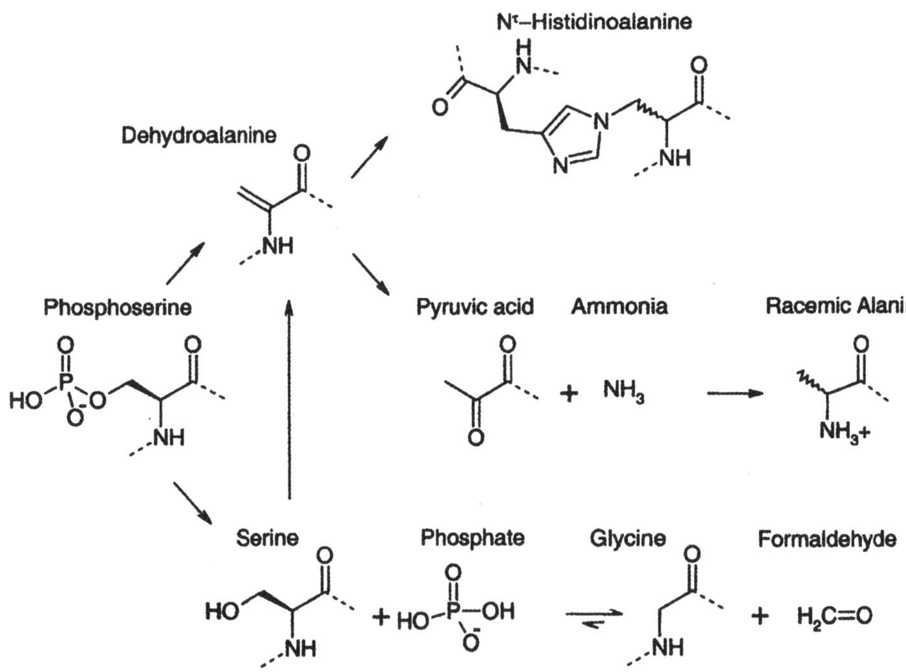

Figure 3. *Instability of phosphoserine residues: intermediates and possible products. Dotted lines indicate polypeptide backbone. (Figure from Cloos and Jensen (2000) Biogerontology 1: 341–56).*

Can non-oxidative protein damage be modulated?

Protein turnover

Obviously the more rapidly proteins are replaced (turned over), the lower the likelihood that any of spontaneous amino acid residue modifications will occur in any one protein. Consequently a high protein turnover rate may be considered an anti-aging strategy and the decline in intracellular proteolytic activity that accompanies aging may be considered as the gradual failure of an anti-senescence function.

One factor that, paradoxically, might control a cell's proteolytic capacity is its growth rate. It is possible that rapidly growing cells possess a higher constitutive proteolytic activity towards aberrant proteins than cells growing slowly; this has been demonstrated in a bacterium at least, where the constitutive proteolytic activity towards an abnormal protein squared as growth rate doubled [126]. The reason for the increased level of proteolytic activity in more rapidly growing cells probably lies in the fact that protein synthesis is never perfect, errors are always being made, hence the greater the rate of protein synthesis the more altered proteins are synthesized per cell per unit time, assuming that error rate remains unaltered by growth rate. Consequently a cell rapidly synthesizing protein would have to deal with more abnormal polypeptide molecules than a cell slowly synthesizing its proteins. Should mammalian cells behave similarly, then one would predict that cells synthesizing proteins rapidly would have a higher intrinsic ability to eliminate aberrant proteins than cells carrying out a slower rate of protein synthesis [127]. Consistent with this idea is the fact that ubiquitin is co-ordinately expressed with a ribosomal protein [128], thereby constitutively providing at least one component of the apparatus for the selective elimination of aberrant polypeptides. Additionally it has been shown that synthesis of stress proteins and chaperone proteins, including those that bind to altered and nascent protein forms on ribosomal subunits, also appears to correlate with growth rate in transformed cells [129–133]. Whilst the constitutive ability to degrade abnormal polypeptides has not been extensively investigated in relationship to development and aging in animals, there is suggestive evidence that longevity is related to intrinsic high levels of stress proteins that presumably improve the organism's ability to deal with aberrant proteins [134, 135]. The effects of hormesis where repeated sublethal heat stress enhances survival of cultured cells and improves resistance to oxidative stress is most probably due to enhanced stress protein expression [136–138].

Recent studies suggest that intramitochondrial proteolysis could play a role in the relationship between mitochondrial dysfunction and aging [139]. A major mitochondrial protease is the Lon protease which is analogous to protease La found in *Escherichia coli*. This enzyme degrades both miscoded and oxidized proteins in mitochondria [140] and loss of this enzyme is associated with the accumulation of oxidized protein aggregates [140, 141], at least in yeast. The activity of the Lon protease has been found to be lower in skeletal muscle from old mice compared to young animals [43], the enzyme being responsible, at least in part, for degradation of oxidized (carbonylated) mitochondrial proteins such as aconitase. It is not known why the Lon protease shows the age-related decline but any of the modifications described here could play a role. Continued synthesis of Lon could however help maintain mitochondrial protein homeostasis. It is therefore interesting to note that epidermal growth factor (EGF) upregulates Lon expression [142]. This reinforces the idea that continued rapid growth is an effective anti-aging strategy, as proposed above. Hence it is suggested that when growth slows during hormonally-controlled decreased protein synthesis, there is an accompanying decline in constitutive proteolytic potential (cytosolic and intramitochondrial) to remove aberrant mitochondria and mitochondrial and cytosolic proteins of whatever origin. Furthermore altered proteins that are formed as consequences of postsynthetic damage would represent an increasing proportion of the abnormal protein load as protein synthesis rates decline. When the constitutive ability to deal with the aberrant polypeptide load is insufficient, then induction of the stress response occurs which includes synthesis of the ubiquitin, stress and chaperone proteins. However the stress response also involves suppression of the synthesis of normal housekeeping proteins. Hence it is theoretically possible that where excess aberrant proteins are continuously generated and/or persist due to very low constitutive proteolytic activity, the stress response could also persist with the continued shut down of housekeeping protein production. Such a condition could eventually compromise viability perhaps due to chaperone overload [143]. In addition, postsynthetic modification of chaperone proteins themselves [144] may also contribute to decreased stress tolerance [145, 146]. In contrast, the beneficial effects of transient non-toxic stress (hormesis) [136–138] and possibly exercise training [147], may, by increasing expression of chaperone proteins, improve a cell's ability to deal with the aberrant polypeptides that otherwise may initiate senescence.

These ideas predict that the threshold at which the stress reponse is induced is low in cells in which protein synthesis is occurring slowly, consequently they would be relatively vulnerable to the presence of aberrant protein forms of whatever origin. It is possible that differences in the constitutive proteolytic activity could explain some of the age-related changes in vulnerability to aberrant protein forms observed in inherited forms of Alzheimer's disease and possibly other late onset pathologies where the defective genes are apparently present from conception. Hence maintenance of a juvenile proteolytic phenotype may be important for suppressing aging and onset of age-related pathology. An effective proteasome function may additionally help in maintenance of DNA integrity too.

Caloric restriction

Caloric restriction is the only almost universal process that is known to suppress the onset of aging and extend maximum lifespan. It is possible that the anti-aging effects of caloric restriction might be at least partly explained in terms of non-oxidative events. First, caloric restriction induces fasting and periods of relative hypoglycaemia which suppresses both extracellular glycation potential as well as intracellular glycation by glycolytic intermediates and their by-products such as methylglyoxal [106]. Indeed the recent observation [148] indicating that cellular defenses are improved (increased reduced glutathione [GSH] content and raised catalase and glutathione peroxidase activities) in glycolysis-depleted cells, again suggests that decreased glycolysis when calorically restricted, compared to the *ad libitum* condition, could be important for enhanced survival. Secondly, periodic fasting induced in the restricted animals would also lower circulating insulin levels [149] which would have the effect of increasing proteasome activity; it is known that insulin suppresses proteasome function [150]. The stimulated protein turnover in the restricted animals, necessary for gluconeogenesis [151], should also help to prevent the accumulation of aberrant protein forms. Third, assuming that ATP demand remains unaltered, less glycolysis would occur in the caloric restricted condition, and mitochondrial ATP generation proportionally increased. This would not only decrease intracellular glycation (as discussed above) but also increase electron supply to mitochondria thereby decreasing the possibility of generating incompletely reduced oxygen molecules, i.e., free-radicals, due to an inbalance between oxygen supply and electron availability. Furthermore caloric restriction would suppress the conversion of excess glucose in *ad libitum* cells to fatty acids whose synthesis requires NADPH. A lowered NADPH availability in the *ad libitum* animals due to fatty acid synthesis could compromise the NADPH-dependent maintenance of glutathione in its reduced state. Any compensating increase in NADPH generation would decrease NADH production by one sixth thereby decreasing electron availability to mitochondria so increasing the possibility of incomplete reduction of oxygen molecules. Any decrease in fatty acid synthesis induced by caloric restriction should therefore increase electron supply to mitochondria and ensure complete oxygen reduction i.e., there would be no reduction in respiration, as recently demonstrated in calorically-restricted yeast [152] and *C. elegans* [153].

Naturally-occurring anti-glycating agents

Despite the fact that glycation has been long recognized as a major age-related source of postsynthetic protein modification, little is known if or how this process is regulated *in vivo*. A possible naturally-occurring anti-glycating molecule is carnosine (β-alanyl-L-histidine) [154, 155] which is found in innervated tissues of many species and sometimes at surprisingly high concentrations (up to 20 mM in humans). The dipeptide appears to contain all the features required for preferential glycation, a target amino group, and proximal imidazole and carboxyl groups; test-tube experiments show that carnosine can indeed protect proteins against glycation by a variety of glycating agents including methylglyoxal, malondialdehyde, deoxyribose, glucose and AGEs, and it can prevent formation of protein carbonyls and protein-protein cross-links [154, 156–158]. The dipeptide also protects cells against AGEs [157], malondialdehyde [159] and the toxicity of an amyloid peptide fragment [160]. Some of the protective activity of carnosine may be related to its ability to behave as an aldehyde/ketone scavenger including lipid peroxidation products [161–163] as well as peroxinitrite damage [164], although its chelating activity towards copper ions [165, 166] and its anti-oxidant and hydroxyl radical scavenging ability [167] must also contribute. Carnosine has also been demonstrated to inhibit aggregation of amyloid peptides [168] and α-synuclein [169], delay senescence in cultured human fibroblasts and rejuvenate senescent cells [89, 90] as well have beneficial effects on hyperoxic cultured cells [170, 171] and in senescence-accelerated mice such as lowering brain malondialdehyde content [172]. The mechanisms by which the rejuvenation occurs are unknown but the process must involve something more than prophylactic events that merely suppress glycation or ROS reactivity [127, 173]. Possible explanations involve carnosine-induced alterations in gene expression [174] which could improve the cells' ability to deal with aberrant proteins, and/or its reaction with protein carbonyls [175] which could facilitate their proteolytic elimination. Whilst the latter idea remains speculative at present, there is some supporting evidence. An increase in intracellular proteolysis has been shown in senescent fibroblasts when cultured with carnosine [158, 176] and the dipeptide appears to facilitate the proteolytic elimination of hypoxia inducible factor 1 (HIF-1 alpha) in cardiomyoblasts [177], a protein that is normally very readily oxidized and then degraded by proteasomes. Molecules such as carnosine and related peptides have been proposed as naturally-occurring anti-glycating agents that might also control the secondary complications of diabetes [173, 178], for example cataracts [179], where carbonyl group reactivity might be involved.

Another naturally-occurring candidate anti-glycating and general aldehyde scavenger is pyridoxamine [180–182]. Pyridoxamine inhibits formation of AGEs and ALEs *in vivo* and in model systems [180], reacts directly with methylglyoxal [181] and decreases formation of protein carbonyls in plasma proteins of diabetic rats, though not in control animals. The product of pyridoxamine's reaction with methylglyoxal, a methylglyoxal-pyridoxamine dimer (MOPD) has been described [181]. Pyridoxamine may also react with protein carbonyls inhibiting conversion of Amadori products to AGEs [183].

Kinetin, which is a naturally-occurring modified form adenine, has been found to inhibit both glycation and oxidative phenomena [138, 184] and furthermore helps cultured human fibroblasts retain their juvenile morphology when approaching the end of their proliferative life-span [185]. It is possible that kinetin's apparent pluripotency, similar to carnosine's [157, 158], is important for its anti-senescence activity.

Other possible antiglycating agents or aldehyde scavengers that could suppress glycation include D-penicillamine which also protects keratinocytes and fibroblasts against glyoxal and methylglyoxal-induced toxicity, most probably by directly and irreversibly reacting with the dicarbonyls forming thiazolidine derivaties [186]. No effects on cellular senescence, however, have been reported.

Aminoguanidine has long been proposed as an anti-glycating agent [187] and it can scavenge aldehydes. It can inhibit both the initial glycation and subsequent AGE formation in model systems and *in vivo*, suppressing the development of secondary complications of diabetes [104 and refs therein]. However its reported toxicity may prevent its widespread therapeutic application [186].

Hydroxylamines have been reported to delay senescence of cultured human fibroblasts [188], although the mechanism(s) involved is/are uncertain. Because hydroxylamines readily react with aldehydes and ketones, it is possible that carbonyl scavenging could be involved, including reaction with those on proteins, similar to the proposed explanation of carnosine's rejuvenating action on senescent cells [189]. Hydroxylamines may also possess anti-oxidant activity [188], again suggesting that pluripotency could be important for controlling aging [127, 189].

Interrelationships between aberrant proteins
Evidence is emerging which suggests that the various forms of protein modification that appear to be associated with senescence should not be considered independently. As already mentioned, erroneously synthesized proteins show an increased potential towards oxidative damage [42], which in turn can increase deamidation phenomena [74], while denaturation increases protein glycation potential [190]. Cross-linked products of oxidation and glycation can inhibit proteasome function [191, 192] thereby compromsing the elimination of aberrant protein forms in general. An inactive proteasome apparatus may inhibit the elimination of inactive/dysfunctional mitochondria, possibly suppressing their replacement while promoting a compensating increase in glycolysis to maintain ATP production. This would increase glycation potential and the generation of more glycoxidized and cross-linked polypeptides capable of inhibiting proteasome function even further [193]. That oxidized proteins may alter DNA structure [194] provides further evidence of the importance for effective and rapid elimination of aberrant polypeptides.

It has been known for some time that proteasome function is essential for cell cycling and DNA repair. There are examples of DNA repair mutations where the relevant proteins have roles in ubiquitin metabolism and/or proteasome function [7]. Should the ubiquitin system or proteasome function become compromised in any way (by synthesis of erroneous ubiquitin or inactivation by cross-linked proteins as outlined above, or even suppression as a result of permanent hyperinsulinaemia in the well-fed) then DNA repair would show decreased efficacy with consequent effects on DNA composition and gene expression.

Conclusions

In this chapter I have tried to show that both proteins and DNA are subject to deleterious alteration via processes that do not directly involve reactive oxygen or nitrogen species. Whilst not denying the role of these reactive free-radical species in senescence and age-related pathologies it is important to remember that macromolecules are intrinsically unstable and that oxygen is not the sole source of their instability. Much emphasis has also been given to the role of reactive aldehydes in modification of proteins and DNA and subsequent formation of cross-linked species, and evidence is accumulating that the intransigence of protein-protein cross-links, DNA-protein cross-links and protein-lipid cross-links (lipofuscin) to destruction are major contributors to the senescent state. It appears that while various mechanisms have evolved to repair or eliminate the immediate precursors of the cross-linked species, there is no obvious process that successfully destroys them. Perhaps the induction of highly reactive oxygen free-radicals by AGEs binding to RAGEs on macrophages (see Chapter 15, this volume) is a "last ditch" attempt at their elimination, all other mechanisms being unsuccessful.

It is possible that the ability of our ancestors to evolve has sewn the seeds of the ultimate demise of the individual. Evolution relies on errors in gene replication and some inefficiency in the elimination of altered protein forms; perfection in these homeostatic process is not only energetically highly demanding but also inhibitory to evolutionary adaptation. Hence senescence may be non-deterministic consequence of our evolution.

References

1. Lindahl T (1993). Instability and decay of the primary structure of DNA. *Nature* 362: 709–15.
2. Setlow RB (1987). Theory presentation and background summary. In: Warner HR, Sprott RL, Butler RN, Schneider EL, eds. *Modern Biological Theories of Aging*. New York: Raven Press, pp. 177–82.
3. Atamna H, Cheung I, Ames BN (2000). A method for detecting abasic sites in living cells: age-dependent changes in base excision repair. *Proc Natl Acad Sci USA* 97: 686–91.
4. Kow YW (2002). Repair of deaminated bases in DNA. *Free Radic Biol Med.* 33: 886–93.
5. Lindahl T, Demple B, Robins P (1982). Suicide inactivation of the *E coli* O^6-methylguanine DNA-methyltransferase. *EMBO J.* 1: 1359–63.
6. Srivenugopal KS, Yuan XH, Bigner DD, Fredman HS, Ali-Osman F (1996). Ubiquitination-dependent proteolysis of O^6-methylguanine-DNA methyltransferase in human and murine tumor cells following inactivation with O^6-benzylguanine or 1,3-bi(2-chloroethyl)-1-nitrosourea. *Biochemistry* 35: 1328–34.
7. van Laar T, van der Eb AJ, Terleth C (2002). A role for RAD23 proteins in 26S proteasome-dependent protein degradation? *Mutat Res.* 499: 53–61.
8. Kleibl K (2002). Molecular mechanisms of adaptive response to alkylating agents in *Escherichia coli* and some remarks on O^6-methylguanine DNA-methyltransferase in other organisms. *Mutat Res.* 512: 67–84.
9. Buccala R, Model P, Cerami A (1984). A possible mechanism for nucleic acid aging and age-related dysfunction in gene expression. *Proc Natl Acad Sci USA* 81: 105–9.
10. Shires TK, Tresnak J, Kaminsky M, Herzog SL, Truc-Pham B (1990). DNA modification *in vivo* by derivatives of glucose: enhancement by glutathione depletion. *FASEB J.* 4: 3340–6.
11. Papoulis A, Al-Abed Y, Bucala R (1995). Identification of N^2-(1-Carboxyethyl)guanine (CEG) as a guanine advanced glycosylation end product. *Biochemistry* 34: 648–55.
12. Baynes JW (2001). The role of AGEs in aging: causation or correlation. *Exp Gerontol.* 36: 1527–37.
13. Baynes JW (2002). The Maillard hypothesis on aging: time to focus on DNA. *Ann NY Acad Sci.* 959: 360–7.
14. Pischetsrieder M, Seidel W, Münch G, Schinzel R (1999). N^2-(-1-Carboxyethyl)deoxyguanosine, a nonenzymatic glycation adduct of DNA, induces single-strand breaks and increases mutation frequencies. *Biochem Biophys Res Commun.* 264: 544–9.
15. Blair IA (2001). Lipid hydroperoxide-mediated DNA damage. *Exp Gerontol.* 36: 1473–81.
16. Brambilla G, Sciaba L, Faggin P, *et al.* (1986). Cytotoxicity, DNA fragmentation and sister-chromatid exchange in Chinese hamster ovary cells exposed to the lipid peroxidation product 4-hydroxynonenal and homologous aldehydes. *Mutat Res.* 171: 169–76.
17. Navasumrit P, Ward TH, O'Conner PJ, Nair J, Frank N, Bartsch H (2001). Ethanol enhances the formation of endogenously and exogenously derived adducts in rat hepatic DNA. *Mutat Res.* 479: 81–94.

18. Roberts MJ, Wondrak GT, Laurean DC, Jacobson MK, Jacobson EL (2002). DNA damage by carbonyl stress in human skin cells. *Mutat Res.* 420: 1–12.
19. Izzotti A, Cartiglia C, Taningher M, De Flora S, Balansky R (1999). Age-related increase of 8-hydroxy-2′-deoxyguanosine and DNA-protein crosslinks in mouse organs. *Mutat Res.* 446: 215–23.
20. Kuykendall JR, Bogdanffy SM (1992). Reaction kinetics of DNA-histone crosslinking by vinyl acetate and acetaldehyde. *Carcinogenesis* 13: 2095–100.
21. Kuykendall JR, Bogdanffy MS (1994). Formation and stability of acetaldehyde-induced crosslinks between poly-lysine and poly-deoxyguanosine. *Mutat Res.* 311: 49–56.
22. Dizdaroglu M, Jaruga P, Birincioglu M, Rodriguez H (2002). Free radical-induced damage to DNA: mechanisms and measurement. *Free Radic Biol Med.* 32: 1102–15.
23. Ullrich O, Reinheckel T, Sitte N, Hass R, Grune T, Davies KJ (1999). Poly-ADP ribose polymerase activates nuclear proteasomes to degrade oxidatively damaged histone. *Proc Natl Acad Sci USA* 96: 6223–8.
24. Rosenberger RF (1991). Senescence and the accumulation of abnormal proteins. *Mutat Res.* 256: 255–62.
25. Rattan SIS, Derventzi A, Clark BFC (1992). Protein synthesis, posttranslational modifications and aging. *Ann NY Acad Sci.* 663: 48–62.
26. Holliday (1995). *Understanding Ageing*. Cambridge, UK: Cambridge University Press.
27. Orgel LE (1963). The maintenance of the accuracy of protein biosynthesis and its relevance to aging. *Proc Natl Acad Sci USA* 49: 517–21.
28. Orgel LE (1970). The maintenance of the accuracy of protein biosynthesis and its relevance to aging: a correction. *Proc Natl Acad Sci USA* 67: 1476.
29. van Leeuwen FW, Gerez L, Benne R, Hol EM (2002). +1 Proteins and aging. *Int J Biochem Cell Biol.* 34: 1502–5.
30. van den Hurk WH, Willems HJ, Bloemen M, Martens GJ (2001). Novel frameshift mutations near short simple repeats. *J Biol Chem.* 276: 11496–8.
31. Evans DAP, van der Kleij AAM, Sonnemans MAF, Burbach JPH, van Leeuwen FW (1994). Frameshift mutations at two hotspots in vasopressin transcripts in postmitotic neurons. *Proc Natl Acad Sci USA* 91: 6059–63.
32. van Leeuwen FW, de Kleijn DPV, van den Hurk HH, *et al.* (1998). Frameshift mutations of β-amyloid precursor protein and ubiquitin B in Alzheimer's and Down patients. *Science* 279: 242–7.
33. van Leeuwen FW, Hol EM, Hermanussen RWH, *et al.* (2000). Molecular misreading in non-neuronal cells. *FASEB J.* 14:1595–602.
34. Lam YA, Pickart CM, Alabn A, *et al.* (2000). Inhibition of the ubiquitin proteasome in Alzheimer's disease. *Proc Natl Acad Sci USA* 97: 9902–6.
35. de Vrij FMS, Sluijs JA, Gregori L, *et al.* (2001). Mutant ubiquitin expressed in Alzheimer's disease causes neuronal death. *FASEB J.* 15: 2680–8.
36. Kirkwood TBL, Holliday R, Rosenberger RF (1984). The stability of the cellular translation process. *Int Rev Cytol.* 92: 93–132.
37. Harley CB, Pollard JW, Chamberlain JW, Stanners CP, Goldstein S (1980). Protein synthetic errors do not increase during the aging cultured human fibroblasts. *Proc Natl Acad Sci USA* 77: 1885–9.
38. Branscombe EW, Galas DJ (1975). Progressive decrease in protein synthesis accuracy induced by streptomycin in *Escherichia coli*. *Nature* 254: 161–3.
39. Rosenberger RF (1982). Streptomycin-induced protein error propagation appears to lead to cell death in *Escherichia coli*. *IRCS Med Sci.* 10: 874–5.

40. Carrier MJ, Kogut M, Hipkiss AR (1984). Changes in intracellular proteolysis in *Escherichia coli* during prolonged growth with a low concentration of dihydrostreptomycin. *FEMS Lett.* 22: 223–7.
41. Holbrook MA, Menninger JR (2002). Erythromycin slows aging of *Saccharomyces cerevisiae*. *J Gerontol.* 57: B29–36.
42. Dukan S, Farewell A, Ballesteros M, Taddei F, Radman M, Nystrom T (2000). Protein oxidation in response to increased transcriptional or translational errors. *Proc Natl Acad Sci USA* 97: 5746–9.
43. Bota DA, Van Remmen H, Davies KJA (2002). Modulation of Lon protease activity and aconitase turnover during aging and oxidative stress. *FEBS Lett.* 532: 103–6.
44. Fisk HA, Yaffe MP (1999). A role for ubiquitination in mitochondrial inheritance in *Saccharomyces cerevisiae*. *J Cell Biol.* 145: 1199–208.
45. Rapoport S, Dubiel W, Muller M (1985). Proteolysis of mitochondria in reticulocytes during maturation is ubiquitin-dependent and is accompanied by a high rate of ATP hydrolysis. *FEBS Lett.* 180: 249–52.
46. Sutovsky P, Moreno RD, Ramalho-Santos J, Dominko T, Simerly C, Schatten G (2000). Ubiquitinated sperm mitochondria, selective proteolysis and the regulation of mitochondrial inheritance in mammalian embryos. *Biol Reprod.* 63: 582–90.
47. Sutovsky P, Moreno R, Ramalho-Santos J, Dominko T, Winston WE, Schatten G (2001). A putative, ubiquitin-dependent mechanism for the recognition and elimination of defective spermatozoa in the mammalian epididymis. *J Cell Sci.* 114: 1665–75.
48. Brunk UT, Terman A (2002). The mitochondrial-lysosomal axis theory of aging – accumulation of damaged mitochondria as a result of imperfect autophagocytosis. *Eur J Biochem.* 269: 1996–2002.
49. Ritz-Timme S, Collins MJ (2002). Racemization of aspartic acid in human proteins. *Ageing Res Rev.* 1: 43–59.
50. Geiger T, Clarke S (1987). Deamidation, isomerization and racemization at asparaginyl and aspartyl residues in peptides. *J Biol Chem.* 262: 785–94.
51. Luthra M, Ranganathan D, Ranganathan S, Balasubramanian D (1994). Racemization of tyrosine in the insoluble protein fraction of brunescent aging human lenses. *J Biol Chem.* 269: 22678–82.
52. Kreil G (1994). Conversion of L-to D-amino acids: a posttranslational reaction. *Science* 266: 996–7.
53. Brunauer LS, Clarke S (1986). Age-dependent accumulation of protein residues which can be hydrolyzed to D-aspartic acid in human erythrocytes. *J Biol Chem.* 261: 12538–43.
54. Fujii N, Satoh K, Harada K, Ishibashi Y (1994). Simultaneous stereoinversion and isomerization at specific aspartic acid residues in alpha A-crystallin from human lens. *J Biochem (Tokyo)* 116: 663–9.
55. Lindner H, Helliger W (2001). Age-dependent deamidation of asparagine residues in proteins. *Exp Gerontol.* 36: 1551–63.
56. Robinson NE, Robinson AB (2001). Deamidation of human proteins. *Proc Natl Acad Sci USA* 98: 12409–13.
57. Hsu YR, Chang WC, Mendiaz ER, *et al.* (1998). Selective deamidation of recombinant human stem cell factor during *in vitro* aging: isolation and characterization of the aspartyl and isoaspartyl homodimers and heterodimers. *Biochemistry* 37: 2251–62.
58. Buchanan DL, Haley EE, Markiw RT (1962). Occurrence of β-aspartyl and γ-glutamyl oligopeptides in human urine. *Biochemistry* 4: 612–23.

59. Pisano JJ, Prado E, Freedman J (1966). β-Aspartylglycine in urine and enzymic hyrolyzates of proteins. *Arch Biochem Biophys.* 117: 394–9.
60. Stratton LP, Kelly RM, Rowe J, *et al.* (2001). Controlling deamidation rates in a model peptide: effects of temperature, peptide concentration and additives. *J Pharm Sci.* 90: 2141–8.
61. Athmer L, Kindrachuk J, Georges F, Napper S (2002). The influence of protein structure on the products emerging from succinimide hydrolysis. *J Biol Chem.* 277: 30502–7.
62. Lapko VN, Purkis AG, Smith DL, Smith JB (2002). Deamidation in human gamma S-crystallin from cataractous lenses is influenced by surface exposure. *Biochemistry* 41: 8638–48.
63. Robinson NE (2002). Protein deamidation. *Proc Natl Acad Sci USA* 99: 5283–8.
64. Takemoto L, Boyle D (1998). Deamidation of specific glutamine residues from alpha-A crystallin during aging of the human lens. *Biochemistry* 37: 13681–5.
65. Takemoto L (2001). Deamidation of Asn-143 of gamma S crystallin from protein aggregates of the human lens. *Curr Eye Res.* 22: 148–53.
66. Aswad DW, Paranandi MV, Schurter BT (2000). Isoaspartate in peptides and proteins: formation, significance and analysis. *J Pharm Biomed Anal.* 21: 1129–36.
67. Ghandour H, Lin B-F, Choi S-W, Mason JB, Selhub J (2002). Folate status and age affect the accumulation of L-isoaspartyl residues in rat liver proteins. *J Nutr.* 132: 1357–60.
68. Lindner H, Sarg B, Grunicke H, Helliger W (1999). Age-dependent deamidation of H1(0) histones in chromatin of mammalian tissues. *J Cancer Res Clin Oncol.* 125: 182–6.
69. Fujii N, Tajima S, Tanaka N, Fujimoto N, Takata T, Shimo-Oka T (2002). The presence of D-β-aspartic acid-containing peptides in elastic fibers of sun-damaged skin: a potent marker for ultraviolet-induced skin aging. *Biochem Biophys Res Commun.* 294: 1047–51.
70. Fonseca MI, Head E, Velazquez P, Cotman CW, Tenner AJ (1999). The presence of isoaspartic acid in β-amyloid plaques indicates plaque age. *Exp Neurol.* 157: 277–88.
71. Watanabe A, Takio K, Ihara Y (1999). Deamidation and isoaspartate formation in smeared tau in paired helical filaments. *J Biol Chem.* 274: 7368–78.
72. Sandmeier E, Hunziker P, Kunz B, Sack R, Christen P (1999). Spontaneous deamidation and isomerization of Asn108 in prion peptide 106-126 and in full length prion protein. *Biochem Biophys Res Commun.* 261: 578–83.
73. Nilsson MR, Driscoll M, Raleigh DP (2002). Low levels of asparagine deamidation can have a dramatic effect on aggregation of amyloidogenic peptides: implications for the study of amyloid formation. *Protein Sci.* 11: 342–9.
74. Ingrosso D, D'Angelo S, Di Carlo E (2000). Increased methyl esterification of altered aspartyl residues in erythrocyte membranes in response to oxidative stress. *Eur J Biochem.* 267: 4397–405.
75. Stadtman ER (1992). Protein oxidation and aging. *Science* 257: 1220–4.
76. Syntichaki P, Xu K, Driscoll M, Tavernarakis N (2002). Specific aspartyl and calpain proteases are required for neurodegeneration in *C. elegans*. *Nature* 419: 939–44.
77. Clarke S (1985). Protein carboxyl methyltransferase: two distinct classes of enzymes. *Annu Rev Biochem.* 54: 479–506.
78. Kim E, Lowenson JD, MacLaren DC, Clarke S, Young SG (1997). Deficiency of a protein-repair enzyme results in the accumulation of altered proteins, retardation of growth and fatal seizures in mice. *Proc Natl Acad Sci USA* 94: 6132–7.
79. Yamamoto A, Takagi H, Kitamura D, *et al.* (1998). Deficiency in protein L-isoaspartyl methyltransferase results in a fatal progressive epilepsy. *J Neurosci.* 18: 2063–74.

80. Lowenson JD, Kim E, Young SG, Clarke S (2001). Limited accumulation of damaged proteins in L-isoaspartyl (D-aspartyl) O-methyltransfearse-deficient mice. *J Biol Chem.* 276: 20695–702.
81. Young AL, Carter WG, Doyle HA, Mamula MJ, Aswad DW (2001). Structural integrity of histone H2B *in vivo* requires the activity of protein L-isoaspartate O-methyltransferase, a putative protein repair enzyme. *J Biol Chem.* 276: 37161–5.
82. Lanthier J, Bouthillier A, Lapointe M, Demeule M, Beliveau R, Desrosiers RR (2002). Down-regulation of protein L-isoaspartyl methyltransferase in human epileptic hippocampus contributes to generation of damaged tubulin. *J Neurochem.* 83: 581–91.
83. Lindquist JA, McFadden PN (1994). Automethylation of protein (D-aspartyl/L-isoaspartyl) carboxyl methyltransferase, a response to enzyme aging. *J Protein Chem.* 13: 23–30.
84. Chavous DA, Jackson FR, O'Connor CM (2001). Extension of the *Drosophila* lifespan by overexpression of a protein repair methyltransferase. *Proc Natl Acad Sci USA* 98: 14814–18.
85. Dyer RB, McMurray CT (2001). Mutant protein in Huntington disease is resistant to proteolysis in affected brain. *Nature Genet* 29: 270–8.
86. Perutz MF, Windle AH (2001). Cause of neural death in neurodegenerative diseases attributable to expansion of glutamine repeats. *Nature* 412: 143–4.
87. Morley, JF, Brignull H, Weyers JJ, Morimoto RJ (2002). The threshold for polyglutamine-expansion protein aggregation and cellular toxicity is dynamic and influenced by aging in *Caenorhabditis elegans. Proc Natl Acad Sci USA* 99: 10417–22.
88. Onodera O, Roses AD, Tsuji S, Vance JM, Strittmatter WJ, Burke JR (1996). Toxicity of expanded polyglutamine-domain proteins in *Escherichia coli. FEBS Lett.* 399: 135–9.
89. McFarland GA, Holliday R (1994). Retardation of the senescence of cultured human diploid fibroblasts by carnosine. *Exp Cell Res.* 212: 167–75.
90. McFarland GA, Holliday R (1999). Further evidence for the rejuvenating effects of the dipeptide L-carnosine on cultured human diploid fibroblasts. *Exp Gerontol.* 34: 35–45.
91. Kuroda M, Harada T (2000). Incorporation of histidine and β-alanine into the macromolecular fraction of beef soup stock solution. *J Food Sci.* 65: 596–603.
92. Kuroda M, Harada T (2002). Distribution of γ-glutamyl-β-alanylhistidine isopeptide in the macromolecular fractions of commerical meat extracts and correlation with the color of the macromolecular fractions. *J Agric Food Chem.* 50: 2088–93.
93. Kuroda M, Ohtake R, Suzuki E, Harada T (2000). Investigation on the formation and the determination of γ-glutamyl-β-alanylhistidine and related isopeptide in the macromolecular fraction of beef soup stock. *J Agric Food Chem.* 48: 6317–24.
94. Morgan PE, Dean RT, Davies MJ (2002). Inactivation of cellular enzymes by carbonyls and protein-bound glycation/glycoxidation products. *Arch Biochem Biophys.* 403: 259–69.
95. Baynes JW, Monnier VM (1989). *The Maillard Reaction in Aging, Diabetes and Nutrition.* New York: Alan R. Liss, Inc.
96. Ahmed, N, Argirov O, Minhas HS, Cordeiro CAA, Thornally PJ (2002). Assay of advanced glycation endproducts (AGEs): surveying AGEs by chromatographic assay with derivatization by 6-aminoquinolyl-N-hydroxysuccinimidyl-carbamate and application to N_ε-carboxymethyl-lysine-and N_ε-(1-carboxyethyl)lysine-modified albumin. *Biochem J.* 364: 1–14.
97. Brownlee, M (1995). Advanced protein glycation in diabetes and aging. *Ann Rev Med.* 46: 223–34.

98. Sell DR, Primc M, Schafer IA, Kovach M, Weiss MA, Monnier VM (1998). Cell-associated pentosidine as a marker of aging in human diploid cells *in vitro* and *in vivo*. *Mech Ageing Dev.* 105: 221–40.
99. Thornally PJ (1999). Clinical significance of glycation. *Clin Lab.* 45: 263–73.
100. Koschinsky T, He C-J, Mitsuhashi T, *et al.* (1997). Orally absorbed reactive glycation products (glycotoxins): an environmental risk factor in diabetic nephropathy. *Proc Natl Acad Sci USA* 94: 6474–9.
101. Yan SD, Chen X, Schmidt AM, *et al.* (1994). Glycated tau protein in Alzheimer disease: a mechanism for induction of oxidant stress. *Proc Natl Acad Sci USA* 91: 7787–91.
102. Shuvaev VV, Laffont I, Serot J-M, Fujii J, Taniguchi N, Siest G (2001). Increased protein glycation in cerebrospinal fluid of Alzheimer's disease. *Neurobiol Aging* 22: 397–402.
103. Bar KJ, Franke S, Wenda B, Muller S, Kientsch-Engel, Stein G, Sauer H (2002). Pentosidine and N_ε-(carboxymethyl)-lysine in Alzheimer's disease and vascular dementia. *Neurobiol Aging* 5749: 1–6.
104. Brownlee M (2001). Biochemistry and molecular cell biology of diabetic complications. *Nature* 414: 813–20.
105. Mathys KC, Ponnampalam SN, Padival S, Nagaraj RH (2002). Semicarbazide-sensitive amino oxidase in aortic smooth muscle cells mediates synthesis of methylglyoxal-AGE: implications for vascular complications in diabetes. *Biochem Biophys Res Commun.* 297: 863–9.
106. Chaplen FWR (1998). Incidence and potential implications of the toxic metabolite methylglyoxal in cell culture: a review. *Cytotechnology* 26: 173–83.
107. Amicarelli F, Colafarina S, Cesare P, *et al.* (2001). Morphofunctional mitochondrial response to methylglyoxal toxicity in *Bufo bufo* embryos. *Int J Biochem Cell Biol.* 33: 1129–39.
108. Nohara Y, Usui T, Kinoshita T, Watanabe M (2002). Generation of superoxide anions during the reaction of guanidino compounds with methylglyoxal. *Chem Pharm Bull.* 50: 179–84.
109. Matsunaga T, Iguchi K, Nakajima T, *et al.* (2001). Glycated high-density lipoprotein induces apoptosis in endothelial cells via a mitochondrial dysfunction. *Biochem Biophys Res Commun* 287: 714–20.
110. Wolff SP, Dean RT (1987). Glucose autoxidation and protein modification: a potential source of "autoxidative glycosylation" in diabetes mellitus. *Biochem J.* 245: 243–50.
111. Uchida K, Stadtman ER (1992). Modification of histidine residues in proteins by reaction with 4-hydroxynonenal. *Proc Natl Acad Sci USA* 89: 4544–8.
112. Uchida K, Kanematsu M, Sakai K, *et al.* (1998). Protein-bound acrolein: potential markers for oxidative stress. *Proc Natl Acad Sci USA* 95: 4882–7.
113. Calingasan NY, Uchida K, Gibson GE (1999). Protein-bound acrolein: a novel marker of oxidative stress in Alzheimer's disease. *J Neurochem.* 72: 751–6.
114. Lovell MA, Xie C, Markesbery WR (2001). Acrolein is increased in Alzheimer's disease brain and is toxic to primary hippocampal cultures. *Neurobiol Aging* 22: 187–94.
115. Baynes JW (2000). From life to death – the struggle between chemistry and biology during aging: the Maillard reaction as an amplifier of genomic damage. *Biogerontology* 1: 235–46.
116. Davies SS, Amarnath V, Montine K, *et al.* (2002). Effects of reactive γ-ketoaldehydes formed by the isoprostane pathway (isoketals) and cyclooxygenase pathway (levuglandins) on proteasome function. *FASEB J.* 16: 715–17.
117. Yim MB, Yim HS, Lee C, Kang SO, Chock PB (2001). Protein glycation: creation of catalytic sites for free radical generation. *Ann NY Acad Sci.* 928: 48–53.

118. Burkle A (2000). Poly(ADP-ribosyl)ation: a posttranslational protein modification linked with genome protection and mammalian longevity. *Biogerontology* 1: 41–6.
119. Burkle A, Beneke S, Brabeck C, *et al.* (2002). Poly(ADP-ribose) polymerase-1, DNA repair and mammalian longevity. *Exp Gerontol.* 37:1203–5.
120. Cervantes-Laurean D, Jacobson EL, Jacobson MK (1996). Glycation and glycoxidation of histones by ADP-ribose. *J Biol Chem.* 271: 10461–9.
121. Wondrak GT, Cervantes-Laurean D, Jacobson EL, Jacobson MK (2000). Histone carbonylation *in vivo* and *in vitro*. *Biochem J.* 351: 769–77.
122. Tanner KG, Landry J, Sternglanz R, Denu JM (2000). Silent information regulator 2 family of NAD-dependent histone/protein deacetylases generates a unique product, 1-O-acetyl-ADP-ribose. *Proc Natl Acad Sci USA* 97:14178–82.
123. Ullrich O, Sitte N, Sommerburg O, Sandig V, Davies KJA, Grune T (1999). Influence of DNA binding on the degradation of oxidized histones by the 20S proteasome. *Arch Biochem Biophys.* 362: 211–16.
124. Ullrich O, Grune T (2001). Proteasomal degradation of oxidatively damaged endogenous histones in K562 human leukemic cells. *Free Radic Biol Med.* 31: 887–93.
125. Cloos PAC, Jensen AL (2000). Age-related de-phosphorylation of proteins in dentin: a biological tool for assessment of protein age. *Biogerontology* 1: 341–56.
126. Rosenberger RS, Carr AJ, Hipkiss AR (1990). Regulation of breakdown of canavanyl proteins in *Escherichia coli* by growth conditions in Lon^+ and Lon-cells. *FEMS Lett.* 68: 19–25.
127. Hipkiss AR (2001). On the "struggle between chemistry and biology during aging" implications for DNA repair, apoptosis and proteolysis, and a novel route of intervention. *Biogerontology* 2: 173–8.
128. Findlay D, Bartlet B, Varshavsky A (1989). The tails of ubiquitin precursors are ribosomal proteins whose fusion facilitates ribosome biogenesis. *Nature* 338: 394–401.
129. Musch MW, Kaplan B, Chang EB (2001). Role of increased basal expressionof heat shock protein 72 in colonic epithelial c2BBE adenocarcinoma cells. *Cell Growth Differ.* 12: 419–26.
130. Pirkkala L, Nykanen P, Sistonen L (2001). Roles of heat shock transcription factors in regulation of heat shock response and beyond. *FASEB J.* 15:1118–31.
131. Powers SK, Locke M, Demirel HA (2001). Exercise, heat shock proteins, and myocardial protection from I-R injury. *Med Sci Sports Exerc.* 33: 386–92.
132. Vercoutter-Edouart AS, Czesszak X, Crepin M, *et al.* (2001). Proteomic detection of changes in protein synthesis induced by fibroblast growth factor-2 in MCF-7 human breast cancer cells. *Exp Cell Res.* 262: 59–68.
133. Wataba K, Saito T, Fukunaka K, Ashihara K, Nushimura M, Kudo R (2001). Overexpression of heat shock proteins in carcinogenic endometrium. *Int J Cancer* 91: 448–56.
134. Lithgow GJ, White TM, Melov S, Johnson TE (1995). Thermotolerance and extended life-span conferred by single gene mutations and induced by thermal stress. *Proc Natl Acad Sci USA* 92: 7540–4.
135. Tatar M, Khazaeli AA, Curtsinger JW (1997). Chaperoning extended life. *Nature* 390: 30.
136. Minois N (2000). Longevity and aging: beneficial effects of exposure to mild stress. *Biogerontology* 1: 15–29.
137. Fonager J, Beedholm R, Clarke BFC, Rattan SIS (2002). Mild stress-induced stimulation of heat-shock protein synthesis and improved functional ability of human fibroblasts undergoing aging *in vitro*. *Exp Gerontol.* 37:1223–8.

138. Verbeke P, Siboska GE Clarke BFC, Rattan SIS (2002). Kinetin inhibits protein oxidation and glycoxidation *in vitro*. *Biochem Biophys Res Commun.* 276: 1265–70.
139. Bota DA, Davies KJA (2001). Protein degradation in mitochondria: implications for oxidative stress, aging and disease: a novel etiological classification of mitochondrial proteolytic disorders. *Mitochondrion* 1: 33–49.
140. Bota DA, Davies KJA (2002). Lon protease preferentially degrades oxidized mitochondrial aconitase by an ATP-stimulated mechanism. *Nature Cell Biol.* 4: 674–80.
141. Suzuki CK, Suda K, Schatz G (1994). Requirement for the yeast gene LON in intramitochondrial proteolysis and maintenance of respiration. *Science* 264: 273–6.
142. Zhu Y, Wang M, Lin H, Huang C, Shi X, Luo J (2002). Epidermal growth factor upregulates the transcription of mouse lon homology ATP-dependent protease through extracellular signal-regulated protein kinase – and phosphatidylinositol-3-kinase-dependent pathways. *Exp Cell Res.* 280: 97–106.
143. Nardai G, Csermely P, Söti C (2002). Chaperone function and chaperone overload in the aged. A preliminary analysis. *Exp Gerontol.* 37: 1257–62.
144. Soti C, Csermely P (2000). Molecular chaperones and the aging process. *Biogerontology* 1: 225–33.
145. Cherian M, Abraham EC (1995). Decreased molecular chaperone property of α-crystallin due to posttranslational modifications. *Biochem Biophys Res Commun.* 208: 675–9.
146. Derham BK, Harding JJ (1999). α-Crystallin as molecular chaperone. *Prog Retinal Eye Res.* 18: 463–509.
147. Radak Z, Kaneko T, Tahara T, *et al.* (1999). The effect of exercise training on oxidative damage of lipids, proteins and DNA in rat skeletal muscle: evidence for beneficial outcomes. *Free Radic Biol Med.* 27: 69–74.
148. Boada J, Cuesta E, Foig T, *et al.*J (2002). Enhanced antioxidant defenses and resistance to TNF-α in a glycolysis-depleted lung epithelial cell line. *Free Radic Biol Med.* 33: 1409–18.
149. Facchini FS, Hua NW, Reaven GM, Stoohs RA (2000). Hyperinsulinemia: The missing link among oxidative stress and age-related diseases? *Free Radic Biol Med.* 29: 1302–6.
150. Hamel FG, Bennet RG, Harmon KS, Duckworth WC (1997). Insulin inhibition of proteasome activity in intact cells. *Biochem Biophys Res Commun.* 234: 671–4.
151. Hagopian K, Ramsey JJ, Weindruch R (2003). Caloric restriction increases gluconeogenic and transaminase enzyme activities in mouse liver. *Exp Gerontol.* 38: 267–78.
152. Lin S-J, Kaeberlein M, Andalis AA, *et al.* (2002). Calorie restriction extends *Saccharomyces cerevisiae* lifespan by increasing respiration. *Nature* 418: 344–8.
153. Houthoofd K, Braeckman BP, Lenaerts I, *et al.* (2002). No reduction of metablolic rate in food restricted *Caenorhabditis elegans*. *Exp Gerontol.* 38: 267–78.
154. Hipkiss AR, Michaelis J, Syrris P (1995). Non-enzymatic glycosylation of the dipeptide L-carnosine, a potential anti-protein-cross-linking agent. *FEBS Lett.* 371: 81–5.
155. Vinson JA, Howard III TB (1996). Inhibition of protein glycation and advanced glycation end products by ascorbic acid and other vitamins and nutrients. *Nutr Biochem.* 7: 659–63.
156. Hipkiss AR, Chana H (1998). Carnosine protects proteins against methylglyoxal-mediated modifications. *Biochem Biophys Res Commun.* 248: 28–32.
157. Hipkiss AR, Preston JE, Himsworth DTM, *et al.* (1998). Pluripotent protective effects of carnosine, a naturally-occurring dipeptide. *Ann NY Acad Sci.* 854: 37–53.
158. Hipkiss AR, Brownson C, Bertani MF, Ruiz E, Ferro A (2002). Reaction of carnosine with aged proteins: another protective process? *Ann NY Acad Sci.* 959: 285–94.

159. Hipkiss AR, Preston JE, Himsworth DTM, Worthington VC, Abbot NJ (1997). Protective effects of carnosine against malondialdehyde-induced toxicity towards cultured rat brain endothelial cells. *Neurosci Lett.* 238: 135–8.
160. Preston JE, Hipkiss AR, Himsworth DTJ, Romero IA, Abbott JN (1998). Toxic effects of β-amyloid (25-35) on immortalised rat brain endothelial cell: protection by carnosine, homocarnosine and β-alanine. *Neurosci Lett.* 242: 105–8.
161. Nagasawa T, Yonekura T, Nishizawa N, Kitts DD (2001) *In vitro* and *in vivo* inhibition of muscle lipid and protein oxidation by carnosine. *Mol Cell Biochem.* 225: 29–34.
162. Yen WJ, Chang LW, Lee CP, Duh PD (2002). Inhibition of lipid peroxidation and nonlipid oxidative damage by carnosine. *J Am Oil Chem Soc.* 79: 329–33.
163. Aldini G, Carini M, Beretta G. Bradamante S, Facino R M (2002). Carnosine is a quencher of 4-hydroxy-nonenal: through what mechanism of reaction? *Biochem Biophys Res Commun.* 298: 699–706.
164. Fontana M, Pinnen F, Lucente G, Pecci L (2002). Prevention of peroxynitrite-dependent damage by carnosine and related sulphonamido pseudodipeptides. *Cell Mol Life Sci.* 59: 546–51.
165. Decker EA, Ivanov V, Zhu B-Z, Frei B (2001). Inhibition of low-density lipoprotein oxidation by carnosine, histidine. *J Agric Food Chem.* 49: 511–16.
166. Price DL, Rhett PM, Thorpe SR, Baynes JW (2001). Chelating activity of advanced glycation end-product inhibitors. *J Biol Chem.* 276: 48967–72.
167. Quinn PJ, Boldyrev AA, Formazuyk VE (1992). Carnosine: Its properties, functions and potential therapeutic applications. *Mol Aspects Med.* 13: 379–444.
168. Münch G, Mayer S, Michaelis J, *et al.* (1997). Influence of advanced glycation end-products and AGE-inhibitors on nucleation-dependent polymerization of β-amyloid peptide. *Biochim Biophys Acta* 1360: 17–29.
169. Kim KS, Choi SY, Kwon HY, Won MH, Kang T-C, Kang JH (2002). The ceruloplasmin and hydrogen peroxide system induces α-synuclein aggregation *in vitro*. *Biochimie* 84: 625–31.
170. Gille JJP, Pasman P, van Berkel CGM, Joenje H (1991). Effect of antioxidants on hyperoxia-induced chromosomal breakage in Chinese hamster ovary cell: protection by carnosine. *Mutagenesis* 6: 313–18.
171. Kantha SS, Wada S, Tanaka H, Takeuchi M, Watabe S, Ochi H (1996). Carnosine sustains the retention of cell morphology in continuous fibroblast culture subjected to nutritional insult. *Biochem Biophys Res Commun.* 223: 278–82.
172. Yuneva MO, Bulygina ER, Gallant SC, *et al.* (1999). Effect of carnosine on age-induced changes in senescence-acccelerated mice. *J Anti-Aging Med.* 2: 337–42.
173. Hipkiss AR (1998). Carnosine, a protective, anti-ageing peptide? *Int J Biochem Cell Biol.* 30: 863–8.
174. Ikeda D, Wada S, Yoneda C (1999). Carnosine stimulates vimentin expression in cultured rat fibroblasts. *Cell Struct Function* 24: 79–87.
175. Brownson C, Hipkiss AR (2000). Carnosine reacts with a glycated protein. *Free Radic Biol Med.* 28: 1564–70.
176. Hipkiss AR, Michaelis, J, Syrris, P, Dreimanis M (1995). Strategies for the extension of human lifespan. *Perspect Hum Biol.* 1: 59–70.
177. Bharadwaj LA, Davies GF, Xavier IJ, Ovsenek N (2002). L-carnosine and verapamil inhibit hypoxia-induced expression of hypoxia inducible factor (HIF-1 alpha) in H9c2 cardiomyoblasts. *Pharmacol Res.* 45: 175–81.

178. Yamano T, Niijima A, Limori S, Tsuruoka N, Kiso Y, Nagai K (2001). Effect of L-carnosine on the hyperglycemia caused by intracranial injection of 2-deoxy-D-glucose in rats. *Neurosci Lett*. 313: 78–82.
179. Babizhayev MA, Deyev AI, Yermakova VN, *et al.* (2001). N-Acetylcarnosine, a natural histidine-containing dipeptide, as a potent ophthalmic drug in treatment of human cataracts. *Peptides* 22: 979–94.
180. Onorato JM, Jenkins AJ, Thorpe SR, Baynes JW (2000). Pyridoxamine, an inhibitor of advanced glycation reactions, also inhibits advanced lipoxidation reactions – mechnism of action of pyridoxamine. *J Biol Chem*. 275: 21177–84.
181. Nagaraj RH, Sarkar P, Mally A, Biemel KM, Lederer MO, Padayatti PS (2002). Effect of pyridoxamine on chemical modification of proteins by carbonyls in diabetic rats: characterization of a major product from the reaction of pyridoxamine and methylglyoxal. *Arch Biochem Biophys*. 402: 110–19.
182. Burcham PC, Kaminskas LM, Fontaine FR, Petersen DR, Pyke SM (2002). Aldehyde-sequestering drugs: tools for studying protein damage by lipid peroxidation products. *Toxicology* 181–182: 229–36.
183. Khalifah RG, Baynes JW, Hudson BG (1999). Amadorins: novel post-Amadori inhibitors of advanced glycation reactions. *Biochem Biophys Res Commun*. 257: 251–8.
184. Olsen A, Siboska GE, Clarke BFC, Rattan SIS (1999). N^6-furfuryladenine, kinetin, protects against Fenton reaction-mediated oxidative damage to DNA. *Biochem Biophys Res Commun*. 265: 499–502.
185. Rattan SIS, Clarke BFC (1994). Kinetin delays the onset of aging characteristics in human fibroblasts. *Biochem Biophys Res Commun*. 201: 665–72.
186. Wondrak GT, Cervantes-Laurean D, Roberts MJ, *et al.* (2002). Identification of a-dicarbonyl scavengers for cellular protection against carbonyl stress. *Biochem Pharmacol*. 63: 361–73.
187. Brownlee M (1989). Pharmacological modulation of the advanced glycosylation reaction. In: JW Baynes, Monnier VM, eds. *The Maillard Reaction in Aging, Diabetes and Nutrition*. New York: Alan R. Liss, Inc, pp. 235–48.
188. Atamna H, Paler-Martinez A, Ames BN (2000). N-t-butyl hydroxylamine, a hydrolysis product of α-phenyl-N-t-butyl nitrone, is more potent in delaying senescence in human lung fibroblasts. *J Biol Chem*. 275: 6741–8.
189. Hipkiss AR (2001). On the anti-aging activities of aminoguanidine and N-t-butylhydroxylamine. *Mech Ageing Dev*. 122: 169–71.
190. Seidler NW, Yeargans GS (2002). Effects of thermal denaturation on protein glycation. *Life Sci*. 70: 1789–99.
191. Friguet B, Szweda LI (1997). Inhibition of the multicatalytic proteinase (proteasome) by 4-hydroxynonenal cross-linked protein. *FEBS Lett*. 405: 21–5.
192. Bulteau A-L, Verbeke P, Petropoulos I, Chaffotte A-F, Friguet B (2001). Proteasome inhibition in glyoxal-treated fibroblasts and resistance of glycated glucose-6-phosphate dehydrogenase to 20S proteasome degradation *in vitro*. *J Biol Chem*. 276: 45662–8.
193. Takizawa N, Takada K, Ohkawa K (1993). Inhibitory effect of nonenzymatic glycation on ubiquitination and ubiquitin-mediated degradation of lysosome. *Biochem Biophys Res Commun*. 192: 700–6.
194. Luxford C, Dean RT, Davies MJ (2002). Induction of DNA damage by oxidised amino acids and proteins. *Biogerontology* 3: 95–102.

Transcriptional and Translational Dysregulation During Aging

Suresh I.S. Rattan
Laboratory of Cellular Aging, Danish Centre for Molecular Gerontology, Department of Molecular Biology, University of Aarhus, DK-8000 Aarhus – C, Denmark

Introduction

Each cell type in an organism comes to acquire a unique pattern of gene expression through differentiation during development. It is obvious that this pattern of gene expression must be maintained for the normal functioning of cells and for the survival of the organism. Since aging is considered to be a result of failure of maintenance at all levels, attempts have been made to look for an age-related drifting-away of cells in terms of changes in the pattern of gene expression during aging. In most cases, mRNA levels of different genes have been estimated by RNA-DNA hybridization, using cDNA or genomic probes for specific genes. More recently, the availability of gene array technology has made it possible to compare the expression of thousands of genes. The results obtained show that during aging the expression of some genes increases, of some it decreases and of others it remains constant [1–7]. In all such studies on measuring the levels of mRNA in young and old cells and tissues, it is assumed that this estimate is a direct measure of gene activity. This is a simplistic notion, because it is well known that post-transcriptional changes, such as the processing, transport and turnover of RNA, change significantly the levels of mRNA.

A better way to find out if the stability of the genome decreases and if the pattern of gene expression changes during aging, is to study the expression of a cell-type-specific gene. For example, the globin gene is normally repressed in fibroblasts, and no globin-like RNA could be detected either in young or in old human fibroblasts showing thereby that the stability of globin gene expression remains unchanged during aging [8]. Similar results have been reported for five tissue-specific genes (myelin basic protein in brain, atrial natriuretic factor in heart, albumin in liver, kappa immunoglobulin in spleen and a skin-specific keratin), the fidelity of whose

T. von Zglinicki (ed.), Aging at the Molecular Level, 179–191.

expression was maintained during aging of Wistar rats [9]. In contrast to this, some relaxation of the expression of endogenous murine leukemia virus-related RNA and globin RNA was found in mouse brain and liver [10]. However, no reactivation of repressed α-fetoprotein genes was observed in adult rat livers during aging [11]. More studies of a similar nature and with a wide range of cell-type-specific genes will be required in order to resolve this important issue of the stability of gene expression during aging.

Transcriptional alterations

Studies on the synthesis of RNA and on its processing during aging have been few. There are three types of RNA in a cell, of which about 70–80% is rRNA, 10–15% is tRNA and 5–7% is mRNA. The level of total transcription is generally reduced by 15–30% during aging, but the proportion of different RNAs does not change significantly. Although the endogenous nucleotide pool and the activities of the enzymes involved in RNA synthesis are also reduced during aging, it does not appear to affect the transcription of all RNAs equally [12]. For each type of RNA some age-related changes have been observed in various aging systems. For example, a significant decline in the content and synthesis of rRNAs has been reported in aging beagles, rodent organs and cultured cells [13]. This decline in rRNAs was previously thought to be associated with the loss of ribosomal genes. However, no age-related decline in the gene copy number of rRNA has been observed in human fibroblasts or in mouse myocytes [14]. Furthermore, the number of rRNA genes in a cell is already in great excess (between 200 and 1000 copies), and a small loss with age may not have any serious consequences for cell function and survival. However, whether there is a differential loss of various rRNA species during aging, and what effects such a loss might have, is not known.

tRNA and aminoacylation

Other major RNAs and their partner components of protein synthetic machinery are tRNAs and aminoacyl-tRNA synthetases (aaRS). There is at least one tRNA for each codon which is translatable into an amino acid, but there is no tRNA for the stop codons. For several amino acids, for example glycine, alanine, valine, leucine, serine and arginine, there are 4 to 6 isoacceptor tRNA species. The function of a tRNA in transferring the amino acid to the ribosome-mRNA complex is dependent upon the enzymes aaRS. Levels of tRNAs and aaRS have been considered to be rate limiting for protein synthesis. According to one of the earlier molecular theories of aging, the codon restriction theory, a random loss of various isoaccepting tRNAs will progressively restrict the readability of codons resulting in the inefficiency and inaccuracy of protein synthesis [15]. There is some evidence that a shift in the pattern of isoaccepting tRNAs occurs during development and aging in some plants, nematodes, insects and rat liver and skeletal muscle [16, 17]. Similarly, a 30- to 60-fold increase in the amount of UAG suppressor tRNA has been reported in the brain, spleen and liver of old mice, and has been related to increased expression of Moloney murine leukemia virus in fibroblasts [18]. Other characteristics of tRNAs that have been studied during aging include the rate of synthesis, total levels, aminoacylation capacity and nucleoside composition. There is no generalized pattern that emerges from these studies, and the reported changes vary significantly among different species.

The aminoacylation capacity of different tRNAs varies to different extents during aging, but the reasons for such variability are not known. However, the fidelity of aminoacylation did not differ significantly in cell-free extracts prepared from young and old rat livers [19]. In the case of aaRS, an increase or decrease in the specific activities of almost all of them has been reported in various organs of aging mice without any apparent correlation with tissue/cell type and its protein synthetic activity. A significant decline in the specific activities of 17 aaRs has been reported in the liver, lung, heart, spleen, kidney, small intestine and skeletal muscle of aging female mice, and during development and aging of nematodes [20]. An increase in the proportions of the heat-labile fraction of several of these enzymes has been reported in the liver, kidney and brain of old rats [21]. However, no universal pattern can be seen for the changes in the activities of various synthetases in different organs and in different animals. Although an age-related decrease in the efficiency of aaRS can be crucial in determining the rate and accuracy of protein synthesis, direct evidence in this respect is lacking at present.

mRNA processing and stability

Another step in the transfer of genetic information that can be rate-limiting is the availability of mRNAs for translation. Post-transcriptional processing of eukaryotic mRNA is highly complex and is still not very well understood, particularly with respect to aging. However, a few reports are available on changes in various aspects of mRNA-processing during aging. For example, a decrease in the total poly(A^+) mRNA has been reported in aging rat brain and liver, in rabbit liver, in the liver, heart and oviduct of quails [22], and in the post-mortem brain tissues obtained from patients with Alzheimer's disease [23]. This is thought to be due to both a decrease in the length of the poly(A) tail and the rate of polyadenylation of mRNA during aging. However, an application of more sensitive methods for measuring the age-related changes in the length of poly(A) tail have shown no significant differences in rat hepatocytes, and in the liver, kidney and brain samples obtained from calorie-restricted and freely-fed rats of different ages [24]. Similarly, no age-related differences in mRNA cap structure and its translatability *in vitro* have been observed. Thus although it appears that at a gross level there are no major alterations in mRNA characteristics, it is possible that individual mRNA species do undergo changes, including splicing, transport from nucleus to cytoplasm, binding to ribosomes, stability and turnover during aging [12, 25]. For example, there is some evidence that altered splicing of fibronectin mRNA during cellular aging may be related to alterations in trans-acting factors that bind to cAMP-responsive elements (CRE) at the 5′ end of the fibronectin gene and regulate its expression [25]

Translational alterations

Although the genomic instructions of life are written in the language of nucleic acids, the life is actually "lived" in the language of proteins. The genetic information encoded in DNA becomes functionally meaningful only when it is accurately transcribed and translated into RNA and proteins, respectively. Whereas two types of RNA, transfer RNA and ribosomal RNA, are themselves functional molecules, the genetic information transcribed into mRNA has to be generally translated from a language of nucleic acids into a language of amino acids in order to produce proteins which are the functional gene products. Previously it was estimated that in a human cell there were about 80 000 genes per haploid genome, of which about 22 000 were housekeeping genes and the rest were tissue-specific [26]. However, since the near-completion of the human genome sequencing, the number of human genes is estimated to be around 30 000 only, but which may give rise to more than 100 000 functional gene products [27]. Furthermore, in order to become a functional protein, a newly synthesized polypeptide chain has to undergo a wide variety of post-translational modifications that determine its activity, stability, specificity and transportability (see Chapter 11, this volume). Misregulation of genetic information transfer at any of these steps can be critical for the failure of homeostasis.

Proteins are the most versatile macromolecules necessary for the organization of internal cellular structures, for the formation of the energy-creating and metabolic utilizing systems in the cell, for the transport of ions and larger molecules over the cell membranes and for maintaining intra- and intercellular communication pathways. Proteins interact with all other macromolecules including DNA, RNA, carbohydrates and lipids, and are required for maintenance and repair at all levels of biological organization. Protein synthesis is thus crucial for the survival of a living system, and any disturbance at this level can cause large imbalances and deficiencies.

A decline in the rate of total protein synthesis, without making any distinction between the cytoplasmic and mitochondrial protein synthesis, is one of the most common age-associated biochemical changes that has been observed in a wide variety of cells, tissues, organs and organisms, including human beings [28, 29]. Although there is a considerable variability among different tissues and cell types in the extent of decline (varying from 20% to 80%), the fact remains that the bulk protein synthesis slows down during aging. Furthermore, it has been shown that the conditions, such as calorie-restriction, that increase the lifespan and retard the aging process in many organisms, also slow down the age-related decline in protein synthesis [30]. These observations reinforce the view that slowing down of protein synthesis is an integral part of the aging process. The implications and consequences of slower rates of protein synthesis are manifold in the context of aging and age-related pathology.

It should be however pointed out that age-related slowing down of bulk protein synthesis does not mean that the synthesis of each and every protein becomes slower uniformly during aging. A significant increase in the heterogeneity of protein synthesis during aging has been observed. Using two-dimensional gel-electrophoresis (2D-gels) increased or decreased levels of various proteins have been reported, but the reasons for those changes are not always clear [29]. It appears that changes in the protein universe or proteomics during aging are mainly of a quantitative nature and any apparent qualitative change in terms of "new and aging-specific proteins" is mainly due to post-translational modifications, including phosphorylation, oxidation, glycoxidation, methylation and others.

Age-related changes in protein synthesis are regulated both at the transcriptional and pre-translational levels in terms of the availability of individual mRNA species for translation, and at the translational and post-translational levels in terms of alterations in the components of the protein synthetic machinery and the pattern of post-synthetic modifications that determine the activity, specificity and stability of a protein.

Efficiency and accuracy of protein synthesis

Eukaryotic protein synthesis is a highly complex process which requires about 200 small and large components to function effectively and accurately in order to translate one mRNA molecule while using large quantities of cellular energy. There are three major components of the translational apparatus: (1) the translational particle, the ribosome; (2) the amino acid transfer system or charging system; and (3) the translational factors. The protein-synthesizing apparatus is highly organized and its macromolecular components are not freely diffusible within cells.

The rate and accuracy of protein synthesis (as also of DNA and RNA synthesis) have been presumably gone through natural selection and evolved to optimal levels according to the overall life history of an organism. Since, the error frequency of amino acid misincorporation is generally considered to be quite high (10^{-3} to 10^{-4}) as compared with nucleotide misincorporation, the role of protein error feedback in aging has been a widely discussed issue. At present, no direct estimates of protein error levels in any aging system have been made primarily due to the lack of appropriate methods to determine spontaneous levels of errors in a normal situation.

However, several indirect attempts have been made to determine the accuracy of translation in cell-free extracts, using synthetic templates or natural mRNAs. Studies on the accuracy of protein synthesis during aging that have been performed on animal tissues, such as chick brain, mouse liver, and rat brain, liver and kidney, did not reveal any major age-related differences in the capacity and accuracy of ribosomes to translate poly(U) in cell-free extracts [29]. However, these attempts to estimate the error frequencies during translation *in vitro* of poly(U) template were inconclusive because the error frequencies encountered in the assays were several times greater than the estimates of natural error frequencies [31, 32].

Another indirect method that has been used to detect misincorporation of amino acids during aging is by 2D-gel electrophoresis of proteins, by which at least one kind of error, that is the misincorporation of a charged amino acid for an uncharged one (or vice versa) can be demonstrated because of "stuttering" of the protein spot on 2D-gels. Using this method, no age-related increase in amino acid misincorporation affecting the net charge of proteins was observed in histidine-starved human fibroblasts and in nematodes [29, 33]. In contrast to this, using mRNA of CcTMV coat protein for translation by cell extracts prepared from young and old human fibroblasts, a seven-fold increase in cysteine misincorporation during cellular aging has been observed [34, 35]. These studies also showed that an aminoglycoside antibiotic paromomycin (Pm), which is known to reduce ribosomal accuracy during translation *in vivo* and *in vitro* induces more errors in the translation of CcTMV coat protein mRNA by cell extracts prepared from senescent human fibroblasts than those from young cells. Further indirect evidence that indicates the role of protein errors in cellular aging can be drawn from studies on the increase in the sensitivity of human fibroblasts to the life-shortening and aging-inducing effects of Pm and another aminoglycoside antibiotic G418 [36, 37]. Similarly, increased longevity of high-fidelity mutants in *Podospora anserina* indicate the role of protein errors in lifespan [38, 39]. It has also been reported that an increase in the accuracy of protein synthesis in the yeast treated with small doses of erythromycin increases its lifespan [40].

Although a global "error catastrophe" as a cause of aging [41–43], is now considered to be unlikely, it is not ruled out that some kind of errors in various components of protein synthetic machinery including tRNA charging may have long-term effects on cellular stability and survival [44, 45]. Better methods are still required for measuring the basal levels of translational errors in young and old cells, tissues and organisms [46, 47] (see Chapter 11, this volume).

Initiation of protein synthesis

The translational process can be envisaged as proceeding in three steps – initiation, elongation and termination, followed by post-translational modifications, including folding, which give the protein a functional tertiary structure. The translation of an mRNA begins with the formation of a so-called initiation complex between the ribosome and the initiator codon. It is an intricate process, which consumes energy and involves at least seven initiation factors (eIFs) consisting of 24 different subunits, two subunits of ribosomes, and an initiating tRNA called methionyl (Met)-$tRNA_i$. The whole process of the formation of the 80S initiation complex takes about 2 to 3 seconds in cell-free assays and is thought to occur much faster *in vivo* [48, 49]. The initiation step is considered to be the main target for the regulation of protein synthesis during cell cycle, growth, development, hormonal response and under stress conditions including heat shock, irradiation and starvation [50].

With respect to aging, however, the rate of initiation appears to remain unaltered. For example, using *in vitro* assays, the conversion of isolated 40S and 60S ribosomal subunits into the 80S initiation complex has been reported to decrease by less than 15% in old *Drosophila*, rat liver and kidney, and mouse liver and kidney [51–54]. On the other hand, since polysomal fraction of the ribosomes decreases during aging, it implies that the activity of an anti-ribosomal-association factor eIF-3 may increase during aging. The activity of eIF-2, which is required for the formation of the ternary complex of Met-$tRNA_i$, GTP and eIF-2, has been reported to decrease in rat tissues during development and aging [51, 52]. Attenuation of hypusine formation eIF-5A during senescence of human diploid fibroblasts has been reported [55]. Similar studies on other eIFs and in other aging systems are yet to be performed and it is necessary that detailed studies on eIFs are also undertaken in the context of aging and the question of the regulation of protein synthesis at the level of initiation is reinvestigated.

Several studies have been performed on the age-related changes in the number of ribosomes, thermal stability, binding to aminoacyl-tRNA, the level of ribosomal proteins and rRNAs, sensitivity to aminoglycoside antibiotics, and the fidelity of ribosomes [16]. Although there is a slight decrease in the number of ribosomes in old animals, this does not appear to be a rate-limiting factor for protein synthesis due to a ribosomal abundance in the cell. Instead, several studies indicate that the biochemical and biophysical changes in ribosomal characteristics may be more important for translational regulation during aging. For example, the ability of aged ribosomes to translate synthetic poly(U) or natural globin mRNA decreases significantly [56]. A decrease in the translational capacity of ribosomes has also been observed in rodent tissues such as muscle, brain, liver, lens, testis and parotid gland and in various organs of *Drosophila* [17, 57].

The reasons for the functional changes observed in aging ribosomes are not known at present. Some attempts have been made to study the effect of aging on rRNAs and ribosomal proteins. Although a three-fold increase in the content of rRNA has been reported in late passage senescent human fibroblasts [58], it is not clear if the quantity and quality of individual rRNA species undergo alterations during aging, and what effect such a change might have on the functioning of ribosomes. Similarly, although an increase in the levels of mRNA for ribosomal protein L7 has been reported in aged human fibroblasts, and rat preadipocytes [59], there are no differences in the electrophoretic patterns of the ribosomal proteins in young and old *Drosophila* and mouse liver [16].

Elongation and termination of protein synthesis

The formation of the 80S initiation complex is followed by the repetitive cyclic event of peptide chain elongation, which is a series of reactions catalysed by elongation factors (eEFs). Various estimates of the elongation rates in eukaryotic cells give a value in the range of 3 to 6 amino acids incorporated per ribosome per second, which is several times slower than the prokaryotic elongation rate of 15 to 18 amino acids incorporated per second [48]. With regard to aging, a slowing-down of the elongation phase of protein synthesis has been suggested to be crucial in bringing about the age-related decline in total protein synthesis. A decline of up to 80% in the rate of protein elongation has been reported by estimating the rate of phenylalanyl-tRNA binding to ribosomes in poly(U)-translating cell-free extracts from old *Drosophila*, nematodes and rodent organs [53, 60–63]. *In vivo*, a two-fold decrease in the rate of polypeptide chain elongation in old WAG albino rat liver and brain cortex has been reported. Similarly, a decline of 31% in the rate of protein elongation in the livers of male Sprague-Dawley rats has been reported by measuring the rate of polypeptide chain assembly, which was 5.7 amino acids per second in young animals and was 4.5 amino acids per second in 2 year old animals [64]. However, these estimates of protein elongation rates have been made for "average" size proteins. It will be important to see if there is differential regulation of protein elongation rates for different proteins during aging.

The elongation of polypeptide chain is mediated by 2 elongation factors, eEF-1 and eEF-2 in eukaryotes (a third factor, eEF-3, is reported only in yeast), which are highly conserved during evolution [65]. There is a high abundance (between 3 and 10% of the soluble protein), of eEF-1A, previously abbreviated as EF-1alpha. There are multiple copies or isoforms of the eEF-1 gene that undergo cell type and/or developmental stage-specific expression reported in yeast, fungi, brine-shrimp, *Drosophila*, toad and mammalian cells and tissues [66]. eEF-1A has several other functions in addition to its requirement in protein synthesis. For example, it has been reported to bind to cytoskeletal elements; it is associated with endoplasmic reticulum; it is part of the valyl-tRNA-synthetase complex; it is associated with mitotic apparatus; it is involved in maintaining the accuracy of protein synthesis and protein degradation; it binds calmodulin in protozoan parasites; it induces rapid fragmentation of cytoplasmic microtubule arrays in fibroblasts; and its overexpression increases the susceptibility of mammalian cells to transformation [67–69].

The activity of eEF-1 declines with age in rat livers and *Drosophila*, and the drop parallels the decrease in protein synthesis [16, 17]. This decline in the activity of eEF-1 has been correlated only to eEF-1A. Using more specific cell-free stoichiometric and catalytic assays, a 35–45% decrease in the activity and amounts of active eEF-1A has been reported for serially passaged senescent human fibroblasts [70], old mouse and rat livers and brains [71, 72]. The germ line insertion of an extra copy of eEF-1A gene under the regulation of a heat shock promoter resulted in a better survival of transgenic *Drosophila* at high temperature [73]. However, this relative increase in the lifespan of transgenic insects at high temperature was not accompanied by any increase in the levels of mRNA, amount and activity of eEF-1A [74]. Similarly, no increased expression of eEF-1A genes was observed in *Drosophila* with extended longevity phenotype in a long-lived strain [75]. However, the increased longevity of eEF-1A high-fidelity mutants of a fungus *Podospora anserina* suggest that the life prolonging effects of eEF-1A may be due to its role in maintaining the fidelity of protein synthesis [38].

In the case of eEF-2 that catalyses the translocation of peptidyl-tRNA on the ribosome during the elongation cycle, conflicting data are available regarding the changes during aging. For example, a lack of difference in the rate of translocation has been observed during the translation of poly(U) by cell-free extracts prepared from young and old *Drosophila* and from rodent organs [53, 76]. Similarly, although the proportion of heat-labile eEF-2 increases during aging, the specific activity of eEF-2 purified from old rat and mouse liver remains unchanged [77]. In contrast, a decline of more than 60% in the amount of active eEF-2 has been reported during aging of human fibroblasts in culture, measured by determining the content of diphtheria toxin-mediated ADP-ribosylatable eEF-2 in cell lysates [78]. However, using the same assay, no age-related change in the amount of ADP-ribosylatable eEF-2 was detected in rat livers [72]. But, increased fragmentation of eEF-2 due to oxidation has also been reported in old rat livers [79]. A change towards acidic subtypes and increased phosphorylation of eEF-2 variants have been reported in the aging rat heart [80].

The cycle of peptide chain elongation continues until one of the three stop codons (UAA, UAG, UGA) is reached. There is no aa-tRNA complementary to these codons, and instead a termination factor or a release factor (RF) binds to the ribosome and induces the hydrolysis of both the aminoacyl linkage and the GTP releasing the completed polypeptide chain from the ribosome. Studies on aging *Drosophila* and old rat livers and kidneys have shown that the release of ribosome bound N-formylmethionine, a measure of the rate of termination, was not affected with age [53]. Direct estimates of the activity of the termination factor during aging have not been yet made.

Conclusion

Although a decline in bulk transcription and translation during aging is a widely observed change, it is not reflected at the level of individual RNA species and proteins. Whereas the levels of several mRNAs and proteins decrease with age, those of an equally large number either increase or remain unaltered during aging in a cell-, tissue-, organ- and species-specific manner. This indicates that there is no global dysregulation of transcription and translation which can affect the transfer of genetic information from genes to gene products equally and universally.

Cataloguing gene expression profiles during aging, mostly at the level of mRNAs and in some cases at the level of proteins, have not led to any universal patterns except that generally a decrease in the expression of genes involved in the basic metabolic processes and an increase in the expression of stress response pathway genes have been observed [1, 3, 4]. Even such changes may be variable depending on the organism and the cell-type being studied [2]. Thus, the data collected so far indicate that any age-related alterations in the regulation of transcription and translation of a gene may have a whole range of specific regulatory steps, without any global impairments in the transcriptional and translational machinery of the cells.

References

1. Goyns MH, Charlton MA, Dunford JE, *et al.* (1998) Differential display analysis of gene expression indicates that age-related changes are restricted to a small cohort of genes. *Mech Ageing Dev.* 101: 73–90.
2. Shelton DN, Chang E, Whittier PS, Choi D, Funk WD (1999). Microarray analysis of replicative senescence. *Curr Biol.* 9: 939–45.
3. Lee CK, Klopp RG, Weindruch R, Prolla TA (1999). Gene expression profile of aging and its retardation by caloric restriction. *Science* 285: 1390–3.
4. Ly DH, Lockhart DJ, Lerner RA, Schultz PG (2000). Mitotic misregulation and human aging. *Science* 287: 2486–92.
5. Jiang CH, Tsien JZ, Schultz PG, Hu Y (2001). The effects of aging on gene expression in the hypothalamus and cortex of mice. *Proc Natl Acad Sci USA* 98: 1930–4.
6. Miller RA, Galecki A, Shmookler-Reis RJ (2001). Interpretation, design, and analysis of gene array expression experiments. *J Gerontol Biol Sci.* 56A: B52–7.
7. Welle S (2002). Gene transcript profiling in aging research. *Exp Gerontol.* 37: 583–90.

8. Kator K, Cristofalo V, Charpentier R, Cutler RG (1985). Dysdifferentiative nature of aging: passage number dependency of globin gene expression in normal human diploid cells grown in tissue culture. *Gerontology* 31: 355–61.
9. Sato AI, Schneider EL, Danner DB (1990). Aberrant gene expression and aging: examination of tissue-specific mRNAs in young and old rats. *Mech Ageing Dev.* 54: 1–12.
10. Ono T, Shinya K, Uehara Y, Okada S (1989). Endogenous virus genomes become hypomethylated tissue-specifically during aging process of C57BL mice. *Mech Ageing Dev.* 50: 27–36.
11. Richardson A, Rutherford MS, Birchenall-Sparks MC, Roberts MS, Wu WT, Cheung HT (1985). Levels of specific messenger RNA species as a function of age. In: Sohal RS, Birnbaum LS, Cutler RG, eds. *Molecular Biology of Aging: Gene Stability and Gene Expression*. New York: Raven Press, pp. 228–41.
12. Müller WEG, Agutter PS, Schröder, HC (1995). Transport of mRNA into the cytoplasm. In: Macieira-Coelho A, ed. *Molecular Basis of Aging*. Boca Raton: CRC Press, pp. 353–88.
13. Medvedev ZA (1986). Age-related changes of transcription and RNA processing. In: Platt D, ed. *Drugs and Aging*. Berlin: Springer-Verlag, pp. 1–19.
14. Be Miller PM, Baker UA, Schmit JC (1985). Cellular aging and ribosomal RNA. In: Sohal RS, Birnbaum LS, Cutler RG, eds. *Molecular Biology of Aging: Gene Stability and Gene Expression*. New York: Raven Press, pp. 223–8.
15. Strehler BL, Hirsch G, Gusseck D, Johnson R, Bick M (1971). Codon restriction theory of ageing and development. *J Theor Biol.* 33: 429–74.
16. Rattan SIS (1995). Translation and post-translational modifications during aging. In: Macieira-Coelho A, ed. *Molecular Basis of Aging*. Boca Raton: CRC Press, pp. 389–420.
17. Van Remmen H, Ward WF, Sabia RV, Richardson A (1995). Gene expression and protein degradation. In: Masoro E, ed. *Handbook of Physiology: Aging*. Oxford: Oxford University Press, pp. 171–234.
18. Schröder HC, Ugarkovic D, Müller WEG, Mizushima H, Nemoto F, Kuchino Y (1992). Increased expression of UAG suppressor tRNA in aged mice: consequences for retroviral gene expression. *Eur J Gerontol.* 1: 452–7.
19. Takahashi R, Goto S (1988). Fidelity of aminoacylation by rat-liver tyrosyl-tRNA synthetase. Effect of age. *Eur J Biochem.* 178: 381–6.
20. Gabius H-J, Graupner G, Cramer F (1983). Activity patterns of aminoacyl-tRNA synthetases, tRNA methylases, arginyltransferase and tubulin:tyrosine ligase during development and ageing of *Caenorhabditis elegans*. *Eur J Biochem.* 131: 231–4.
21. Takahashi R, Mori N, Goto S (1985). Alteration of aminoacyl tRNA synthetases with age: accumulation of heat-labile moleculaes in rat liver, kidney and brain. *Mech Ageing Dev.* 33: 67–75.
22. Bernd A, Batke E, Zahn RK, Müller WEG (1982). Age-dependent gene induction in quail oviduct. XV. Alterations of the poly(A) associated protein pattern and of the poly(A) chain length of mRNA. *Mech Ageing Dev.* 19: 361–77.
23. Langstrom NS, Anderson JP, Lindroos HG, Winblad B, Wallace WC (1989). Alzheimer's disease-associated reduction of polysomal mRNA translation. *Mol Brain Res.* 5: 259–69.
24. Kristal BS, Conrad CC, Richardson A, Yu BP (1993). Is poly(A) tail length altered by aging or dietary restriction? *Gerontology* 39: 152–62.
25. Brewer G (2002). Messenger RNA decay during aging and development. *Ageing Res.* Rev. 1: 607–25.
26. Antequera F, Bird A (1993). Number of CpG islands and genes in human and mouse. *Proc Natl Acad Sci USA* 90: 11995–9.

27. Goodstadt L, Ponting CP (2001). Sequence variation and disease in the wake of the draft human genome. *Hum Mol Genet.* 10: 2209–14.
28. Rattan SIS (1995). Ageing – a biological perspective. *Mol Aspects Med.* 16: 439–508.
29. Rattan SIS (1996). Synthesis, modifications and turnover of proteins during aging. *Exp Gerontol.* 31: 33–47.
30. Ramsey JJ, Harper ME, Weindruch R (2000). Restriction of energy intake, energy expenditure, and aging. *Free Rad Biol Med.* 29: 946–68.
31. Kirkwood TBL, Holliday R, Rosenberger RF (1984). Stability of the cellular translation process. *Int Rev Cytol.* 92: 93–132.
32. Holliday R (1995). *Understanding Ageing*. Cambridge: Cambridge University Press.
33. Wojtyk RI, Goldstein S (1980). Fidelity of protein synthesis does not decline during aging of cultured human fibroblasts. *J Cell Physiol.* 103: 299–303.
34. Luce MC, Bunn CL (1987). Altered sensitivity of protein synthesis to paromomycin in extracts from aging human diploid fibroblasts. *Exp Gerontol.* 22: 165–77.
35. Luce MC, Bunn CL (1989). Decreased accuracy of protein synthesis in extracts from aging human diploid fibroblasts. *Exp Gerontol.* 24: 113–25.
36. Holliday R, Rattan SIS (1984). Evidence that paromomycin induces premature ageing in human fibroblasts. *Monogr Dev Biol.* 17: 221–33.
37. Buchanan JH, Stevens A, Sidhu J (1987). Aminoglycoside antibiotic treatment of human fibroblasts: intracellular accumulation, molecular changes and the loss of ribosomal accuracy. *Eur J Cell Biol.* 43: 141–7.
38. Silar P, Picard M (1994). Increased longevity of *EEF*-1a high-fidelity mutants in *Podospora anserina. J Mol Biol.* 235: 231–6.
39. Silar P, Rossignol M, Haedens V, Derhy Z, Mazabraud A (2000). Deletion and dosage modulation of the eEF1A gene in *Podospora anserina:* effect on the life cycle. *Biogerontology* 1: 47–54.
40. Holbrook MA, Menninger JR (2002). Erythromycin slows aging of *Saccharomyces cerevisiae. J Gerontol Biol Sci.* 57A: B29–36.
41. Orgel LE (1963). The maintenance of the accuracy of protein synthesis and its relevance to ageing. *Proc Natl Acad Sci USA* 49: 517–21.
42. Orgel LE (1973). The maintenance of the accuracy of protein synthesis and its relevance to ageing: a correction. *Proc Natl Acad Sci USA* 67: 1476.
43. Medvedev ZA (1990). An attempt at a rational classification of theories of ageing. *Biol Rev.* 65: 375–98.
44. Kowald A, Kirkwood TBL (1993). Accuracy of tRNA charging and codon:anticodon recognition; relative importance for cellular stability. *J Theor Biol.* 160: 493–508.
45. Silar P (1994). Is translational accuracy an out-dated topic? *Trends Genet.* 10: 73–4.
46. Holliday R (1996). The current status of the protein error theory of aging. *Exp Gerontol.* 31: 449–52.
47. Nyström T (2002). Translational fidelity, protein oxidation, and senescence: lessons from bacteria. *Ageing Res Rev.* 1: 693–703.
48. Merrick WC (1992). Mechanism and regulation of eukaryotic protein synthesis. *Microbiol Rev.* 56: 291–315.
49. Hershey JWB, Merrick WC (2000). The pathway and mechanism of inititation of protein synthesis. In: Sonenberg N, Hershey JWB, Mathews MB, eds. *Translational Control of Gene Expression.* New York: Cold Spring Harbor Laboratory Press, pp. 33–88.
50. Schneider RJ (2000). Translational control during heat shock. In: Sonenberg N, Hershey JWB, Mathews MB, eds. *Translational Control of Gene Expression.* New York: Cold Spring Harbor Laboratory Press, pp. 581–93.

51. Vargas R, Castaneda M (1983). Age dependent decrease in the activity of protein synthesis initiation factors in rat brain. *Mech Ageing Dev.* 21: 183–91.
52. Vargas R, Castañeda M (1984). Heterogeneity of protein synthesis initiation factors in developing and aging rat brain. *Mech Ageing Dev.* 26: 371–8.
53. Webster GC (1986). Effect of aging on the components of the protein synthesis system. In: Collatz KG, Sohal RS, eds. *Insect Aging*. Berlin: Springer-Verlag, pp. 207–16.
54. Webster GC, Webster SL, Landis WA (1981). The effect of age on the initiation of protein synthesis in *Drosophila* melanogaster. *Mech Ageing Dev.* 16: 71–9.
55. Chen ZP, Chen KY (1997). Dramatic attenuation of hypusine formation on eukaryotic initiation factor 5A during senescence of IMR-90 human diploid fibroblasts. *J Cell Physiol.* 170: 248–54.
56. Nokazawa T, Mori N, Goto S (1984). Functional deterioration of mouse liver ribosomes during aging: translational activity and the activity for formation of the 47S initiation complex. *Mech Ageing Dev.* 26: 241–51.
57. Ward W, Richardson A (1991)/ Effect of age on liver protein synthesis and degradation. *Hepatology* 14: 935–48.
58. Adam G, Simm A, Braun F (1987). Levels of ribosomal RNA required for stimulation from quiescence increase during cellular aging *in vitro* of mammalian fibroblasts. *Exp Cell Res.* 169: 345–56.
59. Kirkland JL, Hollenberg CH, Gillon WS (1993). Effects of aging on ribosomal protein L7 messenger RNA levels in cultured rat preadipocytes. *Exp Gerontol.* 28: 557–63.
60. Vargas R, Castaneda M (1981). Role of elongation factor 1 in the translational control of rodent protein synthesis. *J Neurochem.* 37: 687–94.
61. Webster GC, Webster SL (1983). Decline in synthesis of elongation factor one (EF-1) precedes the decreased synthesis of total protein in aging *Drosophila* melanogaster. *Mech Ageing Dev.* 22: 121–8.
62. Webster GC, Webster SL (1984). Specific disappearance of translatable messenger RNA for elongation factor one in aging *Drosophila* melanogaster. *Mech Ageing Dev.* 24: 335–42.
63. Richardson A, Semsei I (1987). Effect of aging on translation and transcription. *Rev Biol Res Aging* 3: 467–83.
64. Merry BJ, Holehan AM (1991). Effect of age and restricted feeding on polypeptide chain assembly kinetics in liver protein synthesis *in vivo*. *Mech Ageing Dev.* 58: 139–50.
65. Riis B, Rattan SIS, Clark BFC, Merrick WC (1990). Eukaryotic protein elongation factors. *TIBS.* 15: 420–4.
66. Knudsen SM, Frydenberg J, Clark BFC, Leffers H (1993). Tissue-dependent variation in the expression of elongation factor-1alpha isoforms: isolation and characterization of a cDNA encoding a novel variant of human elongation factor 1alpha. *Eur J Biochem.* 215: 549–54.
67. Shiina N, Gotoh Y, Kubomura N, Iwamatsu A, Nishida E (1994). Microtubule severing by elongation factor 1a. *Science* 266: 282–5.
68. Gonen H, Smith CE, Siegel NR, *et al.* (1994) Protein synthesis elongation factor EF-1a is essential for ubiquitin-dependent degradation of certain Na-acetylated proteins and may be substituted for by the bacterial elongation factor Tu. *Proc Natl Acad Sci USA* 91: 7648–52.
69. Rattan SIS (1995). Protein synthesis and regulation in eukaryotes. In: Bittar EE, Bittar N, eds. *Principles of Medical Biology, Volume 4, Cell Chemistry and Physiology.* Greenwich: JAI Press, pp. 247–63.

70. Cavallius J, Rattan SIS, Clark BFC (1986). Changes in activity and amount of active elongation factor-1a in aging and immortal human fibroblast cultures. *Exp Gerontol.* 21: 149–57.
71. Rattan SIS, Cavallius J, Hartvigsen G, Clark BFC (1986). Amounts of active elongation factor-1a and its activity in livers of mice during ageing. *Trends Ageing Res.* 147: 135–40.
72. Rattan SIS, Ward WF, Glenting M, Svendsen L, Riis B, Clark BFC (1991). Dietary calorie restriction does not affect the levels of protein elongation factors in rat livers during ageing. *Mech Ageing Dev.* 58: 85–91.
73. Shepherd JCW, Walldorf U, Hug P, Gehring WJ (1989). Fruitflies with additional expression of the elongation factor EF-1a live longer. *Proc Natl Acad Sci USA* 86: 7520–1.
74. Shikama N, Ackermann R, Brack C (1994). Protein synthesis elongation factor EF-1a expression and longevity in *Drosophila* melanogaster. *Proc Natl Acad Sci USA* 91: 4199–203.
75. Dudas SP, Arking R (1994). The expression of the EF1a genes of *Drosophila* is not associated with the extended longevity phenotype in a selected long-lived strain. *Exp Gerontol.* 29: 645–57.
76. Webster GC (1985). Protein synthesis in aging organisms. In: Sohal RS, Birnbaum LS, Cutler RG, eds. *Molecular Biology of Aging: Gene Stability and Gene Expression.* New York: Raven Press, pp. 263–89.
77. Takahashi R, Mori N, Goto S (1985). Accumulation of heat-labile elongation factor 2 in the liver of mice and rats. *Exp Gerontol.* 20: 325–31.
78. Riis B, Rattan SIS, Derventzi A, Clark BFC (1990). Reduced levels of ADP-ribosylatable elongation factor-2 in aged and SV40-transformed human cells. *FEBS Lett.* 266: 45–7.
79. Parrado J, Bougria M, Ayala A, Castaño A, Machado A (1999). Effects of aging on the various steps of protein synthesis: fragmentation of elongation factor 2. *Free Rad Biol Med.* 26: 362–70.
80. Jäger M, Holtz J, Redpath NT, *et al.* (2002) The ageing heart: influence of cellular and tissue ageing on total content and distribution of the variants of elongation factor-2. *Mech Ageing Dev.* 123: 1305–19.

Metabolic Regulation of Gene Silencing and Life Span

Haim Y. Cohen, Kevin J. Bitterman and David A. Sinclair

The timing of gene expression defines the particular activity of a single cell or tissue and ensures the proper development of the organism. In higher eukaryotes most genes are "silent," meaning that they are transcriptionally inert and that this state is heritable. Many changes in gene expression occur during cellular senescence and organismal aging [1, 2]. Although most of these are considered to be downstream effects rather than causes of aging, there is evidence that gene silencing may be important for determining lifespan. This chapter will provide an introduction to mechanisms of gene silencing and discuss how this (and related mechanisms) may influence longevity. There will be particular focus on how the environment and metabolic activity may regulate gene silencing and lifespan.

Transcriptional silencing

Gene silencing is defined as a heritable chromatin structure within a specific chromosomal region that represses transcription of genes in a promoter non-specific manner [3]. Recent studies have demonstrated that the establishment of silent chromatin shares an impressive mechanical similarity across eukaryotes, from yeast to mammals, despite significant differences in the specific proteins involved. Silencing involves changing the chromatin into a more tightly packaged structure that is thought to be refractory to most DNA-binding proteins [4]. Nucleosomes are the basic unit for packaging DNA. Each unit combines 146 base pairs of DNA wrapped in two loops around an octamer of histones H2A, H2B, H3 and H4.

The extent of silencing at a particular locus is determined by the covalent modification of the nucleosome core and the binding of accessory proteins. In this chapter we introduce the biochemical events that establish silent chromatin and discuss the connection between gene silencing, metabolism and aging in model organisms. Within the context of this system, the possible implications of this process in higher organisms are also presented.

T. von Zglinicki (ed.), Aging at the Molecular Level, 193–211.

"Silencing" differs from "repression" because it is region-specific rather than promoter-specific [3]. Integration of a gene into a silenced area will suppress its transcription, often regardless of its promoter. A second property of a silenced region is that the chromatin architecture is maintained over successive mitotic and meiotic divisions, as a way to ensure the stability of gene expression within cell lineages. In the fruit fly *Drosophila melanogaster* this phenomenon is known as Position Effect Variegation (PEV) [5]. In the budding yeast *Saccharomyces cerevisiae* a position effect is also observed at three known silent loci: telomeres, the mating loci (*HML* and *HMR*) and the ribosomal DNA (rDNA) array. Only the telomeres show variegated expression whereas the other yeast loci are constitutively silent.

Although silent chromatin has a more closed structure, recent findings have shown that it does not fully block accessibility of certain transcription and DNA repair proteins. In *Drosophila*, a bacteriophage T7 RNA polymerase can still transcribe across the silent region of the Ultrabithorax gene [6] and heterochromatin at the yeast *HMR* locus is permissive to the constitutive binding of an activator, HSF, and two components of the pre-initiation complex (PIC), TBP and Pol II [7]. These and other findings suggest that silent chromatin is a dynamic and partially open structure.

Covalent modifications and silent chromatin

The modification of specific amino acid residues within the core histones is key to the establishment of heterochromatin. Specific amino acids in histone tails are targets for modifications including acetylation, phosphorylation, methylation, poly(ADP ribosylation) and ubiquitylation [8]. Of these, acetylation and methylation have been the most well studied in the context of gene silencing.

Acetylation

It has been known for many years that there is an inverse correlation between histone acetylation and silencing. Early work by Broach using budding yeast showed that both the silent mating-type cassettes and telomeres are hypoacetylated [9]. Braunstein (1996) was the first to show that the acetylation pattern of histone H4 at the silent mating locus is identical to the heterochromatin of the silent centromeric loci of *Drosophila* [10]. The primary targets for histone acetylation are specific lysines in the amino terminal tails of histones H3 and H4. In addition, though more infrequent, histones H2A and H2B are targets for acetylation. The extent of histone acetylation is determined by the activities of two opposing classes of enzyme, namely histone acetyltransferases (HATs), which catalyze acetyl transfer reactions, and histone deacetylases (HDACs), which remove acetyl groups. For histone H3, lysines 9, 14, 18 and 23 are known targets for acetylation (Figure 1), whereas for histone H4, lysines 5, 8, 12 and 16 have been found to be acetylated (Figure 1) [4]. In the early 1980's, Cary and colleagues [11] used high-resolution proton NMR to show that the amino-terminal of histones H3 and H4 interact with the DNA backbone of the nucleosome under physiological conditions. A number of findings suggest that acetylation decreases the affinity of DNA for histones and promotes the binding of

Histone H3

NT ARTKQTARK STGGKAPRKQ LATKAARKSA PATGGVKKPH CT

HISTONE H4

NT SGRGKGGKG LGKGGAKRHR ----//---- VTYTEHAKRK LKR CT

Figure 1. *Post-translational acetylation and methylation of histones. Silent heterochromatic regions are associated with decreased levels of DNA recombination and gene expression. Heterochromatin is specified by specific modifications of the core histones, known as the "histone code". The amino terminal tails of histones H3 and H4 are subject to numerous post-translational modifications. For histone H3, lysines 9,14,18 and 23 are known targets for acetylation, whereas on histone H4 lysines 5, 8, 12 and 16 are targets. Methylation occurs on lysines 4, 9, and 27 and arginines 2,17 and 26 of histone H3 as well as lysine 79 and arginine 3 of H4. Amino acid code: white box - acetylated; black box – methylated; grey box – acetylated and methylated.*

accessory proteins, in effect promoting a more open structure that is permissive to transcription [12, 13].

Methylation and ubiquitylation

In addition to acetylation, core histone tails are also subject to methylation [8]. Unlike acetylation, methylation can occur on a variety of residues including glutamate, leucine, cysteine, lysine, arginine, and histidine [14]. Methylation can have dual effects on silencing depending on which residue is methylated. Although histone methylation had been known for many years, the importance of this modification in gene silencing has become apparent only recently. A breakthrough came with the characterization of the methyltransferase complexes. One class of methyltranferases, characterized by a so-called "SET" domain, appears to mediate lysine methylation. The first SET domain was identified in the mammalian Suv39h1, which methylates lysine 9 of histone H3 (Figure 1) [15]. Since then many proteins have been found to contain these domains and to possess methyltransferase activity. The methylation of lysine 4 of histone H3 in *S. cerevisiae* is facilitated by a protein called Set1 [16]. Several lines of evidence show that methylation of lysine 4 is important for silencing at all three silent domains in yeast. In a screen for rDNA silencing mutants, Bryk *et al.* [17] identified a mutation in *SET1* which abolished silencing at the rDNA. ChIP analysis confirmed that SET1 binds to the rDNA and the deletion of *SET1* was shown to affect the methylation level of lysine 4. These results suggested that *SET1* is involved in the silencing of rDNA via methylation of lysine 4.

There is only limited information about the role of chromatin methylation in organismal aging. Recent work showed that the tri-methylated form of lysine 20 of H4 significantly increases between 30 day and 450 day old rats [18]. Similar results have been obtained in growth-inhibited cell lines [18].

Regardless of the great amount of data on ubiquitylation, its role in silencing is not yet clear. Unfortunately, it is too early to know whether changes in the methylation or ubiquitylation status of histones are involved in aging or have relevance to longevity regulation.

SIR2-dependent gene silencing

The first definitive link between silencing and aging was discovered in the mid-1990s in *S. cerevisiae*. In yeast, silencing is mediated via the silent information regulators, the Sir proteins, first described by Ivy *et al.* in 1985 [19]. Four complementation groups (*sir1-4*) were defined in a genetic screen for mutations that relieved silencing at the mating type loci [20]. The Sir2 protein is the only Sir family member required for silencing at all three silent loci. The Sir2/3/4 complex mediates silencing at telomeres and the *HM* loci, whereas the RENT complex (regulator of nucleolar silencing and telophase exit) comprised of Net1, Cdc14 and Sir2, mediates silencing at the rDNA [21]. Silencing at the rDNA is also required to suppress recombination between the tandemly repeated rDNA units and can repress transcription of PolII genes integrated at this locus [3]. Sir2 does not appear to regulate transcription of native rDNA genes because the levels of rRNA in a *sir2* null mutant are not appreciably different than in wild-type cells [22].

Sir2 is the founding member of a large family of proteins known as "sirtuins" that extends from prokaryotes to humans [23]. Studies in *Salmonella typhimurium* showed that the Sir2 homologue CobB rescues the lethality of a *CobT* mutant [24]. Since CobT acts as pyrimidine nucleotide transferase, this raised the possibility that CobB might perform a similar reaction. Based on this finding, it was originally suggested that sirtuins function as mono ADP-ribosylases and such activity could be shown *in vitro* [25]. Today however, this activity is not considered physiologically relevant.

A breakthrough came in 2000 from the Guarente and Sternglanz laboratories [26, 27], who showed that Sir2 possesses deacetylase activity *in vitro*, targeting lysines 9 and 16 of histones H3 and H4, respectively. The second striking finding was that sirtuins require nicotinamide adenine dinucleotide (NAD^+) as a co-substrate. Because NAD^+ is essential for many metabolic reactions, it was proposed that intracellular NAD^+ levels might directly influence Sir2 activity and silencing, thus linking metabolism to gene expression.

Yeast Sir2 has become the founding member of Class III histone deacetylases (HDACs). Unlike Class I or II HDACs, Sir2-like deacetylases are not inhibited by trichostatin A (TSA) and have the unique characteristic of being NAD^+-dependent. While many Sir2-like enzymes have been shown to readily deacetylate histone substrates *in vitro,* at least two Sir2 homologues, yeast Hst2 and human SIRT2, are localized to the cytoplasm [28] and human SIRT1, a nuclear protein, has recently been shown to target p53 for deacetylation [29–31]. In addition, eubacterial species such as *Salmonella* lack histones entirely, yet their genomes still encode Sir2 homologues [24]. While yeast Sir2 itself is known to have specificity for lysine 16 of histone H4 and lysines 9 and 14 of histone H3 [27], the above results suggest that the Sir2 family of deacetylases may act on a broad range of substrates and that only a subset of these enzymes are likely to target histones for deacetylation *in vivo*.

Sir proteins and the regulation of longevity

Although the caloric intake and metabolic rate of every species seems to impact on its rate of aging, the connection between metabolism and aging is best understood for *S. cerevisiae*. In fact, the mechanism of lifespan extension by caloric restriction is almost completely elucidated. For this reason, this chapter will focus on the *S. cerevisiae* system, with reference to its relevance potential in higher organisms.

Many researchers have proposed that aging in dividing human cells may be fundamentally different from aging in those that remain in a post-mitotic state. In *S. cerevisiae* a distinction is also made between the aging of mitotic cells and those that are quiescent [32]. Yeast "replicative life span" is defined as the number of divisions an individual yeast cell undergoes before dying. The alternative measure, "chronological life span," also referred to as "post-diauxic survival," is the length of time a population of cells remains viable in a non-dividing state following nutrient deprivation [32]. One attractive feature of *S. cerevisiae* is that the progenitor cell is easily distinguished from its descendants. Cell division is asymmetric: a newly formed "daughter" cell is almost always smaller than the "mother" cell that gave rise to it. Yeast mother cells divide about 20 times before dying and, as described below, undergo characteristic structural and metabolic changes as they age.

In 1950, Andrew Barton showed that individual mother cells are mortal, but it took another forty years before we had a molecular understanding of the mechanism [33]. Two papers in the mid 1990's reported that the daughters from old mother cells inherit the characteristics of old age and, correspondingly, experience a shortened lifespan [34, 35]. This finding suggested that yeast aging is due to the accumulation of a "senescence factor" that appears stochastically during the life span of the cell [36]. In old cells, this factor may diffuse from the mother to the daughter cell, thus explaining how such phenotypes can be inherited. The second intriguing finding came from a screen for long-lived yeast mutants [37]. Four mutations called *uth1-4* (pronounced *youth*) were isolated in this screen, and three of them were involved in silencing. One of the mutations resulted in a C-terminal truncation of the Sir4 protein and was called *Sir4-42*. This semi-dominant mutation extended yeast life span by 45% [37] and caused the redistribution of the Sir complex to the rDNA [38]. The redistribution of Sir proteins to the rDNA was also shown to occur in very old wild-type yeast cells, implying that changes at the rDNA were a cause of yeast aging and *SIR4-42* extends life because it counters this process earlier than wild-type cells [37].

Since this finding, researchers have come to the consensus that the primary cause of yeast replicative aging stems from changes within the nucleolus, the distinct nuclear region responsible for ribosomal RNA (rRNA) transcription and ribosome assembly.

The yeast rDNA locus (*RDN1*) consists of 100–200 tandemly-repeated, 9 kb units encoding the ribosomal RNAs. In 1997, Sinclair and Guarente [39] showed that the enlarged and fragmented nucleolus of old wild-type cells is the result of accumulation of extrachromosomal rDNA circles (ERCs). They showed that recombination between two adjacent rDNA repeats leads to an ERC that replicates each cell division and is preferentially segregated to the mother cell. Thus, once formed, ERCs exponentially accumulate, reaching a copy number of over 1000 in old cells.

One prediction of this model was that any factor that affects the rate of ERC formation will affect life span, and multiple lines of evidence have shown this to be the case. First, the ectopic release of an ERC into a virgin cell shortens life span and causes premature aging, demonstrating that ERCs are sufficient to cause aging [39, 40]. Second, a variety of genetic manipulations that decrease either ERC formation or ERC replication extend life span, demonstrating that ERCs are a primary cause of aging.

One of the largest lifespan extensions in yeast is afforded by loss-of-function mutations in the *FOB1* gene. FOB1 does not apparently play a role in silencing but it strongly promotes rDNA recombination [41, 42]. Strains lacking *FOB1* have a greatly reduced rate of ERC formation and a life span over twice that of wild-type cells [43]. There are now many mutations and manipulations that are known to reduce ERC formation and extend life span [44]. All of these reduce ERC formation either by increasing the extent of heterochromatin at the rDNA (which stabilizes the locus) or by directly suppressing homologous recombination at the rDNA [39, 40, 42, 43, 45, 46].

In 1999, Kaeberlein and colleagues [43] showed that a single extra copy of the *SIR2* gene extends yeast lifespan by 30% by suppressing rDNA recombination, and this effect is strictly dependent on Sir2 enzymatic activity [47]. A year later, Tissenbaum and Guarente reported that in the nematode *C. elegans,* extra copies of the *SIR2* homologue sir-2.1 also extend lifespan by 50% [48]. Genetic analysis indicates that sir-2.1 functions upstream of daf-16 in the insulin-like signaling pathway, suggesting that Sir2 proteins may be part of the conserved IGF-1 pathway that regulates worm longevity in response to environmental conditions such as nutrient availability and crowding [49]. Recently, the closest human homologue of *SIR2*, SIRT1 (Figure 2), has been shown to inhibit apoptosis through deacetylation of p53 [29–31]. These findings suggest that Sir2 and its homologues may have a conserved role in the regulation of survival at both the cellular and organismal level.

The Rpd3 histone deacetylase family influences longevity

Though the literature devoted to Sir2 is vast, and continues to grow, it is not the only histone deacetylase which has been implicated in the aging process. Yeast *RPD3* encodes a class I histone deacetylase with specificity for lysines 5 and 12 of histone H4 [50]. In yeast, Rpd3 is required for the proper timing of replication origin firing [51] and the heterochromatic structure of rDNA during stationary phase [52]. Rpd3 also represses the transcription of a number of critical genes including *HO, TRK2, STE6, PHO5* and *SPO13* [53, 54].

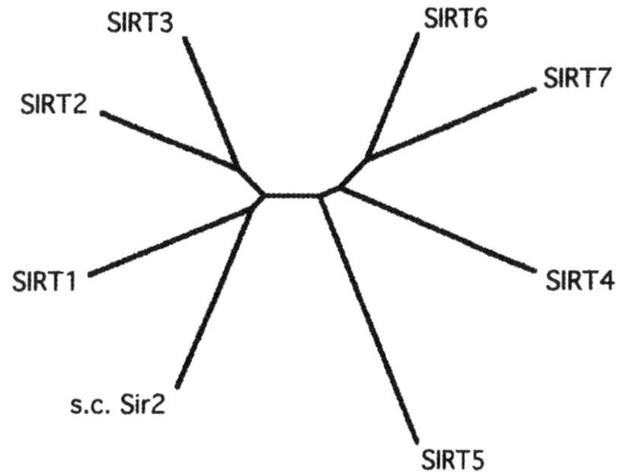

Figure 2. *Phylogenetic tree of human Sirtuins and yeast Sir2.*

Deletion of *RPD3* leads to hyperacetylation of histones, similar to that seen following deletion of *SIR2* but, paradoxically, this leads to an *increase* in rDNA silencing at all three silent loci [50, 55, 56]. Notably, this difference between *RPD3* and *SIR2* extends to lifespan as well: an *rpd3* mutant lives significantly longer than wild type cells [55]. These findings have recently been extended to the fruit fly *Drosophila melanogaster* [57]. While complete loss of Rpd3 function is lethal in this organism, Rogina and colleagues showed that a partial loss-of-function extends lifespan in both male and female flies. The fact that levels of this protein are reduced under nutrient limiting conditions prompted speculation that Rpd3 functions in the calorie restriction pathway [57, 58]. A similar result has also been observed for yeast *RPD3* [59].

Because Rpd3 and Sir2 act on different lysine residues on histones H3 and H4, and because they have opposing effects on silencing and lifespan, it is likely that the overall pattern of core histone acetylation governs the degree of silencing at these loci. In addition to changes in the state of silent heterochromatin, overall changes in gene expression patterns resulting from manipulation of these genes may also regulate yeast lifespan [60, 61], although this remains speculation.

Sir2 links silencing and genome stability to metabolism

Sir2 catalyzed deacetylation is an atypical reaction from both a thermodynamic and mechanistic standpoint. Although TSA-sensitive HDACs catalyze deacetylation without the need of a cofactor, Sir2 requires NAD^+. This is in spite of the fact that deacetylation is a simple and energetically favorable hydrolysis reaction. Furthermore, NAD^+ is not required catalytically for this reaction, but is actually consumed by it. Hydrolysis of the glycosidic bond between the ribose and nicotinamide moieties of NAD^+ liberates roughly 8.2 kcal/mol of free energy [62].

This leads one to question why the cell would couple cleavage of a high-energy bond in a metabolically valuable molecule to an already exothermic reaction? One possible explanation for these findings is that the NAD^+-dependence of Sir2 may allow for regulation of its activity through changes in the availability of this co-substrate [63], which, in turn, are dependent on the energetic and redox state of the cell. An additional possibility is that the products formed, nicotinamide and *O*-acetyl-ADP-ribose, may initiate a signal transduction pathway or facilitate feedback regulation [62, 64]. These possibilities will be discussed in more detail below after taking a closer look at the reaction mechanism of Sir2.

The molecular mechanism of NAD^+-dependent deacetylation

Sir2 couples NAD^+ hydrolysis to the deacetylation of the ε amine of lysine residues with a 1:1 stoichiometry [26, 27, 63, 65–67]. The overall reaction actually consists of two hydrolysis steps, which are thought to be coupled [68]. The first step is the cleavage of the high-energy glycosidic bond that joins the ADP-ribose moiety of NAD^+ to nicotinamide. This is followed by cleavage of the C-N bond between the acetyl group and lysine. Upon cleavage, Sir2 then catalyzes the transfer of the acetyl group to ADP-ribose [65–67].

Although the exact mechanism of the reaction is not yet known, the crystal structures of two archaeal Sir2 homologs (Sir2-Af1 and 2), as well as the catalytic core of human SIRT2, have provided information [68–71]. The archaeal structures have been particularly revealing as both the wild type and several catalytically deficient varieties of Sir2-Af1 have been crystalized with an (albeit incomplete) NAD^+ molecule [68, 69] and with an acetylated p53 peptide substrate [71]. The conservation is immediately apparent when the structures of these distantly related homologous proteins are compared. The conserved core domain of these enzymes forms two regions that bind an NAD^+ molecule within a deep pocket between them. The larger of these two domains is reminiscent of the so-called Rossmann fold, a motif found in many NAD(H)/NADP(H) binding proteins [72]. The smaller domain coordinates a structural zinc atom [68].

The NAD^+ binding pocket itself is divided into three distinct regions, termed sites A, B and C. Site A appears to bind the adenine-ribose moiety of NAD^+, while site B contacts the nicotinamide-ribose [68]. Many of the mutations that diminish or abolish the activity of Sir2-like enzymes map to site B [25, 47, 68, 69], consistent with this site being involved in catalysis. Site C forms a deep core within the pocket and does not appear to make direct contact with NAD^+ in any of the solved structures. Mutation of conserved residues within this pocket – specifically, residues equivalent to Ser24, Asn99 and Asp101 of Sir2-Af1 – either severely diminish or abolish catalytic activity [68, 69].

Min and colleagues have suggested that because these residues are on the surface of the NAD^+ binding pocket they do not likely contribute to protein stability, but more likely are involved in catalysis [68]. They proposed that the first step in deacetylation may be reminiscent of a serine protease reaction. A conformational change in NAD^+, due to a rotation around the ester bond joining the adenine ribose to the pyrophosphate and/or the phosphodiester bonds within the pyrophosphate moiety, would position the nicotinamide in proximity to site C. In this conformation, Ser24 may act as a catalytic base in the cleavage of the glycosidic linkage between nicotinamide and ribose [68]. Arguing against this model though, Chang *et al.* find that mutation of Ser24 in Sir2-Af1 decreases activity by only 6-fold suggesting that this residue is not essential for activity. From examination of the structures of these mutants, they conclude instead that site C residues play a role in NAD^+ binding and positioning, rather than in catalysis [69]. In either case, hydrolysis of the acetyl group from N-acetyl lysine in the second step of the reaction, seems to require conserved residues in site B. Specifically, His116 (His118 in Sir2-Af2) is proposed to be directly involved in catalyzing this step [69].

A cleft between the "Rossmann fold" domain and zinc-binding domain serves as a protein-substrate binding site [71]. The acetyl-lysine side chain appears to insert into a tunnel in this cleft, bringing it in proximity to site B of the NAD^+ binding region. In particular, the structure of Sir2-Af2 bound to a p53 peptide shows the acetyl-lysine in close proximity to the putative catalytic His118. This is proposed to make the acetyl group of the acetyl-lysine a better nucleophile during deacetylation [71]. Based on the available data, a current model for the mechanism of deacetylation proposes a nucleophilic attack by the carbonyl oxygen of the N-acetyl group of the substrate on the C1′ of the nicotinamide ribose [63, 73]. Several models have been proposed though detailing the exact chemistry of the two hydrolysis steps [68, 69, 71], and further investigation will be needed to elucidate the precise mechanism.

The Sir2-Af2-peptide structure indicates that all of the contacts that anchor the protein-substrate to its binding site on the enzyme are simple hydrogen bonds between the two peptide backbones [71]. According to the authors, these interactions form an enzyme-substrate β sheet which they refer to as a "β staple," as the substrate appears to link together two distinct regions of Sir2-Af2. As Tanny and Moazed point out, this means that any peptide with a stretch of amino acids containing an acetylated lysine and which are flexible enough to form a β staple, is a putative substrate for Sir2-like enzymes [73]. This idea is consistent with the broad specificity displayed by many of these enzymes *in vitro*. Indeed, Avalos *et al.* tested the ability of archaeal Sir2-Af1, Sir2-Af2, human SIRT2, and yeast Sir2p to act on both acetylated histones and two monoacetylated p53 peptides. The authors found that all four enzymes were able to deacetylate each of the four substrates [71]. While mutation of several non-conserved residues did alter the specificity of Sir2-Af2, it is likely that *in vivo* substrate specificity will be primarily determined through protein-protein interactions outside the catalytic domain of the enzyme [73].

While a complete understanding of the chemistry underlying the reaction requires further investigation, the products of Sir2-catalyzed deacetylation are known [26, 65, 66]. The first step, the cleavage of the glycosidic bond joining the nicotinamide moiety of NAD^+ to ADP-ribose, results in the release of free nicotinamide. This important regulatory molecule, which is a precursor of nicotinic acid in the cell and a form of vitamin B3, is discussed in detail below.

The other reaction product, namely a regioisomer of 2′- and 3′-*O*-acetyl ADP ribose, results from cleavage of the C-N bond between the acetyl group and lysine followed by transfer of the acetyl to ADP-ribose [63, 74]. The initial product of the transfer reaction appears to be the 2′ form of this molecule, which quickly converts to the 3′-*O*-acetyl ADP ribose, eventually reaching an equilibrium between the two [69].

It has been proposed that *O*-acetyl ADP ribose may initiate a signal transduction cascade, as the metabolic instability of these molecules is reminiscent of the initiators of other signaling pathways [63]. This in turn is cited as one possible rationale for the consumption of NAD^+ during deacetylation. In support of this idea, injection of *O*-acetyl ADP ribose into living cells has been shown to cause a delay/block in the cell cycle and maturation of oocytes, although it must be cautioned that ADP ribose has a similar effect [75]. Furthermore, it has recently been demonstrated that *O*-acetyl ADP ribose is metabolized by ADP-ribose hydrolases and, predominately, by an as yet unidentified enzyme [76].

NAD synthesis and Sir2 activity

While production of *O*-acetyl ADP ribose provides an attractive hypothesis to explain the seemingly wasteful consumption of NAD^+ in the Sir2 reaction, others have been posited as well. As mentioned above, it has been proposed that the strict NAD^+-dependence of Sir2-catalyzed deacetylation may allow for regulation of enzymatic activity through availability of the co-substrate itself.

Although there are many ways that the availability of NAD^+ may be altered, via alterations in redox or respiration for example [77, 78], recent work indicates that NAD^+ biosynthesis is an important player. NAD^+ is essential for many key metabolic reactions and in most organisms there are two parallel pathways for its biosynthesis. NAD is synthesized *de novo* from tryptophan and recycled from nicotinamide via the NAD^+ salvage pathway [79] (Figure 3).

The *de novo* NAD^+ synthesis pathway, also known as the kynurenine pathway, is catalyzed by the *BNA* (**b**iosynthesis of **n**icotinic **a**cid) genes, which convert tryptophan to quinolinic acid and subsequently to nicotinic acid mononucleotide (NaMN). At this point the *de novo* and the salvage pathways converge [80, 81]. The yeast *de novo* pathway does not appear to be involved in silencing or lifespan extension by calorie restriction (see below) [45, 81].

In yeast and many lower eukaryotes, the salvage pathway for NAD^+ synthesis consists of four steps, whereas in mammals it consists of two (K. Bitterman, personal communication). As shown in Figure 3, nicotinamide produced from NAD^+ cleavage is first converted to nicotinic acid by Pnc1, a nicotinamidase [82, 83] and is also utilized from the medium [81, 84]. Nicotinic acid is subsequently converted into nicotinic acid mononucleotide (NaMN) by a phosphoribosyltransferase encoded by *NPT1* [85] (Figure 3). At this point, the NAD^+ salvage pathway and the *de novo* NAD^+ synthesis pathway converge, and NaMN is converted to desamido-NAD^+ (NaAD) by a nicotinate mononucleotide adenylyltransferase (NaMNAT). In *S. cerevisiae*, there are two ORFs with homology to bacterial NaMNAT genes [67, 86], which have been named *NMA1* and *NMA2*, respectively [85]. In *Salmonella*, the final step in the regeneration of NAD^+ is catalyzed by an NAD synthetase [87]. An as yet uncharacterized ORF, *QNS1*, is predicted to encode the NAD synthetase of this organism.

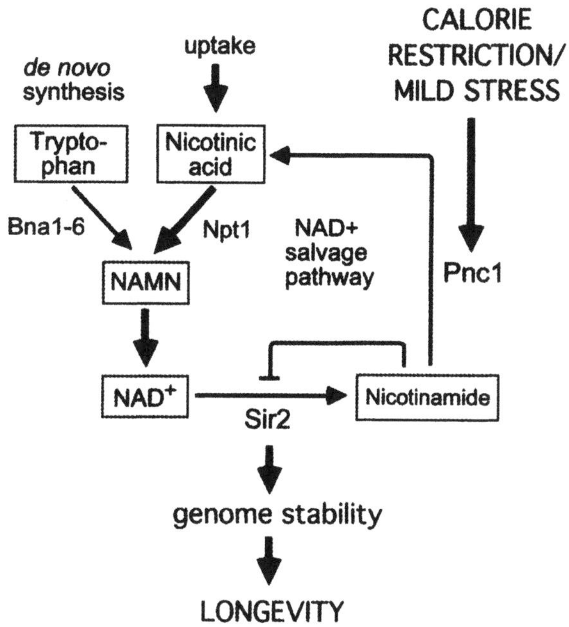

Figure 3. Model for the regulation of silencing and lifespan in S. cerevisiae by Pnc1 and nicotinamide. Disparate environmental stimuli including calorie restriction, heat and osmotic stress serve as inputs to a common pathway of longevity, mediated by the Sir2 NAD$^+$-dependent deacetylase. Cells coordinate a response to these inputs by inducing transcription of PNC1, which encodes an enzyme that converts nicotinamide (a potent Sir2 inhibitor) to nicotinic acid. Nicotinic acid is used as a precursor for NAD$^+$ synthesis via the NAD$^+$ salvage pathway. Together, these events increase Sir2 histone deacetylase activity, which promotes silencing at the rDNA, stabilizes the locus, and increases replicative lifespan.

Cells lacking *NPT1* or *PNC1* show a loss of silencing at telomeres and the rDNA, reminiscent of a *sir2* mutant [81]. Furthermore, *NPT1* is required for lifespan extension by the caloric restriction mimic, *cdc25-10* [45]. Conversely, it was demonstrated that increased dosage of NAD$^+$ salvage pathway genes increase silencing at the above mentioned loci, and a single extra copy of the *NPT1* gene extends life span by 60% [85].

Calorie restriction and the regulation of Sir2 enzymatic activity
As for most organisms, yeast life span is extended by dietary or "calorie" restriction. Yeast provided with limiting glucose or amino acids live up to 50% longer [45, 88]. Importantly, the lifespan extension by glucose restriction is *SIR2*-dependent, suggesting that Sir2 is part of the caloric restriction pathway [45]. Importantly, because Sir2

protein levels do not change in response to calorie restriction, the increase in silencing associated with this treatment is thought to be due to the stimulation of Sir2 activity [45].

Guarente and colleagues have proposed that calorie restriction increases lifespan by increasing available NAD^+ and a recent paper by Lin *et al.* [78] offers an explanation. *S. cerevisiae* is a facultative anaerobe and can generate energy via either fermentation or, more efficiently, via respiration. When glucose levels are high, cells preferentially utilize fermentation but when glucose becomes limiting, respiration is preferred and carbon is shunted towards the mitochondrial tricarboxylic acid (TCA) cycle and the electron transport chain [89]. By measuring oxygen consumption, Lin *et al.* showed that yeast grown on low glucose respire at a threefold higher rate [78]. Furthermore, yeast that cannot respire due to a *cyt1* mutation that inactivates the electron-transport chain, are not longer lived when calorie restricted. The artificial induction of genes involved in respiration (by overexpressing the *HAP4* gene) causes a Sir2-dependent extension of life span, even in the absence of glucose restriction [78]. The authors propose a model whereby increased respiration leads to increased oxidation of NADH to NAD^+ in the mitochondria. This change in the NAD^+/NADH ratio would be transmitted outside the mitochondria where it would effectively stimulate Sir2 activity.

While this is an attractive hypothesis, the overall increase in the cellular NAD^+/NADH predicted by this model has yet to be observed. Measurements of total NAD/NADH in calorie restricted cells showed no increase under glucose restriction [90]. Similarly, there was no increase in NAD^+ levels or NAD^+/NADH ratios detected in long-lived cells over expressing *NPT1*, even though Sir2 activity was clearly increased [85]. Furthermore, cells lacking *PNC1* are compromised for Sir2-dependent silencing, though NAD^+ levels are unaltered [81]. Clearly there is not a clear association between NAD^+ levels and Sir2 activity. One explanation is that these experiments measured total NAD^+, so there remains the possibility that the flux through the salvage pathway is increased or "free" (versus enzymatically "bound") NAD levels are altered [85]. With regard the latter hypothesis, measurements of free NAD^+ by ^{13}C-NMR appear to rule out that hypothesis (Anderson, Neves, Sinclair, unpublished result). It has also been postulated that Sir2 may be regulated by local (perhaps nuclear specific) changes in NAD^+ that cannot easily be detected [85]. One intriguing possibility, which may resolve these apparent conflicts, concerns a product of the Sir2-catalyzed reaction, nicotinamide.

The role of nicotinamide metabolism in longevity regulation

Nicotinamide (NAM) was recently demonstrated to be a strong inhibitor of yeast Sir2 and human SIRT1 activity, both *in vitro* and *in vivo* [29, 64]. We found that cells grown in the presence of this compound showed a near complete loss of Sir2-dependent silencing, and had a replicative lifespan identical to that of a *sir2* mutant. *In vitro*, the IC_{50} of nicotinamide for SIRT1 was found to be relatively low ($\sim 50\ \mu M$), leading to speculation that this small molecule may be a physiologically relevant regulator of Sir2-like enzymes [64]. Based on these findings, we proposed that fluctuations in cellular nicotinamide levels may directly control the activity of Sir2 proteins *in vivo*, and that these fluctuations may in turn be regulated by enzymes involved in nicotinamide metabolism, namely Pnc1.

PNC1 encodes a nicotinamidase that is situated in a key position to regulate NAD^+-dependent deacetylases. This enzyme converts nicotinamide into nicotinic acid as part of the NAD^+ salvage pathway (Figure 3). Thus Pnc1 may reduce levels of this inhibitor and simultaneously increase the availability of NAD^+ to Sir2. Importantly, we showed that the product of this reaction, the structurally similar nicotinic acid, has no effect on Sir2 activity either *in vitro* or *in vivo*. This raises the possibility that high levels of Pnc1 induce silencing by removing the inhibitory effects of nicotinamide (Figure 3). In support of this model, we find that overexpression of *PNC1* extends replicative lifespan by 70%, and the lifespan of these cells is not further increased by growth on low glucose. In addition, the lifespan of cells lacking *PNC1* is not augmented by gluose restriction [83].

If nicotinamide turns out to regulate Sir2, why would an organism utilize a single gene and a small ubiquitous molecule to govern such a critical process as lifespan extension? Perhaps this system allows the organism to process inputs from multiple stimuli and facilitate a coordinated defense response. Another intriguing possibility is that this design permits the rapid evolution of strategies to suit a changing environment via changes in the *PNC1* promoter. The Sir2 pathway is ancient and perhaps early life forms used nicotinamide directly as an indicator of nutrient availability. Its role as a signaling molecule regulating longevity and survival may have been conserved to the present day.

Nicotinamide metabolism in higher organisms

It will be interesting to determine whether nicotinamide levels do indeed decrease under conditions of calorie restriction or mild stress, and if so, whether this mechanism is conserved in higher organisms. The possibility that nicotinamide is the major regulator of Sir2-like proteins raises many questions regarding the role of nicotinamide and its targets in such organisms. In mammals, there is already evidence for a link between nicotinamide, metabolism and stress resistance. Poly(-adenosine diphosphate-ribose) polymerase-1 (PARP) is a nuclear enzyme that cleaves NAD^+ to covalently attach poly(ADP-ribose) to acceptor proteins. This two-step reaction generates nicotinamide, which exerts an inhibitory effect on PARP-1 allowing for autoregulation. PARP enzymes have been implicated in numerous cellular functions including DNA break repair, telomere-length regulation, histone modification, and the regulation at the transcriptional level of key proteins including ICAM-1 and nitric oxide synthase [91]. PARP enzymes might be regulated by nicotinamide metabolism as part of a general stress response.

As was mentioned, nicotinamide also inhibits human *SIRT1* both *in vitro* and *in vivo* [29, 64]. *SIRT1* negatively controls p53 activity, indicating that nicotinamide levels may also regulate apoptosis and DNA repair [29, 92]. Consistent with this, the expression of NNMT (a human nicotinamide methyl transferase) in cells and tissues correlates with tumorigenesis [93] and radioresistance [94]. There is no obvious homologue of *PNC1* in model organisms more complex than *Drosophila*, which likely reflects differences in nicotinamide metabolism. It will be important to determine whether nicotinamide regulates multiple Sir2-like proteins and if so, which metabolic pathways in turn regulate nicotinamide levels. Clearly there is still much to be learned about this family of proteins and their potential roles in cell and organismal survival.

Perspective

Despite our success in understanding the relationship between metabolism and aging in simple budding yeast, many key questions remain to be answered. It is now clear that histones are not the only target of the sirtuin family. In fact, histones are proving to be only one class of many that are regulated by acetylation. Because the sirtuins are fairly indiscriminate in their activity *in vitro*, there are likely to be mediators *in vivo* that direct the sirtuins to their target. The identification of the sirtuin targets and the putative mediators is one of the major challenges of the field. Another pertinent area of research regards the regulation of the mammalian deacetylases themselves. Are they regulated indirectly (by metabolism via NAD^+ availability for example) or are there specific regulatory genes like yeast *PNC1* that respond to metabolic stimuli and specifically modulate NAD^+ levels and/or nicotinamide? The recent finding that SIRT1 regulates p53 activity suggests that in mammals, the sirtuins may control the balance between cell survival and apoptosis under stress. Although there is much work to do, work in model organisms has at least proven that longevity regulation is a real biological phenomenon. The promise is that one day these pathways might be manipulated in humans to enhance fitness and, perhaps one day, extend lifespan.

Acknowledgments

We are grateful to members of the Sinclair laboratory for manuscript preparation. Work cited from the Sinclair laboratory was supported by the National Institute on Aging (AG19972-01) and the Harvard-Armenise Foundation. D.A.S. is an Ellison Medical Research Foundation Special Fellow and K.B. is supported by the American Federation of Aging Research.

References

1. Shelton DN, Chang E, Whittier PS, Choi D, Funk WD (1999). Microarray analysis of replicative senescence. *Curr Biol.* 9: 939–45.
2. Prolla TA (2002). DNA microarray analysis of the aging brain. *Chem Senses.* 27: 299–306.
3. Moazed D (2001). Common themes in mechanisms of gene silencing. *Mol Cell.* 8: 489–98.
4. Li E (2002). Chromatin modification and epigenetic reprogramming in mammalian development. *Nat Rev Genet.* 3: 662–73.
5. Henikoff S (1995). Gene silencing in *Drosophila. Curr Top Microbiol Immunol.* 197: 193–208
6. McCall K, Bender W (1996). Probes of chromatin accessibility in the *Drosophila* bithorax complex respond differently to Polycomb-mediated repression. *EMBO* J 15: 569–80.
7. Sekinger EA, Gross DS (2001). Silenced chromatin is permissive to activator binding and PIC recruitment. *Cell* 105: 403–14.

8. Goll MG, Bestor TH (2002). Histone modification and replacement in chromatin activation. *Genes Dev.* 16: 1739–42.
9. Braunstein M, Rose AB, Holmes SG, Allis CD, Broach JR (1993). Transcriptional silencing in yeast is associated with reduced nucleosome acetylation. *Genes Dev* 7: 592–604.
10. Braunstein M, Sobel RE, Allis CD, Turner BM, Broach JR (1996). Efficient transcriptional silencing in Saccharomyces cerevisiae requires a heterochromatin histone acetylation pattern. *Mol Cell Biol.* 16: 4349–56.
11. Cary PD, Crane-Robinson C, Bradbury EM, Dixon GH (1982). Effect of acetylation on the binding of N-terminal peptides of histone H4 to DNA. *Eur J Biochem.* 127: 137–43.
12. Hong L, Schroth GP, Matthews HR, Yau P, Bradbury EM (1993). Studies of the DNA binding properties of histone H4 amino terminus. Thermal denaturation studies reveal that acetylation markedly reduces the binding constant of the H4 "tail" to DNA. *J Biol Chem.* 268: 305–14.
13. Tse C, Sera T, Wolffe AP, Hansen JC (1998). Disruption of higher-order folding by core histone acetylation dramatically enhances transcription of nucleosomal arrays by RNA polymerase III. *Mol Cell Biol.* 18: 4629–38.
14. Zhang Y, Reinberg D (2001). Transcription regulation by histone methylation: interplay between different covalent modifications of the core histone tails. *Genes Dev.* 15: 2343–60.
15. Rea S, Eisenhaber F, O'Carroll D, *et al.* (2000) Regulation of chromatin structure by site-specific histone H3 methyltransferases. *Nature* 406: 593–9.
16. Briggs SD, Bryk M, Strahl BD, *et al.* (2001). Histone H3 lysine 4 methylation is mediated by Set1 and required for cell growth and rDNA silencing in *Saccharomyces cerevisiae. Genes Dev.* 15: 3286–95.
17. Bryk M, Briggs SD, Strahl BD, Curcio MJ, Allis CD, Winston F (2002). Evidence that Set1, a factor required for methylation of histone H3, regulates rDNA silencing in *S. cerevisiae* by a Sir2-independent mechanism. *Curr Biol.* 12: 165–70.
18. Sarg B, Koutzamani E, Helliger W, Rundquist I, Lindner HH (2002). Postsynthetic trimethylation of histone H4 at lysine 20 in mammalian tissues is associated with aging. *J Biol Chem.* 277: 39195–201.
19. Ivy JM, Hicks JB, Klar AJ (1985). Map positions of yeast genes SIR1, SIR3 and SIR4. *Genetics* 111: 735–44.
20. Rine J, Herskowitz I (1987). Four genes responsible for a position effect on expression from HML and HMR in *Saccharomyces cerevisiae. Genetics* 116: 9–22.
21. Straight AF, Shou W, Dowd GJ, *et al.* (1999). Net1, a Sir2-associated nucleolar protein required for rDNA silencing and nucleolar integrity. *Cell* 97: 245–56.
22. Buck SW, Sandmeier JJ, Smith JS (2003). RNA polymerase I propagates unidirectional spreading of rDNA silent chromatin. *Cell* 111: 1003–14
23. Frye RA (2000). Phylogenetic classification of prokaryotic and eukaryotic Sir2-like proteins. *Biochem Biophys Res Commun.* 273: 793–8.
24. Tsang AW, Escalante-Semerena JC (1998). CobB, a new member of the SIR2 family of eucaryotic regulatory proteins, is required to compensate for the lack of nicotinate mononucleotide: 5,6-dimethylbenzimidazole phosphoribosyltransferase activity in cobT mutants during cobalamin biosynthesis in *Salmonella typhimurium* LT2. *J Biol Chem.* 273: 31788–94.
25. Tanny JC, Dowd GJ, Huang J, Hilz H, Moazed D (1999). An enzymatic activity in the yeast Sir2 protein that is essential for gene silencing. *Cell* 99: 735–45.
26. Landry J, Slama JT, Sternglanz R (2000). Role of NAD(+) in the deacetylase activity of the SIR2-like proteins. *Biochem Biophys Res Commun.* 278: 685–90.

27. Imai S, Armstrong CM, Kaeberlein M, Guarente L (2000). Transcriptional silencing and longevity protein Sir2 is an NAD- dependent histone deacetylase. *Nature* 403: 795–800.
28. Perrod S, Cockell MM, Laroche T, *et al.* (2001). A cytosolic NAD-dependent deacetylase, Hst2p, can modulate nucleolar and telomeric silencing in yeast. *EMBO J.* 20: 197–209.
29. Luo J, Nikolaev AY, Imai S, *et al.* (2001). Negative control of p53 by Sir2alpha promotes cell survival under stress. *Cell* 107: 137–48.
30. Langley EPM, Faretta M, Bauer UM, *et al.* (2002). Human SIR2 deacetylates p53 and antagonizes PML/p53-induced cellular senescence. *EMBO J.* 21: 2383–96.
31. Vaziri H, Dessain SK, Ng Eaton E, *et al.* (2001). hSIR2 (SIRT1) functions as an NAD-dependent p53 deacetylase. *Cell* 107: 149–59.
32. Sinclair D, Mills K, Guarente L (1998). Aging in *Saccharomyces cerevisiae. Annu Rev Microbiol.* 52: 533–60.
33. Barton A (1950). Some aspects of cell division in *Saccharomyces cervisiae. J Gen Microbiol.* 4: 84–6.
34. Kennedy BK, Austriaco NR Jr, Guarente L (1994). Daughter cells of *Saccharomyces cerevisiae* from old mothers display a reduced life span. *J Cell Biol.* 127: 1985–93.
35. Johnston JR (1966). Reproductive capacity and mode of death of yeast cells. *Antonie Van Leeuwenhoek* 32: 94–8.
36. Egilmez NK, Jazwinski SM (1989). Evidence for the involvement of a cytoplasmic factor in the aging of the yeast *Saccharomyces cerevisiae. J Bacteriol.* 171: 37–42.
37. Kennedy BK, Austriaco NR Jr, Zhang J, Guarente L (1995). Mutation in the silencing gene SIR4 can delay aging in *S. cerevisiae. Cell* 80: 485–96.
38. Kennedy BK, Gotta M, Sinclair DA, *et al.* (1997). Redistribution of silencing proteins from telomeres to the nucleolus is associated with extension of life span in *S. cerevisiae. Cell* 89: 381–91.
39. Sinclair DA, Guarente L (1997). Extrachromosomal rDNA circles – a cause of aging in yeast. *Cell* 91: 1033–42.
40. McVey M, Kaeberlein M, Tissenbaum HA, Guarente L (2001). The short life span of *Saccharomyces cerevisiae* sgs1 and srs2 mutants is a composite of normal aging processes and mitotic arrest due to defective recombination. *Genetics* 157: 1531–42.
41. Kobayashi T, Heck DJ, Nomura M, Horiuchi T (1998). Expansion and contraction of ribosomal DNA repeats in *Saccharomyces cerevisiae*: requirement of replication fork blocking (Fob1) protein and the role of RNA polymerase I. *Genes Dev.* 12: 3821–30.
42. Defossez PA, Prusty R, Kaeberlein M, *et al.* (1999). Elimination of replication block protein Fob1 extends the life span of yeast mother cells. *Mol Cell.* 3: 447–55.
43. Kaeberlein M, McVey M, Guarente L (1999). The SIR2/3/4 complex and SIR2 alone promote longevity in *Saccharomyces cerevisiae* by two different mechanisms. *Genes Dev.* 13: 2570–80.
44. Sinclair DA (2002). Paradigms and pitfalls of yeast longevity research. *Mech Ageing Dev.* 123: 857–67.
45. Lin SJ, Defossez PA, Guarente L (2000). Requirement of NAD and SIR2 for life-span extension by calorie restriction in *Saccharomyces cerevisiae. Science* 289: 2126–8.
46. Park PU, Defossez PA, Guarente L (1999). Effects of mutations in DNA repair genes on formation of ribosomal DNA circles and life span in *Saccharomyces cerevisiae. Mol Cell Biol.* 19: 3848–56.
47. Armstrong CM, Kaeberlein M, Imai SI, Guarente L (2002). Mutations in *Saccharomyces cerevisiae* gene SIR2 can have differential effects on *in vivo* silencing phenotypes and *in vitro* histone deacetylation activity. *Mol Biol Cell.* 13: 1427–38.

48. Tissenbaum HA, Guarente L (2001). Increased dosage of a sir-2 gene extends lifespan in *Caenorhabditis elegans*. *Nature* 410: 227–30.
49. Kenyon C (2001). A conserved regulatory system for aging. *Cell* 105: 165–8.
50. Rundlett SE, Carmen AA, Kobayashi R, Bavykin S, Turner BM, Grunstein M (1996). HDA1 and RPD3 are members of distinct yeast histone deacetylase complexes that regulate silencing and transcription. *Proc Natl Acad Sci USA* 93: 14503–8.
51. Vogelauer M, Rubbi L, Lucas I, Brewer BJ, Grunstein M (2002). Histone acetylation regulates the time of replication origin firing. *Mol Cell.* 10: 1223–33.
52. Sandmeier JJ, French S, Osheim Y, *et al.* (2002). RPD3 is required for the inactivation of yeast ribosomal DNA genes in stationary phase. *EMBO J.* 21: 4959–68.
53. Grunstein M (1997). Histone acetylation in chromatin structure and transcription. *Nature* 389: 349–52.
54. Struhl K (1998). Histone acetylation and transcriptional regulatory mechanisms. *Genes Dev.* 12: 599–606.
55. Kim S, Benguria A, Lai CY, Jazwinski SM (1999). Modulation of life-span by histone deacetylase genes in *Saccharomyces cerevisiae*. *Mol Biol Cell.* 10: 3125–36.
56. Vannier D, Balderes D, Shore D (1996). Evidence that the transcriptional regulators SIN3 and RPD3, and a novel gene (SDS3) with similar functions, are involved in transcriptional silencing in *S. cerevisiae*. *Genetics* 144: 1343–53.
57. Rogina B, Helfand SL, Frankel S (2002). Longevity regulation by *Drosophila* Rpd3 deacetylase and caloric restriction. *Science* 298: 1745.
58. Pletcher SD, Macdonald SJ, Marguerie R, *et al.* (2002). Genome-wide transcript profiles in aging and calorically restricted *Drosophila melanogaster*. *Curr Biol.* 12: 712–23.
59. Jiang JC, Wawryn J, Shantha Kumara HM, Jazwinski SM (2002). Distinct roles of processes modulated by histone deacetylases Rpd3p, Hda1p, and Sir2p in life extension by caloric restriction in yeast. *Exp Gerontol.* 37: 1023–30.
60. Chang KT, Min KT (2002). Regulation of lifespan by histone deacetylase. *Ageing Res Rev.* 1: 313–26.
61. Bernstein BE, Tong JK, Schreiber SL (2000). Genomewide studies of histone deacetylase function in yeast. *Proc Natl Acad Sci USA* 97: 13708–13.
62. Moazed D (2001). Enzymatic activities of Sir2 and chromatin silencing. *Curr Opin Cell Biol* 13: 232–8.
63. Sauve AA, Celic I, Avalos J, Deng H, Boeke JD, Schramm VL (2001). Chemistry of gene silencing: the mechanism of NAD^+-dependent deacetylation reactions. *Biochemistry* 40: 15456–63.
64. Bitterman KJ, Anderson RM, Cohen HY, Latorre-Esteves M, Sinclair DA (2002). Inhibition of silencing and accelerated aging by nicotinamide, a putative negative regulator of yeast sir2 and human SIRT1. *J Biol Chem.* 277: 45099–107.
65. Tanny JC, Moazed D (2001). Coupling of histone deacetylation to NAD breakdown by the yeast silencing protein Sir2: evidence for acetyl transfer from substrate to an NAD breakdown product. *Proc Natl Acad Sci USA* 98: 415–20.
66. Tanner KG, Landry J, Sternglanz R, Denu JM (2000). Silent information regulator 2 family of NAD-dependent histone/protein deacetylases generates a unique product, 1-O-acetyl-ADP-ribose. *Proc Natl Acad Sci USA* 97: 14178–82.
67. Smith JS, Brachmann CB, Celic I, *et al.* (2000). A phylogenetically conserved NAD^+-dependent protein deacetylase activity in the Sir2 protein family. *Proc Natl Acad Sci USA* 97: 6658–63.
68. Min J, Landry J, Sternglanz R, Xu RM (2001). Crystal structure of a SIR2 homolog-NAD complex. *Cell* 105: 269–79.

69. Chang JH, Kim HC, Hwang KY, *et al.* (2002). Structural basis for the NAD-dependent deacetylase mechanism of Sir2. *J Biol Chem.* 277: 34489–98.
70. Finnin MS, Donigian JR, Pavletich NP (2001). Structure of the histone deacetylase SIRT2. *Nat Struct Biol.* 8: 621–5.
71. Avalos JL, Celic I, Muhammad S, Cosgrove MS, Boeke JD, Wolberger C (2002). Structure of a Sir2 enzyme bound to an acetylated p53 peptide. *Mol Cell.* 10: 523–35.
72. Rossmann MG, Argos P (1978). The taxonomy of binding sites in proteins. *Mol Cell Biochem.* 21: 161–82.
73. Tanny JC, Moazed D (2002). Recognition of acetylated proteins: lessons from an ancient family of enzymes. *Structure (Camb)* 10: 1290–2.
74. Jackson MD, Denu JM (2002). Structural Identification of 2′- and 3′-O-acetyl-ADP-ribose as novel metabolites derived from the Sir2 family of beta-NAD^+-dependent histone/protein deacetylases. *J Biol Chem.* 277: 18535–44.
75. Borra MT, O'Neill FJ, Jackson MD, *et al.* (2002). Conserved enzymatic production and biological effect of O-acetyl-ADP-ribose by silent information regulator 2-like NAD^+-dependent deacetylases. *J Biol Chem.* 277: 12632–41.
76. Rafty LA, Schmidt MT, Perraud AL, Scharenberg AM, Denu JM (2002). Analysis of O-acetyl-ADP-ribose as a target for nudix ADP-ribose hydrolases. *J Biol Chem.* 277: 47114–22.
77. Kaeberlein M, Andalis AA, Fink GR, Guarente L (2002). High osmolarity extends life span in *Saccharomyces cerevisiae* by a mechanism related to calorie restriction. *Mol Cell Biol.* 22: 8056–66.
78. Lin SJ, Kaeberlein M, Andalis AA, *et al.* (2002). Calorie restriction extends *Saccharomyces cerevisiae* lifespan by increasing respiration. *Nature* 418: 344–8.
79. Foster JW, Park YK, Penfound T, Fenger T, Spector MP (1990). Regulation of NAD metabolism in *Salmonella typhimurium*: molecular sequence analysis of the bifunctional nadR regulator and the nadA-pnuC operon. *J Bacteriol.* 172: 4187–96.
80. Grant RS, Passey R, Matanovic G, Smythe G, Kapoor V (1999). Evidence for increased de novo synthesis of NAD in immune-activated RAW264.7 macrophages: a self-protective mechanism? *Arch Biochem Biophys.* 372: 1–7.
81. Sandmeier JJ, Celic I, Boeke JD, Smith JS (2002). Telomeric and rDNA silencing in *Saccharomyces cerevisiae* are dependent on a nuclear NAD(+) salvage pathway. *Genetics* 160: 877–89.
82. Ghislain M, Talla E, Francois JM (2002). Identification and functional analysis of the *Saccharomyces cerevisiae* nicotinamidase gene, PNC1. *Yeast* 19: 215–24.
83. Anderson RM, Bitterman KJ, Wood JG, Medvedik O, Sinclair DA (2002). Nicotinamide and Pnc1 mediate lifespan extension by calorie restriction and stress. *Nature* (Submitted).
84. Llorente B, Dujon B (2000). Transcriptional regulation of the *Saccharomyces cerevisiae* DAL5 gene family and identification of the high affinity nicotinic acid permease TNA1 (YGR260w). *FEBS Lett.* 475: 237–41.
85. Anderson RM, Bitterman KJ, Wood JG, *et al.* (2002) Manipulation of a nuclear NAD^+ salvage pathway delays aging without altering steady-state NAD^+ levels. *J Biol Chem.* 277: 18881–90.
86. Emanuelli M, Carnevali F, Lorenzi M, *et al.* (1999) Identification and characterization of YLR328W, the *Saccharomyces cerevisiae* structural gene encoding NMN adenylyltransferase. Expression and characterization of the recombinant enzyme. *FEBS Lett.* 455: 13–17.
87. Hughes KT, Olivera BM, Roth JR (1988). Structural gene for NAD synthetase in *Salmonella typhimurium*. *J Bacteriol.* 170: 2113–20.

88. Jiang JC, Jaruga E, Repnevskaya MV, Jazwinski SM (2000). An intervention resembling caloric restriction prolongs life span and retards aging in yeast. *FASEB J.* 14: 2135–7.
89. Pronk JT, Yde Steensma H, Van Dijken JP (1996). Pyruvate metabolism in *Saccharomyces cerevisiae. Yeast* 12: 1607–33.
90. Lin SS, Manchester JK, Gordon JI (2001). Enhanced gluconeogenesis and increased energy storage as hallmarks of aging in *Saccharomyces cerevisiae. J Biol Chem.* 276: 36000–7.
91. Virag L, Szabo C (2002). The therapeutic potential of poly(ADP-Ribose) polymerase inhibitors. *Pharmacol Rev.* 54: 375–429.
92. Vaziri H, Dessain SK, Eaton EN, *et al.* (2001) hSIR2(SIRT1) functions as an NAD-dependent p53 deacetylase. *Cell* 107: 149–59.
93. Lal A, Lash AE, Altschul SF, *et al.* (1999). A public database for gene expression in human cancers. *Cancer Res.* 59: 5403–7.
94. Kassem H, Sangar V, Cowan R, Clarke N, Margison GP (2002). A potential role of heat shock proteins and nicotinamide N-methyl transferase in predicting response to radiation in bladder cancer. *Int J Cancer* 101: 454–60.

The Proteasome in Aging

Géraldine Carrard and Bertrand Friguet
Laboratoire de Biologie et Biochimie Cellulaire du Vieillissement (EA 3106 IFR 117/Biologie Systémique), Université Denis Diderot Paris 7, cc 7128, 2 place Jussieu, 75251 Paris Cedex 05, France

Introduction

The proteasome is a major intracellular proteolytic system found in Archaebacteria and Eukaryotes, and degrades the majority of intracellular proteins including oxidized and ubiquitinated proteins [1, 2]. There is a large body of experimental evidence indicating that proteasome function is compromised during aging, a feature that may have important implications in the cellular aging process [3–6]. Indeed, aging is characterized by a progressive and irreversible decline of the different physiological functions of the organism during the last part of its life. Increased modification of macromolecules is a common mark of aging [7, 8] and proteins have been reported to be crucial targets for numerous post-translational damages (e.g., oxidation, glycation, glycoxidation, conjugation with lipid peroxidation products) which have been shown to directly impair their biological functions [9–11]. Interestingly, calorie restriction, the only intervention that slows down aging, modulates the age-related accumulation of altered proteins [12, 13]. The age-related build up of oxidatively modified protein raises the problem of the efficacy of the proteolytic systems in charge of eliminating these damaged proteins, in particular the efficacy of the proteasomal system. Following a detailed description of the proteasomal system, the impact of aging on proteasome structure and function is discussed in the light of studies aimed at characterizing the fate of proteasome during oxidative stress. The possible implication of age-related alterations on the proteasomal system in immune senescence and neurodegeneration is then presented.

T. von Zglinicki (ed.), Aging at the Molecular Level, 213–231.

The proteasome

The proteasome was observed for the first time by Harris [14] in human erythrocytes and is ubiquitous among eukaryotic cells. The presence of the proteasome in Archaebacteria was more recently observed [15] and the 20S proteasome was isolated from Actinomycetes such as Mycobacterium [16] and Rhodococus [17]. The 20S proteasome, a 700 kDa multiproteolytic complex, represents up to 1% of soluble cellular proteins [18]. In the mammalian cell, this complex constitutes the main nonlysosomal proteolytic machinery implicated in protein degradation. The proteasome plays a crucial role in the turnover of cytosolic proteins but is not only a housekeeping enzyme. In fact, beside eliminating abnormal proteins that are either misfolded or altered, it also participates in the activation of essential functions of the cell. Indeed, the proteasome is implicated in a broad range of cellular pathways [19, 20] such as apoptosis [21, 22], cell cycle [23], cell differentiation [24], DNA repair [25, 26] and degradation of many important rate-limiting enzymes in metabolic pathways.

Structure

Organization

The 20S proteasome is located in the cytosol, in the nucleus and is associated to the endoplasmic reticulum [27]. It can be easily visualized by electron microscopy and exhibits a cylinder shape of about 15 nm length and 11 nm diameter [28]. Its quaternary structure consists of four stacked rings containing each seven subunits of 21 to 35 kDa [29]. The main difference between the archaebacterial and eukaryotic 20S proteasomes is the level of complexity. In fact, the archaebacterial proteasomes contain mainly two types of subunits, while eukaryotic proteasome contains 14 different subunits that can be classified in two families α and β according to their sequence alignment similarities [30]. The primary sequences of α subunits show up to 20% homology and the sequences of β subunits are more diverse. However, their tertiary structures are highly homologous and exhibit a sandwich of five anti-parallel β strands surrounded by five to six α helixes [1]. The X-ray diffraction analysis of the yeast proteasome clearly showed three cavities of about 5 mm diameter each [31]. The α subunits form the outer rings of the cylinder while the β subunits form the two inner rings. The two outer chambers are formed by the junction of one α and one β rings and the unique catalytic chamber is surrounded by the two β rings and contains the three types of β subunits carrying the active sites (β1, β2 and β5) [28, 31]. This compartment organization implies that substrate proteins are recognized and unfolded prior to their access to the catalytic sites.

Assembly

Assembly of the four rings of the 20S proteasome is a complex procedure, which involves specific chaperones and needs the processing of the pro-β-subunits [32, 33]. Indeed, five of the seven β subunits are expressed as subunit precursors containing a N-terminus propeptide which is cleaved during assembly [34–36] and the prosequence seems to be essential only in eukaryotic proteasome assembly [37]. The prosequences of β1, β2 and β5 protect the catalytic threonine residues of each subunit from N-α-acetylation [32] and are autocatalytically cleaved [34, 35]. The α subunits allow a tight regulation of the proteasome assembly and more particularly the α7 subunit which is thought to co-ordinate the assembly of the rest of the α subunits into the two outer rings [38].

Post-translational modifications

The proteasome is finely regulated by several physiological regulators. However, biochemical modifications on the proteasome subunits occur and can also modify proteasome function and specificity. Several studies reported that the peptidase activities of the 20S proteasome can be modulated by post-translational modification such as glycosylation in plants [39], or ADP-ribosylations [40], phosphorylations [41, 42] and oxidation [43] of mammalian proteasome subunits. γ-interferon treatment lowers the phosphorylation level of the α7 and α3 subunits on serine residues. This regulation favors the PA28–20S association and lowers the amount of 26S proteasome [44]. Finally, proteasome inhibitors such as lactacystin may promote structural modification of the 20S core proteasome, with increased exposure of cysteine residues, which are prone to S-thiolation. Proteasome glutathiolation is enhanced by inhibitors but still occurs in their absence and proteasome glutathiolation seems to be a mechanism of proteasome regulation *in vivo* [45].

Catalytic activities

The proteasome (E.C. 3.4.25.1) belongs to a particular class of hydrolases and its proteolytic mechanism consists of a nucleophile attack which is conducted by the N-terminus threonine residue of the three β catalytic subunits [46, 47]. The catalytic mechanism was solved using the crystal structure of yeast proteasome and site directed mutagenesis [34, 47, 48]. The highly conserved lysine 33 (a proton acceptor), the aspartic acid 17 and the N-terminus threonine form the catalytic triad. Two copies of each subunit can be visualized in the structure which means that the catalytic chamber contains a total of six active sites [31].. The direct environment of the threonine residues determines the specificity of the three main catalytic activities of the proteasome: chymotrypsin-like (CT-like), trypsin-like (T-like) and peptidylglutamyl peptide hydrolase (PGPH), that cleave peptides respectively on the carboxyl side of hydrophobic, basic and acidic residues [49]. These activities are routinely measured using fluorogenic peptide substrates consisting of a sequence of 3-4 amino acids linked to a fluorogenic group on its C-terminus. The release of the fluorophore is directly related to the amount of protease activity. The most commonly used peptides are succinyl-Leu-Leu-Val-Tyr-amido-4-methylcoumarin (CT-like activity), t-butoxycarbonyl-Leu-Ser-Thr-Arg-amido-4-methylcoumarin (T-like activity) and carbobenzoxy-Leu-Leu-Glu-β-naphthylamide (PGPH activity). The PGPH activity is also called caspase like activity since the hydrolysis of a caspase specific substrate containing an aspartic acid is 50 fold faster than the hydrolysis of the substrate containing the glutamic acid in the last position [49]. Site directed mutagenesis in yeast showed that the subunits β1, β2 and β5 carry the PGPH, T-like and CT-like activity respectively. Two other peptidase activities were also found to cut on the carboxyl side of branched-residues (BrAAP activity) and between small neutral residues (SNAAP activity) [50]. The substrate specificity of the proteasome is rather large and the substrate protein is degraded in a processive manner through the cylinder where the different sites influence each other [51]. In this process, the CT-like activity appears to be the rate-limiting step [34] and the two other activities could be regulated in an allosteric manner as proposed in the bite and chew model [49]. The substrate is degraded in 3 to 20 amino acids long peptides [1, 28], and the peptides are generally hydrolyzed in amino acids by endogenous cellular peptidases such as the tripeptidyl peptidase [52].

Inhibition

Assignment of proteasome functions in the cell would have not been possible without the development and use of specific inhibitors. The first class of inhibitors are natural derivatives such as epoxomycin [53] and lactacystin [54]. The carbon scaffold is closed into a lactone ring in the case of lactacystin and into a morpholino adduct in the case of epoxomycin [53, 55]. These two molecules inhibit the proteasome in an irreversible manner and abolish the three catalytic activities although to different extent. Indeed, lactacystin inhibits preferentially the CT-like activity, and has a greatly reduced effect on the T-like and the PGPH activity. Gliotoxin is also a natural derivative that inhibits reversibly the CT-like activity of the proteasome [56]. The second class of inhibitors consists of aldehyde derivatives of short peptides, analogous to the model substrate. They are all reversible inhibitors and differ by their specificity. The most commonly used is MG132 (carbobenzoxy-Leu-Leu-Leucinal) which inhibits CT-like and PGPH activities [57]. MG101 (acetyl-Leu-Leu-Norleucinal) and leupeptin (N-acetyl-Leu-Leu-Arginal) are inhibitors of the CT-like and the T-like activities respectively [58]. Those aldehydic inhibitors can inhibit other cellular proteases such as calpains. Other nonpeptidic inhibitors have been discovered lately such as boronate acids [59] and vinyl sulfones [60]. Various intracellular proteins were found to inhibit one or several proteasome activities [for review see ref. 61]. For example, the 31 kDa proteasome inhibitor (PI31) is a proline rich protein blocking protease and peptidase activities [62], and can prevent the association of the 20S core particle with its natural activators.

Activators

The 20S proteasome catalytic barrel is most likely closed in its latent form and can switch to an active form under diverse conditions such as heat treatment and addition of detergent at low concentration [18, 31]. Fatty acids can also stimulate the activities of the proteasome. For example, linoleic and linolenic acid can increase the CT-like and PGPH activities but inhibit the T-like activity [63]. The opening of the α rings can be induced by the attachment of regulatory complexes onto the 20S proteasome such as PA700 (19S) or PA28 (11S) or by point mutation of $\alpha 3$ subunit [64]. The 26S proteasome is an essential component of the ubiquitin and ATP-dependent proteolytic pathway. It results from the ATP-dependent association of the 20S proteasome with one ("single-capped") or two ("double-capped") particles of the 19S regulatory complex. The 19S complex is present in the nucleus as well as in the cytoplasm. The axial channel of the 20S proteasome is gated by the Rpt2 ATPase subunit of the 19S complex and controls both substrate entry and product release [65]. The 19S complex is formed of at least 18 subunits with molecular weight ranking from 25 to 110 kDa. Two components form the 19S regulator, the lid and the base. The base is composed of nine subunits, six of them being ATPase subunits, and binds to the 20S particle. The ATPase subunits exhibit a chaperone like activity and are believed to unfold proteins prior to their entry in the proteolytic chamber [66, 67]. The other eight subunits of the 19S regulator form the lid which is responsible for the recognition of the polyubiquitin signal [68, 69]. The 19S regulatory complex is thought to be the mouth of the proteasome [1]. The PA28 or 11S regulator or REG has been isolated in mammalians. Its association with the 20S proteasome is ATP-independent and increases peptidases activities while it does not improve protein degradation [62, 70]. The expression of this complex is induced in cells by treatment with γ interferon. PA28 is composed in the cytosol of two subunit types namely α and β of about 28 kDa [71] forming hexa or heptameric rings type $\alpha3\beta3$ or $\alpha3\beta4$ [72]. The C-terminus end of the PA28α subunit contains a KEKE motif which allows the association with the 20 S proteasome through interaction of the KEKE charges and those present at the surface of the proteasome [73]. The PA28 α and β subunits are related to a nuclear protein, the PA28γ subunit which forms heptameric rings $\gamma7$ and is able to activate proteasome peptidase activities [74]. The role of the PA28 complex is to stimulate 20S proteasome peptidase activity *in vitro* [75] and to facilitate product release *in vivo* [76]. Interestingly the composition of the heptameric rings of the PA28 influences the proteasome function. Indeed, the $\alpha3\beta4$, $\alpha7$ or $\beta7$ rings facilitate the hydrolysis of protein after basic, acid or hydrophobic residues into small peptides while $\gamma7$ rings allow only digestion after basic residues [74]. These findings could be explained by different conformational changes of the proteasome cylinder placing the active site in more accessible position [77]. The PA28-activated proteasome cannot recognize and degrade ubiquitinated substrates. Hybrid proteasome containing one PA28 and one 19S complex associated to the same 20S proteasome has been recently isolated [78]. In those hybrids, the 19S regulator is thought to facilitate the binding and entrance of the substrate protein into the 20S chamber while the 11S regulator controls the exit of the peptides [79].

Immunoproteasome

The assembly of the 20S proteasome with the PA28 activator plays a crucial role in antigen presentation. Indeed, the PA28 complex is thought to regulate the length of peptides produced for further addressing to the type I major histocompatibility complex (MHC class I) [80, 81]. The molecules of the MHC class I complex insure the transport of peptides from the reticulum to the cell surface necessary for antigen presentation [82]. The induction with γ-interferon does not only lead to the expression of PA28 subunits but also generates another kind of proteasome named immunoproteasome. Indeed, beside the 14 already described subunits, the mammalian proteasome contains three additional subunits that can be induced by cytokines such as γ-interferon and TNF-α which are produced by T cells in response to immune stimuli. The cells are then expressing *de novo* β1i (Lmp2), β2i (MECL1) and β5i (Lmp7) that replace β1 (Y), β2 (Z) and β5 (X) respectively [83, 84]. The replacement of subunits occurs during the assembly of the proteasome. This exchange leads to modification of the proteasome substrate specificity during the immune response. In fact, the subunit substitution results in higher CT-like and T-like activities and lower PGPH activity of the immunoproteasome. The decrease of PGPH activity is due to a conformational change of the active site after the replacement of β1 by β1i and leads to a lower affinity of the β1i active site for acidic residues and higher affinity of immunoproteasome for hydrophobic residues [85]. Recent studies showed that β5i plays a structural role and restores epitope presentation in immunoproteasome deficient cells [86, 87]. The resulting peptides with a carboxylic extremity have a higher affinity to the MHC class I complex [83, 88].

Age-related impairment of proteasome function

Since the 26S proteasome is the major cytosolic proteolytic machinery for degradation of abnormal and ubiquitinated proteins, alteration of its activity would therefore contribute to the well-documented accumulation of oxidized proteins with age [9]. Indeed, there is a large body of data indicating that oxidized protein degradation is primarily achieved by 20S proteasome in an ATP- and ubiquitin-independent pathway [2, 89] while other studies in lens have indicated that certain oxidatively modified proteins are degraded by the ubiquitin-proteasome pathway [90]. An impaired function of the proteasomal machinery would be expected to contribute to the age-related general decline of protein turnover, to the decreased ability of the aged cell to cope with stress and also to participate in the aetiology of several age-associated physio-pathological disorders (e.g., immune senescence and cataract formation) and neurodegenerative diseases (e.g., Alzheimer's and Parkinson's diseases).

During the last decade, experimental evidence has been gathered by different groups, including ours, indicating that there is an age-related impairment of proteasome function in a large number of tissues and organs [3, 5, 91–95]. From these and other studies on the fate of proteasome in oxidative and other proteotoxic stress situations, several mechanisms by which proteasome activity is affected have been proposed.

Alteration of proteasome peptidase activities during aging and replicative senescence

As early as 1996, pioneer studies of both Ward and our laboratory have first evidenced alterations of 20S proteasome (or multicatalytic proteinase) activities in rat liver [13, 96]. Both studies were reporting that the PGPH activity was the primary target for inactivation, this activity being reduced by roughly 50% in the 20S proteasome purified from 24-month-old animals as compared with that purified from the young ones. Of particular interest was the finding that dietary restriction of proteins or calories, the only known intervention that slows down aging, restored the PGPH proteasome activity in old animals [13, 97]. Further studies have suggested that T-like and CT-like peptidase activities may also decrease with age in rat liver and that both 20S and 26S proteasome were equally affected [98]. Since no apparent change of proteasome amount with age was observed in rat liver and inactivation was retained in purified proteasome preparations, post-translational modifications of subunits and/or alterations in proteasome assembly have been proposed to account for the reported decline of peptidase activities. This hypothesis was supported by the occurrence of subtle age-related modifications in the 2D gel electrophoresis pattern of 20S proteasome subunits purified from rat liver [97]. Furthermore, the age-related decline of PGPH activity has been correlated with an age-related induction of the proteasome subunit LMP2 [92]. Other investigations on different organs or tissues (e.g., heart, lung, kidney, spinal cord, cerebral cortex and hippocampus) and cell types (lymphocytes, fibroblasts and keratinocytes) have also demonstrated an age-related decline of proteasome activity [99–104] while no change was observed in brain stem and cerebellum [102].

Decreased proteasome activity has been recently associated to a decline of proteasome content in replicative senescence of human fibroblasts and keratinocytes as well as in aging rat myocardiac cells and human epidermis, suggesting that proteasome expression is down-regulated with age [101, 103, 105, 106]. Interestingly, when fibroblasts from healthy centenarians were examined for their proteasome subunit expression levels and peptidase activities, they were behaving rather like cultures from young donors suggesting an hypothetical contribution of a sustained proteasome activity to the successful aging of these individuals [101]. Additional support for a decreased expression of several proteasome subunits has been provided by gene expression profiling in aged mouse skeletal myocytes and human fibroblasts using the oligonucleotide microarray technology [107, 108]. In both situations, less than 2% of the 6347 genes investigated were differentially expressed with age with very few overlap between the two cell types. Still, in both cell types that are either mitotic or post-mitotic, a decreased expression of certain 20S (α2 and α7; β2 and β5i) and 26S (Rpt1, Rpt4, Rpn5 and Rpn6; Rpt5) proteasome subunits was observed in fibroblasts and myocytes, respectively. In mouse myocytes, caloric restriction was found to either maintain or even stimulate the expression of some proteasome subunit (Rpt5) and activator (PA 28 α subunit) [107]. It should be also noted that in a recent study of proteasome from rat muscle, age-dependent increases in proteasome peptidase activities and the expression of certain subunits (α2, α3, β6 and Rpt2) were observed [109]. Finally, not only proteasome content was found lower in aging human epidermis and rat heart, but purified proteasome subunit patterns analyzed by 2D gel electrophoresis were also indicative of modifications of either proteasome subunits or proteasome subunit composition [99, 105]. In addition, 26S proteasome inactivation has been recently correlated with increased modification of the proteasome by carboxymethyl-lysine (CML), a glycoxidation adduct, and by conjugation with 4-hydroxy-2-nonenal (HNE), a lipid peroxidation product in aging human lymphocytes [100].

Alteration of proteasome peptidase activities upon oxidative stress

Both PGPH and T-like peptidase activities have been reported to be inactivated upon metal-catalyzed oxidation of purified rat 20S proteasome *in vitro* [96, 110] while treatment of the proteasome with either 4-hydroxy-2-nonenal (HNE) or nitric oxide

(NO) were found to inactivate the T-like activity or the CT-like activity, respectively [111, 112]. Moreover, the 26S proteasome has been shown to be more sensitive to oxidative inactivation than the 20S proteasome [113]. Exposure of FAO rat hepatoma cells to metal-catalyzed oxidation upon treatment with iron and ascorbate resulted in inactivation of both PGPH and T-like activities of the proteasome [110]. The T-like activity was found to be protected from oxidative inactivation *in vitro* by HSP 90 and α-crystallin (a member of the HSP 27 family) and in FAO cells by overexpressing HSP 90, suggesting a targeted protection of the proteasome by these chaperone proteins. Based on the protective mechanism generated by the expression of the small stress protein Hsp27 in cells exposed to oxidative stress [114] and on the reduction of the load of oxidized and glycoxidized proteins in fibroblasts that have been subjected to repeated mild stress [115], the accumulation of oxidized protein appears to be also dependent on the heat shock response of the cell. In addition, mild oxidative stress of myotubes has also been shown to stimulate the proteasome CT-like activity and to induce an increased expression of the 20S proteasome α-subunits, the 19S p42 ATPase and the ubiquitin-conjugating enzyme $E2_{14k}$ [116]. Using the BJ human fibroblast cell line under replicative senescence, a decline of proteasome activity was observed in senescent cells together with a loss of their ability to induce an increase of oxidized protein degradation in response to mild oxidative stress induced by exposure to low concentration of H_2O_2 [117]. These results indicate that, in addition to age-related alterations in basal proteasome activity, aging is also characterized by an impaired ability of the proteasome to cope with stress.

Treatment by ferric nitriloacetate that induces oxidative stress in kidney and ischemia-reperfusion in brain were found to result in proteasome inactivation that was associated with an increased covalent modification of proteasome subunits by HNE [118, 119]. Moreover, inactivation of the T-like activity upon cardiac ischemia-reperfusion has been recently correlated with specific modification of three 20S proteasome subunits (α1, α2 and α4) by HNE [120]. In contrast, loss in CT-like and PGPH activities observed in crude extracts were relieved upon purification of the 20S proteasome suggesting that proteasome inhibition may also be due to the formation of endogenous inhibitors and/or substrates such as inhibitory damaged proteins. The occurrence of endogenous proteasome inhibitors is most likely relevant for aging since upon 20S proteasome purification from rat heart, the observed age-dependent decline of the CT-like activity in crude extracts was almost completely relieved while T-like and PGPH activities activities were also partially relieved [105]. In fact, HNE treatment of the model protein G6PDH has been previously shown to result in the formation of intra-molecular cross-links and to decrease its susceptibility to proteolysis by the proteasome [121]. Moreover, the HNE modified protein was shown to act as a non-competitive inhibitor of the proteasome for degradation of mildly oxidized protein [122]. Recent data showing that proteasome inhibition observed after UV irradiation of human keratinocytes was partially relieved after immunodepletion of HNE modified proteins from the cellular extracts strongly suggest a contribution of these modified proteins to both age- and oxidant-associated proteasome inhibition [123]. Moreover, intracellular accumulation of lipofuscin upon loading of WI 38 fibroblasts with ceroid was found associated with a decline of proteasome activity, suggesting that lipofuscin may be implicated in proteasome inhibition in living cells [124].

Proteasome in immune senescence

It is commonly accepted that older individuals fail to generate a vigorous immune response, particularly to antigens not previously encountered [125, 126]. This decline in immune responsiveness with age is due to loss of Th cell function which affects both cellular and humoral immunity [127, 128]. Thus, the decreased B cell responses to antigenic stimulation is due to Th cell deficiency and to alterations in B cell development [129]. In addition to lower antigenic response, an increase in autoantigenic response is observed with advancing age [128, 130]. The overall decline of the immune system is linked to several pathologies such as higher susceptibility to infections, autoimmunity and cancer [131–133].

Several studies have demonstrated the crucial role of the transcription factor NFκB in the activation of T cells through the activation of IL-2 and IL-2R genes [134]. The expression of the two latter have been shown to decline with age suggesting a default in their transcriptional activation. In the cytosol, NFκB is in a dormant form attached to its inhibitor IκB. The activation of NFκB occurs after stimulation by numerous agents such as cytokines (IL-1 and TNF-α), bacterial and viral infection [135]. Induced-degradation of IκB by the proteasome decreases with advancing age and results in the decreased induction of NFκB, and thus in the immune decline observed in the elderly [104]. Examination of activation-induced phophorylation and ubiquitination of IκB did not demonstrate any significant age-related alterations [104]. The lowered degradation of IκB was then associated to a decrease in proteasome function in the elderly. Indeed, proteasome CT-like activity was shown to decrease in T cell proteasome enriched fractions [136] and in purified 26S proteasome from human lymphocytes [100]. Neither 20 S proteasome [104] nor 26S proteasome content [100] seemed to decrease in the elderly samples. Since the observed lower activity was not due to a default in proteasome expression, we investigated the integrity of the proteasome structure during aging. The 19S complex subunits were poorly altered upon aging since only two of its subunits, the ATPase subunits S4 and S7, were glycated (CML) and/or conjugated with lipid peroxidation product (HNE). S4 subunit is thought to participate in 26S proteasome assembly in human cells [137]. However, glycation of this subunit did not appear to affect the stability of the 26S proteasome complex, since no dissociation into 20S and 19S was observed with age. In contrast, the 20S core was more prone to post-translational modification during aging. Indeed, α and β subunits were overall more affected by glycation, conjugation with lipid peroxidation product and ubiquitination in the elderly. Those modifications could have a direct impact on proteasome stability or activity. Indeed, modifications of α subunits could interfere with the accessibility of the substrate to the catalytic chamber and/or impact catalytic activities by destabilizing the interaction between regulatory α and catalytic β subunits. For example, the α7 subunit is thought to coordinate the assembly of the rest of the α subunits in human proteasome [38] and was more severely modified by glycation, conjugation with HNE and was ubiquitinated. The most interesting finding regarding lowered protease activity was the modification by both glycation and HNE conjugation of the β5i catalytic subunit which carries out the CT-like hydrolysis. Interestingly, despite glycation of β5i in early ages, the CT-like specific activity was not affected. However, this does not rule out the possibility that glycation occurring in samples from elderly donors may target more crucial lysine residues involved in the catalytic activity. In contrast, conjugation of β5i with HNE resulted in a concomitant decreased CT-like activity of the proteasome complex. Increased ubiquitination of β5i with age may also contribute to proteasome inactivation. The specific modification of 26S proteasome subunits could be central in the defect of activation of transcriptional factors implicated in the immune response and in antigen processing. A lower degradation of infectious protein agents by the 26S proteasome can then result directly in a higher infection level and indirectly in a lowered immune response of the elderly.

Impairment of proteasome function in brain aging and neurodegenerative diseases
Normal brain aging as well as most neurodegenerative disorders are characterized by an increased oxidative stress and the accumulation of damaged and/or aggregated proteins. In addition, proteasome inhibition has been demonstrated to elevate intracellular levels of protein oxidation, to increase neural vulnerability to subsequent injuries and to induce neuronal cell death *in vitro* [138]. It was first demonstrated by Keller and colleagues that there is an age-related decline of proteasome CT-like activity throughout the central nervous system of rats although no variation was detected in the brain stem and the cerebellum [102]. In the spinal cord, the loss of proteasome activity was further followed by a huge decline of 20S proteasome subunit expression [139]. Therefore, decline of proteasome activity in brain aging has been proposed to play a major role in both cell survival, neural vulnerability to subsequent injuries and neurodegeneration.

Evidence for an impaired proteasome function in both Alzheimer's and Parkinson's disease has been recently provided [140, 141]. Indeed, when proteasome activity was analyzed in short-post-mortem-interval autopsied brains from Alzheimer's disease and age-matched controls, a significant decrease was observed in the hippocampus and parahippocampal gyrus, superior and middle temporal gyri and inferior parietal lobule. However, proteasome activity was not decreased in the occipital lobe and in the cerebellum. No decrease of proteasome expression was associated with the loss of proteasome activity, favoring the possibility of proteasome subunit modification at the post-translational level [140]. A decline of all three proteasome peptidase activities was also observed in the substantia nigra of brains from Parkinson's patients as compared with age-matched controls [141]. In sporadic Parkinson's disease, this decline in proteasome peptidase activities was further correlated with a specific loss of 20S proteasome α-subunits in dopaminergic neurons but not in other brain regions [142]. Furthermore, infusion of the proteasome inhibitor lactacystin into the subtantia nigra pars compacta of rats resulted in a dose-dependent degeneration of dopaminergic cell bodies with the cytoplasmic accumulation and aggregation of α-synuclein to form inclusion bodies [143]. Taken together, these findings strongly suggest that the failure of the ubiquitin-proteasome system is critical for the accumulation of aggregated proteins in cytoplasmic inclusions referred to as Lewis bodies and appears as an important aetiopathogenic factor in Parkinson's disease [144].

The pathogenicity of protein aggregates in neurodegenerative disorders has been questioned for a long time. Recent studies report on a direct implication of protein aggregation on the impairment of the ubiquitin-proteasome system [145]. Indeed, transient expression of a huntingtin fragment containing a polyglutamine repeat and a folding mutant of the cystic fibrosis conductance regulator was accompanied by protein aggregation. The presence of intracellular aggregated protein was found to induce accumulation of ubiquitin conjugates and cell cycle arrest by causing an almost complete inhibition of the ubiquitin-proteasome system. Therefore, protein aggregates appear to be simultaneously inhibitors of the ubiquitin-proteasome pathway while being at the same time the products resulting from its inhibition. This process would imply an amplification phenomenon upon accumulation of aggregated proteins as well as threshold effect on the cell capacity to handle by increased amounts of ubiquitin conjugates and abnormal/damaged proteins. This last feature would provide an explanation for the rapid loss of neuronal function that characterizes the progression of most neurodegenerative diseases. In addition, this mechanism would parallel the induced proteasome inhibition reported upon accumulation of oxidized protein aggregates (artificial lipofuscin/ceroid) during post-mitotic aging of fibroblasts [124]. Finally, it should be pointed out that using SH-SY5Y neural cells stably transfected with polyglutamine-green fluorescent protein, polyglutamine expansion has been recently shown not to impair proteasome nor to elevate protein aggregate formation under basal conditions [146]. However, poly-glutamine-green fluorescent protein overexpressing cells exhibited a decreased ability of the proteasome to respond to stress as well as an increased stress-induced protein aggregation.

Conclusion

It is becoming clear that aging is accompanied by alterations of proteasome structure and function which are the result of different factors including decreased expression of proteasome components, production of inhibitory elements and inactivation by reactive oxygen species derived processes. Since proteasome mediated proteolytic processing is implicated in a variety of cellular events such as quality control of proteins, signal transduction, stress and immune response, cell proliferation and programmed cell death, functional impairment of the proteasomal system is expected to contribute to the deregulation of homeostasis and to the decreased capacity to cope with stress of the aged cell. Further studies are required to better establish the molecular mechanisms and pathophysiological implications of the age-dependent alterations of the proteasomal system. The ultimate goal will be to define strategies aimed at preserving the proteasome crucial function that may turn out to be useful for potential therapeutic interventions.

References

1. Coux O, Tanaka K, Goldberg AL (1996). Structure and functions of the 20S and 26S proteasomes. *Annu Rev Biochem.* 65: 801–47.
2. Davies KJ (2001). Degradation of oxidized proteins by the 20S proteasome. *Biochimie* 83: 301–10.
3. Carrard G, Bulteau AL, Petropoulos I, Friguet B (2002). Impairment of proteasome structure and function in aging. *Int J Biochem Cell Biol.* 34: 1461–74.
4. Dunlop RA, Rodgers KJ, Dean RT (2002). Recent developments in the intracellular degradation of oxidized proteins. *Free Radic Biol Med.* 33: 894–906.
5. Shringarpure R, Davies KJ (2002). Protein turnover by the proteasome in aging and disease. *Free Radic Biol Med.* 32: 1084–9.
6. Szweda PA, Friguet B, Szweda LI (2002). Proteolysis, free radicals, and aging. *Free Radic Biol Med.* 33: 29–36.
7. Beckman KB, Ames BN (1998). The free radical theory of aging matures. *Physiol Rev.* 78: 547–81.
8. Stadtman ER (1992). Protein oxidation and aging. *Science* 257: 1220–4.
9. Berlett BS, Stadtman ER (1997). Protein oxidation in aging, disease, and oxidative stress. *J Biol Chem.* 272: 20313–16.
10. Davies KJ (1993). Protein modification by oxidants and the role of proteolytic enzymes. *Biochem Soc Trans.* 21: 346–53.
11. Stadtman ER (1990). Metal ion-catalyzed oxidation of proteins: biochemical mechanism and biological consequences. *Free Radic Biol Med.* 9: 315–25.
12. Goto S, Takahashi R, Araki S, Nakamoto H (2002). Dietary restriction initiated in late adulthood can reverse age-related alterations of protein and protein metabolism. *Ann NY Acad Sci.* 959: 50–6.
13. Shibatani T, Nazir M, Ward WF (1996). Alterations of rat liver 20 S proteasome activities by age and food restriction. *J Gerontol Biol Sci.* 51: 316–322.
14. Harris JR (1968). Release of a macromolecular protein component from human erythrocyte ghosts. *Biochim Biophys Acta* 150: 534–7.
15. Bochtler M, Ditzel L, Groll M, Hartmann C, Huber R (1999). The proteasome. *Annu Rev Biophys Biomol Struct.* 28: 295–317.
16. Lupas A, Zwickl P, Baumeister W (1994). Proteasome sequences in eubacteria. *Trends Biochem Sci.* 19: 533–4.
17. Tamura T, Nagy I, Lupas A, *et al.* (1995). The first characterization of a eubacterial proteasome: the 20S complex of Rhodococcus. *Curr Biol.* 5: 766–74.
18. Tanaka K, Ii K, Ichihara A, Waxman L, Goldberg AL (1986). A high molecular weight protease in the cytosol of rat liver. I. Purification, enzymological properties, and tissue distribution. *J Biol Chem.* 261: 15197–203.
19. Ciechanover A, Orian A, Schwartz AL (2000). The ubiquitin-mediated proteolytic pathway: mode of action and clinical implications. *J Cell Biochem.* 34(Suppl.): 40–51.
20. Lecker SH, Solomon V, Mitch WE, Goldberg AL (1999). Muscle protein breakdown and the critical role of the ubiquitin-proteasome pathway in normal and disease states. *J Nutr.* 129: 227S–37S.
21. Chang YC, Lee YS, Tejima T, *et al.* (1998). mdm2 and bax, downstream mediators of the p53 response, are degraded by the ubiquitin-proteasome pathway. *Cell Growth Differ.* 9: 79–84.

22. Drexler HG (1998). Review of alterations of the cyclin-dependent kinase inhibitor INK4 family genes p15, p16, p18 and p19 in human leukemia-lymphoma cells. *Leukemia* 12: 845–59.
23. King RW, Deshaies RJ, Peters JM, Kirschner MW (1996). How proteolysis drives the cell cycle. *Science* 274: 1652–9.
24. Helin K (1998). Regulation of cell proliferation by the E2F transcription factors. *Curr Opin Genet Dev.* 8: 28–35.
25. Gillette TG, Huang W, Russell SJ, Reed SH, Johnston SA, Friedberg EC (2001). The 19S complex of the proteasome regulates nucleotide excision repair in yeast. *Genes Dev.* 15: 1528–39.
26. Pajonk F, McBride WH (2001). The proteasome in cancer biology and treatment. *Radiat Res.* 156: 447–59.
27. Kruger E, Kloetzel PM, Enenkel C (2001). 20S proteasome biogenesis. *Biochimie* 83: 289–93.
28. Voges D, Zwickl P, Baumeister W (1999). The 26S proteasome: a molecular machine designed for controlled proteolysis. *Annu Rev Biochem.* 68: 1015–68.
29. Lowe J, Stock D, Jap B, Zwickl P, Baumeister W, Huber R (1995). Crystal structure of the 20S proteasome from the archaeon *T. acidophilum* at 3.4 A resolution. *Science* 268: 533–9.
30. Coux O, Nothwang HG, Silva Pereira I, Recillas Targa F, Bey F, Scherrer K (1994). Phylogenic relationships of the amino acid sequences of prosome (proteasome, MCP) subunits. *Mol Gen Genet.* 245: 769–80.
31. Groll M, Ditzel L, Lowe J, *et al.* (1997). Structure of 20S proteasome from yeast at 2.4 A resolution [see comments]. *Nature* 386: 463–71.
32. Arendt CS, Hochstrasser M (1999). Eukaryotic 20S proteasome catalytic subunit propeptides prevent active site inactivation by N-terminal acetylation and promote particle assembly. *EMBO J.* 18: 3575–85.
33. Ramos PC, Hockendorff J, Johnson ES, Varshavsky A, Dohmen RJ (1998). Umplp is required for proper maturation of the 20S proteasome and becomes its substrate upon completion of the assembly. *Cell* 92: 489–99.
34. Chen P, Hochstrasser M (1996). Autocatalytic subunit processing couples active site formation in the 20S proteasome to completion of assembly. *Cell* 86: 961–72.
35. Heinemeyer W, Fischer M, Krimmer T, Stachon U, Wolf DH (1997). The active sites of the eukaryotic 20 S proteasome and their involvement in subunit precursor processing. *J Biol Chem.* 272: 25200–9.
36. Schmidtke G, Kraft R, Kostka S, *et al.* (1996). Analysis of mammalian 20S proteasome biogenesis: the maturation of beta-subunits is an ordered two-step mechanism involving autocatalysis. *EMBO J.* 15: 6887–98.
37. Seemuller E, Lupas A, Baumeister W (1996). Autocatalytic processing of the 20S proteasome. *Nature* 382: 468–71.
38. Gerards WL, de Jong WW, Bloemendal H, Boelens W (1998). The human proteasomal subunit HsC8 induces ring formation of other alpha-type subunits. *J Mol Biol.* 275: 113–21.
39. Schliephacke M, Kremp A, Schmid HP, Kohler K, Kull U (1991). Prosomes (proteasomes) of higher plants. *Eur J Cell Biol.* 55: 114–21.
40. Ullrich O, Reinheckel T, Sitte N, Hass R, Grune T, Davies KJ (1999). Poly-ADP ribose polymerase activates nuclear proteasome to degrade oxidatively damaged histones. *Proc Natl Acad Sci USA* 96: 6223–8.

41. Bose S, Mason GG, Rivett AJ (1999). Phosphorylation of proteasomes in mammalian cells. *Mol Biol Rep.* 26: 11–4.
42. Umeda M, Manabe Y, Uchimiya H (1997). Phosphorylation of the C2 subunit of the proteasome in rice (Oryza sativa L.). *FEBS Lett.* 403: 313–7.
43. Friguet B, Bulteau AL, Conconi M, Petropoulos I (2002). Redox control of 20S proteasome. *Methods Enzymol.* 353: 253–62.
44. Bose S, Brooks P, Mason GG, Rivett AJ (2001). γ-Interferon decreases the level of 26 S proteasomes and changes the pattern of phosphorylation. *Biochem J.* 353: 291–7.
45. Demasi M, Shringarpure R, Davies KJ (2001). Glutathiolation of the proteasome is enhanced by proteolytic inhibitors. *Arch Biochem Biophys.* 389: 254–63.
46. Brannigan JA, Dodson G, Duggleby HJ, *et al.* (1995). A protein catalytic framework with an N-terminal nucleophile is capable of self-activation. *Nature* 378: 416–9.
47. Seemuller E, Lupas A, Stock D, Lowe J, Huber R, Baumeister W (1995). Proteasome from Thermoplasma acidophilum: a threonine protease. *Science* 268: 579–82.
48. Brannigan JA, Dodson GG (1997). A short cut for the immune system. *Nat Struct Biol.* 4: 334–8.
49. Kisselev AF, Akopian TN, Castillo V, Goldberg AL (1999). Proteasome active sites allosterically regulate each other, suggesting a cyclical bite-chew mechanism for protein breakdown. *Mol Cell.* 4: 395–402.
50. Orlowski M, Cardozo C, Michaud C (1993). Evidence for the presence of five distinct proteolytic components in the pituitary multicatalytic proteinase complex. Properties of two components cleaving bonds on the carboxyl side of branched chain and small neutral amino acids. *Biochemistry* 32: 1563–72.
51. Akopian TN, Kisselev AF, Goldberg AL (1997). Processive degradation of proteins and other catalytic properties of the proteasome from *Thermoplasma acidophilum. J Biol Chem.* 272: 1791–8.
52. Tomkinson B (1999). Tripeptidyl peptidases: enzymes that count. *Trends Biochem Sci.* 24: 355–9.
53. Meng L, Mohan R, Kwok BH, Elofsson M, Sin N, Crews CM (1999). Epoxomicin, a potent and selective proteasome inhibitor, exhibits *in vivo* antiinflammatory activity. *Proc Natl Acad Sci USA* 96: 10403–8.
54. Dick LR, Cruikshank AA, Grenier L, Melandri FD, Nunes SL, Stein RL (1996). Mechanistic studies on the inactivation of the proteasome by lactacystin: a central role for clasto-lactacystin beta-lactone. *J Biol Chem.* 271: 7273–6.
55. Fenteany G, Standaert RF, Lane WS, Choi S, Corey EJ, Schreiber SL (1995). Inhibition of proteasome activities and subunit-specific amino-terminal threonine modification by lactacystin. *Science* 268: 726–31.
56. Kroll M, Arenzana-Seisdedos F, Bachelerie F, Thomas D, Friguet B, Conconi M (1999). The secondary fungal metabolite gliotoxin targets proteolytic activities of the proteasome. *Chem Biol.* 6: 689–98.
57. Rock KL, Gramm C, Rothstein L, *et al.* (1994). Inhibitors of the proteasome block the degradation of most cell proteins and the generation of peptides presented on MHC class I molecules. *Cell* 78: 761–71.
58. Mellgren RL (1997). Specificities of cell permeant peptidyl inhibitors for the proteinase activities of mu-calpain and the 20 S proteasome. *J Biol Chem.* 272: 29899–903.
59. Adams J, Behnke M, Chen S, *et al.* (1998). Potent and selective inhibitors of the proteasome: dipeptidyl boronic acids. *Bioorg Med Chem Lett.* 8: 333–8.
60. Bogyo M, McMaster JS, Gaczynska M, Tortorella D, Goldberg AL, Ploegh H (1997). Covalent modification of the active site threonine of proteasomal beta subunits and the

Escherichia coli homolog HslV by a new class of inhibitors. *Proc Natl Acad Sci USA* 94: 6629–34.
61. Kisselev AF, Goldberg AL (2001). Proteasome inhibitors: from research tools to drug candidates. *Chem Biol.* 8: 739–58.
62. Ma CP, Slaughter CA, DeMartino GN (1992). Identification, purification, and characterization of a protein activator (PA28) of the 20 S proteasome (macropain). *J Biol Chem.* 267: 10515–23.
63. Watanabe N, Yamada S (1996). Activation of 20S proteasomes from spinach leaves by fatty acids. *Plant Cell Physiol.* 37: 147–51.
64. Groll M, Bajorek M, Kohler A, *et al.* (2000). A gated channel into the proteasome core particle. *Nat Struct Biol.* 7: 1062–7.
65. Kohler A, Cascio P, Leggett DS, Woo KM, Goldberg AL, Finley D (2001). The axial channel of the proteasome core particle is gated by the Rpt2 ATPase and controls both substrate entry and product release. *Mol Cell.* 7: 1143–52.
66. Braun BC, Glickman M, Kraft R, *et al.* (1999). The base of the proteasome regulatory particle exhibits chaperone-like activity. *Nat Cell Biol.* 1: 221–6.
67. Kloetzel PM (2001). Antigen processing by the proteasome. *Nat Rev Mol Cell Biol.* 2: 179–87.
68. Deveraux Q, Ustrell V, Pickart C, Rechsteiner M (1994). A 26 S protease subunit that binds ubiquitin conjugates. *J Biol Chem.* 269: 7059–61.
69. van Nocker S, Sadis S, Rubin DM, *et al.* (1996). The multiubiquitin-chain-binding protein Mcb1 is a component of the 26S proteasome in *Saccharomyces cerevisiae* and plays a nonessential, substrate-specific role in protein turnover. *Mol Cell Biol.* 16: 6020–8.
70. Dubiel W, Pratt G, Ferrell K, Rechsteiner M (1992). Purification of an 11 S regulator of the multicatalytic protease. *J Biol Chem.* 267: 22369–77.
71. Mott JD, Pramanik BC, Moomaw CR, Afendis SJ, DeMartino GN, Slaughter CA (1994). PA28, an activator of the 20 S proteasome, is composed of two nonidentical but homologous subunits. *J Biol Chem.* 269: 31466–71.
72. Knowlton JR, Johnston SC, Whitby FG, *et al.* (1997). Structure of the proteasome activator REGalpha (PA28alpha). *Nature* 390: 639–43.
73. Realini C, Rogers SW, Rechsteiner M (1994). KEKE motifs. Proposed roles in protein-protein association and presentation of peptides by MHC class I receptors. *FEBS Lett.* 348: 109–13.
74. Realini C, Jensen CC, Zhang Z, *et al.* (1997). Characterization of recombinant REGalpha, REGbeta, and REGgamma proteasome activators. *J Biol Chem.* 272: 25483-92.
75. Stohwasser R, Salzmann U, Giesebrecht J, Kloetzel PM, Holzhutter HG (2000). Kinetic evidences for facilitation of peptide channelling by the proteasome activator PA28. *Eur J Biochem.* 267: 6221–30.
76. Whitby FG, Masters EI, Kramer L, *et al.* (2000). Structural basis for the activation of 20S proteasomes by 11S regulators. *Nature* 408: 115–20.
77. Li J, Gao X, Ortega J, *et al.* (2001). Lysine 188 substitutions convert the pattern of proteasome activation by REGgamma to that of REGs alpha and beta. *EMBO J.* 20: 3359–69.
78. Rivett AJ, Bose S, Brooks P, Broadfoot KI (2001). Regulation of proteasome complexes by gamma-interferon and phosphorylation. *Biochimie* 83: 363–6.

79. Tanahashi N, Murakami Y, Minami Y, Shimbara N, Hendil KB, Tanaka K (2000). Hybrid proteasomes. Induction by interferon-gamma and contribution to ATP-dependent proteolysis. *J Biol Chem.* 275: 14336–45.
80. Preckel T, Fung-Leung WP, Cai Z, *et al.* (1999). Impaired immunoproteasome assembly and immune responses in PA28$^{-/-}$ mice. *Science* 286: 2162–5.
81. Rechsteiner M, Realini C, Ustrell V (2000). The proteasome activator 11 S REG (PA28) and class I antigen presentation. *Biochem J.* 345(Pt 1): 1–15.
82. Rock KL, Goldberg AL (1999). Degradation of cell proteins and the generation of MHC class I-presented peptides. *Annu Rev Immunol.* 17: 739–79.
83. Belich MP, Trowsdale J (1995). Proteasome and class I antigen processing and presentation. *Mol Biol Rep.* 21: 53–6.
84. Hisamatsu H, Shimbara N, Saito Y, *et al.* (1996). Newly identified pair of proteasomal subunits regulated reciprocally by interferon gamma. *J Exp Med.* 183: 1807–16.
85. Groettrup M, Khan S, Schwarz K, Schmidtke G. (2001). Interferon-gamma inducible exchanges of 20S proteasome active site subunits: why? *Biochimie* 83: 367–72.
86. Gileadi U, Moins-Teisserenc HT, Correa I, *et al.* (1999). Generation of an immunodominant CTL epitope is affected by proteasome subunit composition and stability of the antigenic protein. *J Immunol.* 163: 6045–52.
87. Sijts AJ, Ruppert T, Rehermann B, Schmidt M, Koszinowski U, Kloetzel PM (2000). Efficient generation of a hepatitis B virus cytotoxic T lymphocyte epitope requires the structural features of immunoproteasomes. *J Exp Med.* 191: 503–14.
88. Gaczynska M, Rock KL, Goldberg AL (1993). Gamma-interferon and expression of MHC genes regulate peptide hydrolysis by proteasomes. *Nature* 365: 264–7.
89. Grune T, Reinheckel T, Davies KJ (1997). Degradation of oxidized proteins in mammalian cells. *FASEB J.* 11: 526–34.
90. Shang F, Nowell TR Jr, Taylor A (2001). Removal of oxidatively damaged proteins from lens cells by the ubiquitin-proteasome pathway. *Exp Eye Res.* 73: 229–38.
91. Friguet B, Bulteau AL, Chondrogianni N, Conconi M, Petropoulos I (2000). Protein degradation by the proteasome and its implications in aging. *Ann NY Acad Sci.* 908: 143–54.
92. Gaczynska M, Osmulski PA, Ward WF (2001). Caretaker or undertaker? The role of the proteasome in aging. *Mech Ageing Dev.* 122: 235–54.
93. Goto S, Takahashi R, Kumiyama AA, *et al.* (2001). Implications of protein degradation in aging. *Ann NY Acad Sci.* 928: 54–64.
94. Grune T (2000). Oxidative stress, aging and the proteasomal system. *Biogerontology* 1: 31–40.
95. Keller JN, Gee J, Ding Q (2002). The proteasome in brain aging. *Ageing Res Rev.* 1: 279–93.
96. Conconi M, Szweda LI, Levine RL, Stadtman ER, Friguet B (1996). Age-related decline of rat liver multicatalytic proteinase activity and protection from oxidative inactivation by heat-shock protein 90. *Arch Biochem Biophys*: 331: 232–40.
97. Anselmi B, Conconi M, Veyrat-Durebex C, *et al.* (1998). Dietary self-selection can compensate an age-related decrease of rat liver 20 S proteasome activity observed with standard diet. *J Gerontol A Biol Sci Med Sci.* 53: B173–9.
98. Hayashi T, Goto S (1998). Age-related changes in the 20S and 26S proteasome activities in the liver of male F344 rats. *Mech Ageing Dev.* 102: 55–66.
99. Bulteau A, Petropoulos I, Friguet B (2000). Age-related alterations of proteasome structure and function in aging epidermis. *Exp Gerontol.* 35: 767–77.

100. Carrard G, Dieu M, Toussaint O, Raes M, Friguet B (2003). Impact of ageing on proteasome structure and function in human lymphocytes. *Int J Biochem Cell Biol.* 35: 728–39.
101. Chondrogianni N, Petropoulos I, Franceschi C, Friguet B, Gonos ES (2000). Fibroblast cultures from healthy centenarians have an active proteasome. *Exp Gerontol.* 35: 721–8.
102. Keller JN, Hanni KB, Markesbery WR (2000). Possible involvement of proteasome inhibition in aging: implications for oxidative stress. *Mech Ageing Dev.* 113: 61–70.
103. Merker K, Sitte N, Grune T (2000). Hydrogen peroxide-mediated protein oxidation in young and old human MRC-5 fibroblasts. *Arch Biochem Biophys.* 375: 50–4.
104. Ponnappan U, Zhong M, Trebilcock GU (1999). Decreased proteasome-mediated degradation in T cells from the elderly: a role in immune senescence. *Cell Immunol.* 192: 167–74.
105. Bulteau AL, Szweda LI, Friguet B (2002). Age-dependent declines in proteasome activity in the heart. *Arch Biochem Biophys.* 397: 298–304.
106. Petropoulos I, Conconi M, Wang X, *et al.* (2000). Increase of oxidatively modified protein is associated with a decrease of proteasome activity and content in aging epidermal cells. *J Gerontol A Biol Sci Med Sci.* 55: B220–7.
107. Lee CK, Klopp RG, Weindruch R, Prolla TA (1999). Gene expression profile of aging and its retardation by caloric restriction. *Science* 285: 1390–3.
108. Ly DH, Lockhart DJ, Lerner RA, Schultz PG (2000). Mitotic misregulation and human aging. *Science* 287: 2486–92.
109. Bardag-Gorce F, Farout L, Veyrat-Durebex C, Briand Y, Briand M (1999). Changes in 20S proteasome activity during ageing of the LOU rat. *Mol Biol Rep.* 26: 89–93.
110. Conconi M, Petropoulos I, Emod I, Turlin E, Biville F, Friguet B (1998). Protection from oxidative inactivation of the 20S proteasome by heat-shock protein 90. *Biochem J.* 333: 407–15.
111. Conconi M, Friguet B (1997). Proteasome inactivation upon aging and on oxidation-effect of HSP 90. *Mol Biol Rep.* 24: 45–50.
112. Glockzin S, von Knethen A, Scheffner M, Brune B (1999). Activation of the cell death program by nitric oxide involves inhibition of the proteasome. *J Biol Chem.* 274: 19581–6.
113. Reinheckel T, Sitte N, Ullrich O, Kuckelkorn U, Davies KJ, Grune T (1998). Comparative resistance of the 20S and 26S proteasome to oxidative stress. *Biochem J.* 335: 637–42.
114. Arrigo AP (2001). Hsp27: novel regulator of intracellular redox state. *IUBMB Life* 52: 303–7.
115. Verbeke P, Clark BF, Rattan SI (2001). Reduced levels of oxidized and glycoxidized proteins in human fibroblasts exposed to repeated mild heat shock during serial passaging *in vitro*. *Free Radic Biol Med.* 31: 1593–602.
116. Gomes-Marcondes MC, Tisdale MJ (2002). Induction of protein catabolism and the ubiquitin-proteasome pathway by mild oxidative stress. *Cancer Lett.* 180: 69–74.
117. Sitte N, Merker K, Von Zglinicki T, Grune T, Davies KJ (2000). Protein oxidation and degradation during cellular senescence of human BJ fibroblasts: part I – effects of proliferative senescence. *FASEB J.* 14: 2495–502.
118. Keller JN, Huang FF, Zhu H, Yu J, Ho YS, Kindy TS (2000). Oxidative stress-associated impairment of proteasome activity during ischemia-reperfusion injury. *J Cereb Blood Flow Metab.* 20: 1467–73.

119. Okada K, Wangpoengtrakul C, Osawa T, Toyokuni S, Tanaka K, Uchida K (1999). 4-Hydroxy-2-nonenal-mediated impairment of intracellular proteolysis during oxidative stress. Identification of proteasomes as target molecules. *J Biol Chem.* 274: 23787–93.
120. Bulteau AL, Lundberg KC, Humphries KM, *et al.* (2001). Oxidative modification and inactivation of the proteasome during coronary occlusion/reperfusion. *J Biol Chem.* 25: 25.
121. Friguet B, Stadtman ER, Szweda LI (1994). Modification of glucose-6-phosphate dehydrogenase by 4-hydroxy-2-nonenal. Formation of cross-linked protein that inhibits the multicatalytic protease. *J Biol Chem.* 269: 21639–43.
122. Friguet B, Szweda LI (1997). Inhibition of the multicatalytic proteinase (proteasome) by 4-hydroxy-2-nonenal cross-linked protein. *FEBS Lett.* 405: 21–5.
123. Bulteau AL, Moreau M, Nizard C, Friguet B (2002). Impairment of proteasome function upon UVA- and UVB-irradiation of human keratinocytes. *Free Radic Biol Med.* 32: 1157–70.
124. Sitte N, Huber M, Grune T, *et al.* (2000). Proteasome inhibition by lipofuscin/ceroid during postmitotic aging of fibroblasts. *FASEB J.* 14: 1490–8.
125. Ginaldi L, Loreto MF, Corsi MP, Modesti M, De Martinis M (2001). Immunosenescence and infectious diseases. *Microbes Infect.* 3: 851–7.
126. Webster RG (2000). Immunity to influenza in the elderly. *Vaccine* 18: 1686–9.
127. Gravekamp C (2001). Tailoring cancer vaccines to the elderly: the importance of suitable mouse models. *Mech Ageing Dev.* 122: 1087–105.
128. Weksler ME, Szabo P (2000). The effect of age on the B-cell repertoire. *J Clin Immunol*: 20 240–9.
129. Kline GH, Hayden TA, Klinman NR (1999). B cell maintenance in aged mice reflects both increased B cell longevity and decreased B cell generation. *J Immunol.* 162: 3342–9.
130. Stacy S, Krolick KA, Infante AJ, Kraig E (2002). Immunological memory and late onset autoimmunity. *Mech Ageing Dev.* 123: 975–85.
131. Ben-Yehuda A, Danenberg HD, Zakay-Rones Z, Gross DJ, Friedman G (1998). The influence of sequential annual vaccination and of DHEA administration on the efficacy of the immune response to influenza vaccine in the elderly. *Mech Ageing Dev.* 102: 299–306.
132. Dunn PL, North RJ (1991). Effect of advanced ageing on the ability of mice to cause tumour regression in response to immunotherapy. *Immunology* 74: 355–9.
133. Miller RA (2000). Effect of aging on T lymphocyte activation. *Vaccine* 18: 1654–60.
134. Pimentel-Muinos FX, Mazana J, Fresno M (1994). Regulation of interleukin-2 receptor alpha chain expression and nuclear factor.kappa B activation by protein kinase C in T lymphocytes. Autocrine role of tumor necrosis factor alpha. *J Biol Chem.* 269: 24424–9.
135. Ponnappan U (1998). Regulation of transcription factor NFkappa B in immune senescence. *Front Biosci.* 3: D152–68.
136. Ponnappan U (2002). Ubiquitin-proteasome pathway is compromised in CD45RO(+) and CD45RA(+) T lymphocyte subsets during aging. *Exp Gerontol.* 37: 359–67.
137. Mason GG, Murray RZ, Pappin D, Rivett AJ (1998). Phosphorylation of ATPase subunits of the 26S proteasome. *FEBS Lett* 430: 269–74.
138. Ding Q, Keller JN (2001). Proteasomes and proteasome inhibition in the central nervous system. *Free Radic Biol Med.* 31: 574–84.
139. Keller JN, Huang FF, Markesbery WR (2000). Decreased levels of proteasome activity and proteasome expression in aging spinal cord. *Neuroscience* 98: 149–56.
140. Keller JN, Hanni KB, Markesbery WR (2000). Impaired proteasome function in Alzheimer's disease. *J Neurochem*: 75: 436–9.

141. McNaught KS, Jenner P (2001). Proteasomal function is impaired in substantia nigra in Parkinson's disease. *Neurosci Lett.* 297: 191–4.
142. McNaught KS, Belizaire R, Jenner P, Olanow,CW, Isacson O (2002). Selective loss of 20S proteasome alpha-subunits in the substantia nigra pars compacta in Parkinson's disease. *Neurosci Lett.* 326: 155–8.
143. McNaught KS, Bjorklund LM, Belizaire R, Isacson O, Jenner P, Olanow CW (2002). Proteasome inhibition causes nigral degeneration with inclusion bodies in rats. *Neuroreport* 13: 1437–41.
144. McNaught KS, Olanow CW, Halliwell B, Isacson O, Jenner P (2001). Failure of the ubiquitin-proteasome system in Parkinson's disease. *Nat Rev Neurosci.* 2: 589–94.
145. Bence NF, Sampat RM, Kopito RR (2001). Impairment of the ubiquitin-proteasome system by protein aggregation. *Science* 292: 1552–5.
146. Ding Q, Lewis JJ, Strum KM, *et al.* (2002). Polyglutamine expansion, protein aggregation, proteasome activity, and neural survival. *J Biol Chem.* 277: 13935–42.

Aging and Lysosomal Degradation of Cellular Constituents

Alexei Terman and Ulf T. Brunk
Division of Pathology II, Faculty of Health Sciences, Linköping University, Linköping, Sweden

Lysosomes and recycling of cellular components

Biological structures are continuously renewed through degradation and resynthesis of their worn-out/damaged constituents. Cells possess a number of mechanisms to degrade their components. Many cytosolic proteins, mainly short-lived ones, are decomposed by calcium-dependent cysteine proteases, calpains [1], as well as by multicatalytic proteinase complexes, proteasomes [2]. Most long-lived proteins, lipids, other biomolecules, and all organelles are, however, degraded by lysosomes, acidic vacuolar organelles containing several dozens lytic enzymes [3, 4]. Initially, the material to be degraded is sequestered and wrapped up in an autophagosome, a vacuolar structure surrounded by a specific membrane. Autophagosomes then fuse with lysosomes [5, 6]. This process is called macroautophagy. Besides, the material can enter lysosomes through invagination of the membrane (microautophagy) [7] or by selective chaperone-mediated autophagy [6]. Within lysosomes, the material is decomposed into simple molecules, such as amino acids, fatty acids and monosaccharides, which are carried into the cytosol and reutilized in anabolic activities. A failure to synthesize even a single lysosomal enzyme results in serious, often fatal, disorders, known as lysosomal storage diseases [8, 9]. Normally, lysosomes efficiently degrade most autophagocytosed macromolecules and organelles, providing for their successful recycling. The role of lysosomes, along with other recycling systems, is particularly important for postmitotic cells, which cannot renew themselves by cell division. In contrast, proliferating cells, such as intestinal epitheliocytes or bone marrow cells, continuously dilute their worn-out/damaged constituents during successive divisions, normally associated with intense de novo formation of macromolecules and organelles [10].

T. von Zglinicki (ed.), Aging at the Molecular Level, 233–242.

Oxidative stress and macromolecular damage

Accumulating evidence suggests that oxygen-derived free radicals are the most important contributors to overall macromolecular damage [11]. Normal oxygen metabolism is associated with ineluctable, although minor, electron leak resulting in the intra-mitochondrial formation of superoxide anion radicals, $O_2^{\cdot-}$. The latter are toxic to mitochondrial enzymes, particularly to aconitase [11]. Other effects of superoxide include the conversion of Fe^{3+} into Fe^{2+} (which catalyzes Fenton reactions; see below), and the formation of $HO_2^{\cdot}$, an uncharged and more reactive protonated form of superoxide, easily crossing biological membranes [11, 12]. Most mitochondrially-produced superoxide is, however, dismutated and reduced into H_2O_2 by manganese superoxide dismutase, SOD2. Less abundant, non-mitochondrial superoxide forms as a result of the activity of cytosolic oxidases, the endoplasmic P_{450} system, and peroxisomes, and is then converted to H_2O_2 by copper-zinc superoxide dismutase, SOD1. H_2O_2 is an uncharged molecule, which readily diffuses throughout the cell. It is then eliminated by glutathione peroxidase and catalase [11]. Unfortunately, in the Fenton reaction, catalyzed by free Fe^{2+} (usually present in cells, especially within lysosomes), a small portion of H_2O_2 is also converted to the highly reactive hydroxyl radical, $HO^{\cdot}$:

$$Fe^{2+} + H_2O_2 \rightarrow Fe^{3+} + OH^- + HO^{\cdot}$$

Formation of $HO^{\cdot}$ makes oxygen toxicity akin to radiation-induced damage, associated with the radiolysis of water. $HO^{\cdot}$ reacts with surrounding biomolecules including nucleic acids, proteins and lipids. In unsaturated fatty acids $HO^{\cdot}$ initiates a chain reaction, yielding organic peroxides:

$$LH + HO^{\cdot} \rightarrow L^{\cdot} + H_2O$$
$$L^{\cdot} + O_2 \rightarrow LOO^{\cdot}$$
$$LOO^{\cdot} + LH \rightarrow LOOH + L^{\cdot}$$

Many organic peroxides are unstable and easily form aldehydes, such as malondialdehyde, $CHO\text{-}CH_2\text{-}CHO$. Aldehydes are known to possess a number of toxic effects. In particular, they contribute to protein-protein cross-linking by forming aldhyde bridges [13]. Oxygen metabolism, therefore, is associated with formation of various toxic substances, commonly called reactive oxygen species, ROS, exerting multiple harmful effects.

Denham Harman was the first to connect aging with oxygen toxicity. According to his free radical theory of aging [14], molecular damage induced by physiologically formed ROS progressively accumulates, leading to eventual functional decline, decreasing adaptability and death. This theory is supported by numerous findings indicating acceleration of aging by increased oxidative stress and its retardation by administration of antioxidants [15]. Oxidative stress seems to be the main, but not the only factor responsible for age-related damage. Glucose and other reducing sugars can react with amino groups of proteins resulting in the formation of advanced glycosylation end products, AGEs, known to mediate protein-protein cross-linking [16]. Moreover, reducing sugars are reported to induce mutations [17]. Theoretically, free radical formation associated with natural radioactivity also might be involved in aging, but in fact, the levels of background radiation are not high enough to substantially contribute to age-related damage [10].

By recognizing the role of free radicals and other possible damaging factors in aging, we, however, cannot explain why the molecular damage accumulates despite the continual renewal process. The only possible explanation for this is that the renewal process does not occur with perfect accuracy. Some concepts of aging, such as the error catastrophe theory [18] and the somatic mutation theory [19], suggest that synthetic errors are responsible for imperfect renewal. This, however, was not proved to be the case [20]. We, therefore, believe that rather than an erroneous synthesis, it is an imperfect removal of damaged biological material that mainly triggers the aging process [10, 21]. Within postmitotic cells, known to be most dramatically affected by age, the proportion of this damaged material progressively increases, preventing synthesis and interfering with functions of normal biological structures.

Imperfect lysosomal function and cellular aging

Damaged and functionally disabled biological structures, which have not been removed by recycling systems, can be referred to as biological "garbage," or waste material [10, 21]. This includes various altered biomolecules, organelles, and lipofuscin, a non-degradable intralysosomal substance (Figure 1). Biological "garbage" starts to accumulate early in life and increases gradually throughout the life span, indicating that the imperfect removal of damaged structures is an inherent characteristic of cells, not acquired with age.

Age-related DNA alterations include adducts of bases and sugar groups, single- and double-strand breaks, and DNA-protein cross-links. Modification of proteins involves oxidation of various parts of the molecules, reactions with aldehydes, peptide-fragmentation and protein-protein cross-linking. Apparently, aberrant proteins may also result from damage to nucleic acids and the ensuing alterations to protein synthesis [13, 22]. Oxidation of lipids, resulting in their decomposition, formation of hydroperoxide- and alkyl radicals, cyclic endoperoxides and aldehydes, may in turn enhance damage to nucleic acids and proteins [23].

Whereas age-related molecular alterations reflect imperfect functions of various recycling mechanisms (including DNA repair system, cytosolic proteases and proteasomes), the accumulation of damaged organelles is associated with insufficient autophagocytosis, normally providing for further intralysosomal degradation. Cellular aging is characterized by progressive damage to ribosomes, peroxisomes, endoplasmic reticulum, and proteasomes [10, 24]. Of all cellular organelles, mitochondria undergo the most dramatic changes with age, probably because these organelles are the main sites of radical generation [12, 25]. Among the possible reasons for the high vulnerability of mitochondria are the properties of mitochondrial DNA (mtDNA) which, unlike nuclear DNA, is not protected by histones,

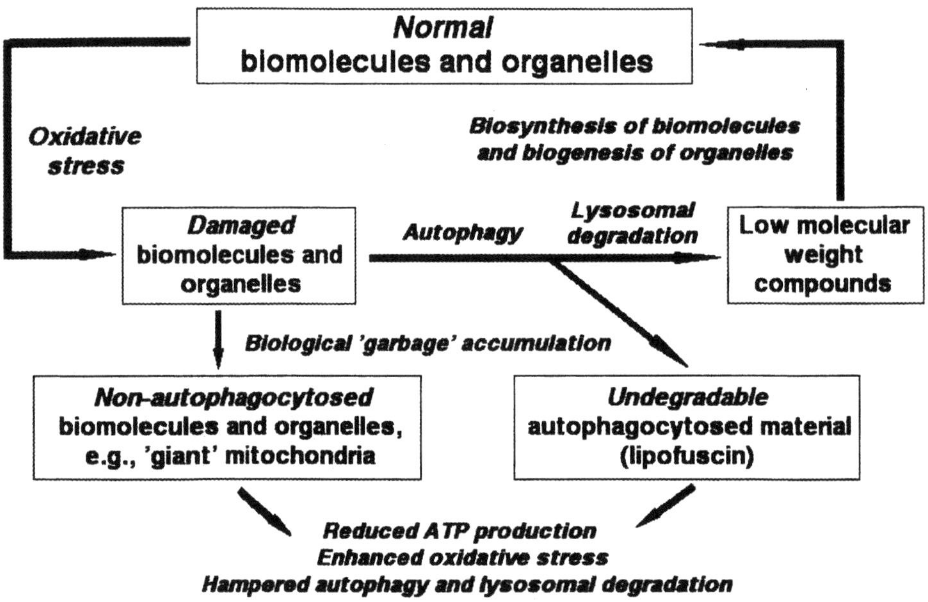

Figure 1. Scheme illustrating the role of lysosomes in aging of postmitotic cells. Names of biological structures are framed; arrows indicate biological processes.

contains a substantially higher proportion of expressed genes, and shows less efficient repair, at least for some types of lesions [26, 27]. Consistent with this, aging primarily affects complexes I and IV of the electron-transport chain, coded by mtDNA, but not proteins which are mainly coded by nuclear genes [26, 27]. Morphologically, many mitochondria of aged cells present an abnormal appearance, with such features as enlargement, swelling, loss of cristae, and partial homogenization of matrix and inner membranes [28, 29]. Some mitochondria become enormously large and, hence, are called "giant" [30]. Giant mitochondria are usually morphologically and functionally defective. ATP production by aging mitochondria progressively declines [31].

It is not understood what interferes with mitochondrial autophagocytosis. One possibility is that some mitochondria initially enlarge (e.g., because of impaired fission associated with DNA/protein damage), and then selectively accumulate, since their autophagocytosis is apparently more difficult than that of small ones [21]. Alternatively, it was proposed that mitochondria are selected for degradation depending on the degree of oxidative damage to their membranes. Because mitochondria with defective DNA are respiratory-deficient, their membranes experience reduced radical attack, resulting in decreased autophagy and preferential survival of such mitochondria [32]. There is also evidence that in some cases mutated mitochondria may have a replicative advantage over normal ones [33, 34].

Lipofuscin, an intralysosomal waste material

Autophagocytosed material is normally degraded within lysosomes, but evidence suggests that lysosomal enzymes fail to degrade all macromolecular substances. Cells also do not exocytose lysosomes with undegraded material to any substantial extent [35]. As a result, undegraded substances gradually accumulate in aging postmitotic cells in the form of brown-yellow, autofluorescent, electron-dense, lysosomal granules called lipofuscin, or age pigment [36]. Lipofuscin represents a mixture of various oxidized biomolecules, mainly of protein and lipid origin (30–70% and 20–50%, respectively) [36, 37]. It also contains small amounts of carbohydrates [37, 38] and metals, including considerable amounts of iron [39, 40]. Besides aging, the undegradable intralysosomal material can be a characteristic of various pathological conditions, including lysosomal storage diseases, vitamin E deficiency, radiation effects, tumors, etc. [36]. This pigment is practically identical to age pigment by its properties and is usually called "ceroid-type lipofuscin" or simply "ceroid." In a number of cell types, such as retinal pigment epithelial cells and macrophages, lipofuscin/ceroid preferentially derives from heterophagocytosed material [41, 42].

An enhancement of lipofuscin formation by oxidative stress and its decrease by administration of antioxidants or iron chelators have been repeatedly demonstrated [36]. These data suggest that oxidation makes biomolecules indigestible by lysosomal enzymes. Macromolecular cross-linking is one of the most typical consequences of ROS attack, and one of the most probable factors responsible for undegradability of autophagocytosed material [43]. Although undegradable material potentially can be formed in any cellular compartment, the most likely site for such formation is the lysosome itself. Lysosomes not only contain autophagocytosed macromolecular substances (including lipidaceous membrane components), but also experience a permanent flux of H_2O_2 and are rich in iron in its reduced Fe^{2+} form, in particular, due to the presence of the reducing agent cysteine. Consequently, many biomolecules become oxidized and undegradable before being degraded by lysosomal enzymes [44]. This hypothesis is supported by the evidence that oxidative stress in combination with protease inhibition (the latter delays degradation of autophagocytosed material thus prolonging time for its oxidation) dramatically enhances lipofuscin formation. This combined effect was almost three times as great as the sum of separate effects of oxidative stress and protease inhibition [45].

Mitochondrial components seem to be a major source for lipofuscin formation. Being affected by ROS more than other cellular compartments, mitochondria contain peroxidized macromolecules and experience extensive autophagocytosis [46]. Furthermore, they also contain iron-rich heme proteins and potentially may contribute to ROS production within lysosomes after being autophagocytosed. Autophagocytosis of mitochondria has been demonstrated electron microscopically [47], and mitochondrial proteins have been found in lipofuscin granules [48, 49].

The consequences of lipofuscin accumulation

By occupying part of the lysosomal system, lipofuscin apparently complicates its functions, forcing it to increase in volume and produce more hydrolytic enzymes [21]. Because of the high adaptability of the lysosomal system, even heavily lipofuscin-loaded postmitotic cells may remain alive. For example, in some human neurons, lipofuscin was reported to occupy up to 75% of the perikaryon [50]. Nevertheless, the adaptability of such cells is obviously limited, which is consistent with profound cellular dysfunctions in lysosomal storage diseases, associated with heavy accumulation of ceroid-type lipofuscin [8, 9]. Furthermore, lipofuscin-loaded human fibroblasts displayed decreased autophagy and shortened survival time when exposed to amino acid starvation [51]. In cultured retinal pigment epithelial (RPE) cells, lipofuscin accumulation has been found to inhibit heterophagocytosis as evaluated by ingestion of latex beads and photoreceptor outer segments [52]. Impaired lysosomal function (as demonstrated by decreased activity of acid phospatase, N-acetyl-β-glucuronidase, and cathepsin D) also has been demonstrated in RPE-cells loaded with exogenous lipofuscin [53]. For aging postmitotic cells, the decreased ability of the lysosomal system to degrade mitochondria may have the most serious consequences, resulting in reduced ATP production, increased formation of ROS and eventual cell death (Figure 1; further discussed in the "mitochondrial-lysosomal axis theory of aging" [21]).

A negative influence of lipofuscin accumulation on cellular degradative processes is also supported by the finding that lipofuscin-loaded fibroblasts have decreased proteasome activity [54]. A possible explanation of this fact is that lipofuscin deposition affects recycling of proteasomes, known to be degraded by lysosomes [55].

Except for interference with intracellular degradation, lipofuscin engenders other pathogenic effects. The increased amounts of iron within lipofuscin granules may promote generation of ROS (see above), sensitizing cells to oxidative injury through lysosomal destabilization. Besides, large numbers of lipofuscin-containing lysosomes with active hydrolases [56, 57] may promote cellular damage when lysosomal membranes are destabilized by pathogenic factors (such as oxidative stress), and the lytic enzymes leak into the cytosol. In support of this, it was demonstrated that lipofuscin-rich fibroblasts are more susceptible to oxidative stress-induced apoptosis than are cells with low lipofuscin content [58]. Furthermore, lipofuscin has been shown to sensitize cells to photodamage [59], a property, which may be of importance for the development of age-related macular degeneration, a common form of blindness in developed countries. In this regard, lipofuscin-loaded RPE cells show increased ROS production and decreased viability when exposed to visible – especially blue – light [59–61].

Both increased sensitivity to oxidative stress and decreased lysosomal degradation – despite the abundance of lysosomal enzymes – would apparently make aged lipofuscin-loaded cells more vulnerable to pathogenic agents and promote the development of various pathologies. In particular, such changes may be involved in age-related neurodegenerative diseases, such as Alzheimer's and Parkinson's diseases [62–64]. In support of this, oxidative stress [65], as well as impaired lysosomal degradation [66], have been shown to induce the formation of beta-amyloid – a hallmark of Alzheimer's disease – under experimental conditions. Furthermore, in aged rat brain neurons, large amounts of lipofuscin and lysosomal enzymes have been found to co-localize with Alzheimer amyloid precursor protein [67].

References

1. Sorimachi H, Ishiura S, Suzuki K (1997). Structure and physiological function of calpains. *Biochem J.* 328(Pt 3): 721–32.
2. Myung J, Kim KB, Crews CM (2001). The ubiquitin-proteasome pathway and proteasome inhibitors. *Med Res Rev.* 21(4): 245–73.
3. Mortimore GE, Miotto G, Venerando R, Kadowaki M (1996). Autophagy. *Subcell Biochem.* 27: 93–135.
4. Klionsky DJ, Emr SD (2000). Autophagy as a regulated pathway of cellular degradation. *Science* 290(5497): 1717–21.
5. Seglen PO, Bohley P (1992). Autophagy and other vacuolar protein degradation mechanisms. *Experientia* 48(2): 158–72.
6. Dice JF (2000). *Lysosomal Pathways of Protein Degradation.* Georgetown, Texas: Eurekah.com/Landes Bioscience.
7. Marzella L, Ahlberg J, Glaumann H (1981). Autophagy, heterophagy, microautophagy and crinophagy as the means for intracellular degradation. *Virchows Archiv B, Cell Pathol Mol Pathol.* 36(2–3): 219–34.
8. Armstrong D, Koppang N (1981). Ceroid-lipofuscinosis, a model for aging. In: RS Sohal, ed. *Age Pigments.* Amsterdam: Elsevier, pp. 355–82.
9. Neufeld EF (1991). Lysosomal storage diseases. *Annu Rev Biochem.* 60: 257–80.
10. Terman A (2001). Garbage catastrophe theory of aging: imperfect removal of oxidative damage? *Redox Rep.* 6(1): 15–26.
11. Halliwell B, Gutteridge JMC (1999). *Free Radicals in Biology and Medicine*, 3rd edn. New York: Oxford University Press.
12. de Grey ADNJ (1999). *The Mitochondrial Free Radical Theory of Aging.* Austin, TX: RG Landes Company.
13. Stadtman ER (2001). Protein oxidation in aging and age-related diseases. *Ann NY Acad Sci.* 928: 22–38.
14. Harman D (1956). Aging: a theory based on free radical and radiation chemistry. *J Gerontol.* 211: 298–300.
15. Harman D (1996). Aging and disease: extending functional life span. *Ann NY Acad Sci.* 786: 321–36.
16. Brownlee M (1995). Advanced protein glycosylation in diabetes and aging. *Annu Rev Med.* 46: 223–34.
17. Lee AT, Cerami A (1992). Role of glycation in aging. *Ann NY Acad Sci.* 663: 63–70.
18. Orgel LE (1973). Ageing of clones of mammalian cells. *Nature* 243(5408): 441–5.
19. Burnet FM (1973). A genetic interpretation of ageing. *Lancet* 2(7827): 480–83.
20. Kirkwood,TB (1989). DNA, mutations and aging. *Mutat Res.* 219(1): 1–7.
21. Brunk UT, Terman A (2002). The mitochondrial-lysosomal axis theory of aging: Accumulation of damaged mitochondria as a result of imperfect autophagocytosis. *Eur J Biochem.* 269: 1996–2002.

22. Kowald A, Kirkwood TB (1996). A network theory of ageing: the interactions of defective mitochondria, aberrant proteins, free radicals and scavengers in the ageing process. *Mutat Res.* 316(5–6): 209–36.
23. Beckman KB, Ames BN (1998). The free radical theory of aging matures. *Physiol Rev.* 78(2): 547–81.
24. Friguet B (2002). Aging of proteins and the proteasome. *Prog Mol Subcell Biol.* 29: 17–33.
25. Cadenas E, Davies KJ (2000). Mitochondrial free radical generation, oxidative stress, and aging. *Free Rad Biol Med.* 29(3–4): 222–30.
26. Richter C (1995). Oxidative damage to mitochondrial DNA and its relationship to ageing. *Int J Biochem Cell Biol.* 27(7): 647–53.
27. Ozawa T (1997). Genetic and functional changes in mitochondria associated with aging. *Physiol Rev.* 77(2): 425–64.
28. Tate EL, Herbener GH (1976). A morphometric study of the density of mitochondrial cristae in heart and liver of aging mice. *J Gerontol.* 31(2): 129–34.
29. Vanneste J, van den Bosch de Aguilar P (1981). Mitochondrial alterations in the spinal ganglion neurons in ageing rats. *Acta Neuropathol.* 54(1): 83–7.
30. Sachs HG, Colgan JA, Lazarus ML (1977). Ultrastructure of the aging myocardium: a morphometric approach. *Am J Anat.* 150(1): 63–71.
31. Yamada K, Sugiyama S, Kosaka K, Hayakawa M, Ozawa T (1995). Early appearance of age-associated deterioration in mitochondrial function of diaphragm and heart in rats treated with doxorubicin. *Exp Gerontol.* 30(6): 581–93.
32. de Grey AD (1997). A proposed refinement of the mitochondrial free radical theory of aging. *BioEssays* 19(2): 161–6.
33. Coller HA, Bodyak ND, Khrapko K (2002). Frequent intracellular clonal expansions of somatic mtDNA mutations: significance and mechanisms. *Ann NY Acad Sci.* 959: 434–47.
34. Aiken J, Bua E, Cao Z, *et al.* (2002). Mitochondrial DNA deletion mutations and sarcopenia. *Ann NY Acad Sci.* 959: 412–23.
35. Terman A, Brunk UT (1998). On the degradability and exocytosis of ceroid/lipofuscin in cultured rat cardiac myocytes. *Mech Ageing Dev.* 100(2): 145–56.
36. Brunk UT, Terman A (2002). Lipofuscin: mechanisms of age-related accumulation and influence on cell functions. *Free Rad Biol Med.* 33(5): 611–19.
37. Porta EA (1991). Advances in age pigment research. *Arch Gerontol Geriat*r. 212(2–3): 303–20.
38. Monserrat AJ, Benavides SH, Berra A, Farina S, Vicario SC, Porta EA (1995). Lectin histochemistry of lipofuscin and certain ceroid pigments. *Histochem Cell Biol.* 103(6): 435–45.
39. Brun A, Brunk U (1970). Histochemical indications for lysosomal localization of heavy metals in normal rat brain and liver. *J Histochem Cytochem.* 18(11): 820–7.
40. Jolly RD, Douglas BV, Davey PM, Roiri JE (1995). Lipofuscin in bovine muscle and brain: a model for studying age pigment. *Gerontology* 41(Suppl 2): 283–95.
41. Glaumann H, Ericsson JL, Marzella L (1981). Mechanisms of intralysosomal degradation with special reference to autophagocytosis and heterophagocytosis of cell organelles. *Int Rev Cytol.* 73: 149–82.
42. Burke JM, Skumatz CM (1998). Autofluorescent inclusions in long-term postconfluent cultures of retinal pigment epithelium. *Invest Ophthalmol Visual Sci.* 39(8): 1478–86.
43. Kikugawa K, Kato T, Beppu M, Hayasaka A (1989). Fluorescent and cross-linked proteins formed by free radical and aldehyde species generated during lipid oxidation. *Adv Exp Med Biol.* 266: 345–57.

44. Brunk UT, Jones CB, Sohal RS (1992). A novel hypothesis of lipofuscinogenesis and cellular aging based on interactions between oxidative stress and autophagocytosis. *Mutat Res.* 275(3-6): 395–403.
45. Terman A, Brunk UT (1998). Ceroid/lipofuscin formation in cultured human fibroblasts: the role of oxidative stress and lysosomal proteolysis. *Mech Ageing Dev.* 104, 277–91.
46. Collins VP, Arborgh B, Brunk U, Schellens JP (1980). Phagocytosis and degradation of rat liver mitochondria by cultivated human glial cells. *Lab Invest.* 42(2): 209–16.
47. Knecht E, Martinez-Ramon A, Grisolia S (1988). Autophagy of mitochondria in rat liver assessed by immunogold procedures. *J Histochem Cytochem.* 36(11): 1433–40.
48. Elleder M, Sokolova J, Hrebicek M (1997). Follow-up study of subunit c of mitochondrial ATP synthase (SCMAS) in Batten disease and in unrelated lysosomal disorders. *Acta Neuropathol.* 93(4): 379–90.
49. Schutt F, Ueberle B, Schnolzer MGHF, Kopitz J (2002). Proteome analysis of lipofuscin in human retinal pigment epithelial cells. *FEBS Lett.* 528(1–3): 217–21.
50. Treff WM (1974). Das involutionsmuster des nucleus dentatus cerebelli. In: D Platt, ed. *Altern.* Stuttgart: Schattauer, pp. 37–54.
51. Terman A, Dalen H, Brunk UT (1999). Ceroid/lipofuscin-loaded human fibroblasts show decreased survival time and diminished autophagocytosis during amino acid starvation. *Exp Gerontol.* 34(8): 943–57.
52. Sundelin S, Wihlmark U, Nilsson SEG, Brunk UT (1998). Lipofuscin accumulation of cultured retinal pigment epithelial cells reduces their phagocytic capacity. *Curr Eye Res.* 17(8): 851–7.
53. Shamsi FA, Boulton M (2001). Inhibition of RPE lysosomal and antioxidant activity by the age pigment lipofuscin. *Invest Ophthalmol Visual Sci.* 42(12): 3041–6.
54. Sitte N, Huber M, Grune T, *et al.* (2000). Proteasome inhibition by lipofuscin/ceroid during postmitotic aging of fibroblasts. *FASEB J.* 14(11): 1490–8.
55. Cuervo AM, Palmer A, Rivett AJ, Knecht E (1995). Degradation of proteasomes by lysosomes in rat liver. *Eur J Biochem.* 227(3): 792–800.
56. Essner E, Novikof, AV (1960). Human hepatocellular pigments and lysosomes. *J Ultrastruct Res.* 3: 374–91.
57. Brunk U, Ericsson JLE (1972). Electron microscopical studies of rat brain neurons. Localization of acid phosphatase and mode of formation of lipofuscin bodies. *J Ultrastruct Res.* 38: 1–15.
58. Terman A, Abrahamsson N, Brunk UT (1999). Ceroid/lipofuscin-loaded human fibroblasts show increased susceptibility to oxidative stress. *Exp Gerontol.* 34(6): 755–70.
59. Wihlmark U, Wrigstad A, Roberg K, Nilsson SE, Brunk UT (1997). Lipofuscin accumulation in cultured retinal pigment epithelial cells causes enhanced sensitivity to blue light irradiation. *Free Rad Biol Med.* 22(7): 1229–34.
60. Rozanowska M, Jarvis-Evans J, Korytowski W, Boulton ME, Burke JM, Sarna T (1995). Blue light-induced reactivity of retinal age pigment. *In vitro* generation of oxygen-reactive species. *J Biol Chem.* 270(32): 18825–30.
61. Davies S, Elliott MH, Floor E, *et al.* (2001). Photocytotoxicity of lipofuscin in human retinal pigment epithelial cells. *Free Rad Biol Med.* 31(2): 256–65.
62. Beal MF (1996). Mitochondria, free radicals, and neurodegeneration. *Curr Opin Neurobiol.* 6(5): 661–6.
63. Busciglio J, Andersen JK, Schipper HM, *et al.* (1998). Stress, aging, and neurodegenerative disorders. Molecular Mechanisms. *Ann NY Acad Sci.* 851: 429–43.

64. Adamec E, Mohan PS, Cataldo AM, Vonsattel JP, Nixon RA (2000). Up-regulation of the lysosomal system in experimental models of neuronal injury: implications for Alzheimer's disease. *Neuroscience* 100(3): 663–75.
65. Misonou H, Morishima-Kawashima M, Ihara Y (2000). Oxidative stress induces intracellular accumulation of amyloid beta-protein (Abeta) in human neuroblastoma cells. *Biochemistry* 39(23): 6951–9.
66. Mielke JG, Murphy MP, Maritz J, Bengualid KM, Ivy GO (1997). Chloroquine administration in mice increases beta-amyloid immunoreactivity and attenuates kainate-induced blood-brain barrier dysfunction. *Neurosci Lett.* 227(3): 169–72.
67. Nakanishi H, Amano T, Sastradipura DF, *et al.* (1997). Increased expression of cathepsins E and D in neurons of the aged rat brain and their colocalization with lipofuscin and carboxy-terminal fragments of Alzheimer amyloid precursor protein. *J Neurochem.* 68(2): 739–49.

Index

MIX
Papier aus verantwortungsvollen Quellen
Paper from responsible sources
FSC® C105338

If you have any concerns about our products,
you can contact us on
ProductSafety@springernature.com

In case Publisher is established outside the EU,
the EU authorized representative is:
Springer Nature Customer Service Center GmbH
Europaplatz 3, 69115 Heidelberg, Germany

Printed by Libri Plureos GmbH
in Hamburg, Germany